MW01627484

Build-to-Order
&
Mass Customization

The Ultimate Supply Chain Management and Lean Manufacturing Strategy for Low-Cost On-Demand Production without Forecasts or Inventory

Dr. David M. Anderson, P.E., fASME, CMC
Build-to-Order Consulting
www.build-to-order-consulting.com

CIM Press
Cambria, California

Build-to-Order & Mass Customization
The Ultimate Supply Chain Management and Lean Manufacturing Strategy for Low-Cost On-Demand Production without Forecasts or Inventory

By Dr. David M. Anderson, P.E., CMC

Published by **CIM Press**
P. O. Box 100
Cambria, California 93428-0100
(805) 924-0200 phone & fax

Library of Congress Cataloging in Publication Data:

Anderson, David M.
Build-to-Order and Mass Customization
The Ultimate Supply Chain Management and Lean Manufacturing Strategy for Low-Cost On-Demand Production without Forecasts or Inventory
Includes index.
1. Build-to-Order, Make-to-Order, etc.
2. Mass Customization, personalization, etc.
3. Lean Production, Lean Manufacturing, etc.
4. Flexible Manufacturing, Agility, etc.
5. Standardization, parts standardization, etc.
6. Product line rationalization, etc.
7. Total cost, cost drivers, etc.

Printed in the United States of America

10 9 8 7 6 5 4

ISBN 1-878072-30-7

ENDORSEMENTS

"An outstanding and practical navigation guide for the complex task of implementing mass customization."

Harold Pinto
President
Fire Rescue Group
Federal Signal Corporation

"Like all of the other books by Dr. Anderson, *Build to Order & Mass Customization* is excellent. I have used it in undergraduate and graduate classes with exceptional results."

Dr. E. Ray Ladd, Professor
Master of Engineering Management Program
Washington State University

"This book captures the essence of what is required for American businesses to transform their business model and drive profitability. *Build-to-Order & Mass Customization* is the roadmap for companies to drive step-level improvements far beyond lean and six-sigma."

Jeff M. Krajacic
President
The Bourton Group

"Build-To-Order & Mass Customization provides a framework of tools for operations improvement that you cannot find anywhere else. This set of tools is indispensable for firms in the U.S. competing with offshore, low labor cost operations."

Dr. W. B. Fredenberger
Department of Management
Langdale College of Business
Valdosta State University

Dedicated to my loving and supportive wife, Lin.

TABLE OF CONTENTS

PART IV: IMPLEMENTATION

PART V: THE BOTTOM LINE

PREFACE

In *craft production,* customers got exactly what they want — later, at a high cost. In *mass production,* customers got low-cost products — now, but not what they wanted. In *mass customization,* customers get exactly what they want — now, and at low cost. Another way at looking at the paradigm shift is from *build-to-forecast* to *build-to-order.*

This book addresses both of these paradigm shifts, hence the title, *Build-to-Order and Mass Customization,* which will be abbreviated throughout the book as *BTO&MC.*

Build-to-order can be defined as the on-demand production of *standard products*, where as mass customization is the on-demand production of *customized products.* They share the same spontaneous supply chain and on-demand lean production capabilities, but with different order-entry scenarios: either standard product designations or customization input. A positive synergy results from combining operations for building both standard and mass-customized products on-demand that may enable both to achieve "critical mass."

The book is based on practical, proven *tactics,* some of which have been refined for decades, for instance, the Toyota Production System that evolved into the *lean production* movement. Other tactics have been developed for this book, but are derived from proven practices and sound original principles. This book combines those proven tactics into an effective *strategy* that can enable companies to build standard or mass-customized products on-demand without forecasts or inventory.

The book includes many innovative, unconventional, and perhaps controversial discussions that challenge the prevailing "wisdom" (or lack thereof) about forecasts and mass production in batches (Ch. 2), outsourcing and offshoring (Ch. 6), the value of internal integration (Ch.6), off-shore manufacturing to save cost (Ch. 6), world trade (Ch. 6), cheap parts (Ch. 7), on-line auction procurement, reverse auctions, and low-bidding (Ch. 7 and 11), supply chain management (Ch. 3 through 7), the buzz about mass customization (Ch. 9), cost cutting (Ch. 11 and 12), and growth (Ch. 14).

The most cited article in this book is David Pringle's Wall Street Journal article, "How Nokia Thrives by Breaking the Rules" (Jan. 3, 2003). The "rules" that Nokia breaks correspond to many of the over-hyped practices that are challenged by this book. The results: Nokia has thrived with 22% operating margins and a dominant position in the highly competitive telecom industry.

One of the most cited books is Dartmouth Professor Sydney Finkelstein's thorough study of business failures, *Why Smart Executives Fail, and What You Can Learn from Their Mistakes* (2003, Portfolio/Penguin), which complements this book's messages about going for profits over market share, arguing the "no side" of decisions, problems with outsourcers becoming competitors, cost pressures alienating suppliers, the costs of realizing merger synergies, severe consequences caused by inadequate order fulfillment rates, executives' inaccurate perceptions of reality, and the real cause of nearly every major business breakdown: pursuing the wrong business model.

The book also includes discussions that may not seem relevant, but really are supportive of BTO&MC such as avoiding the practice of "taking all orders" (Ch. 3), total cost measurements (Ch. 12), downturn strategies, including why not to lay off people (Ch. 13), and the effects of mergers and acquisitions on BTO&MC (Ch. 14).

There are several companies that are implementing many of the principles of this book. The book has many examples from research and the author's extensive consulting experience.

Reading Scenarios

The format of the book allows various reading scenarios. All readers should read the Introduction. Busy executives can obtain an overview of the remainder of the book by reading Chapter 1, plus any other sections of interest. Chapters are cross-referenced for more detailed exploration of any topic. The thorough Table of Contents and Index may also be useful for this purpose. Chapter 14 summarizes the business case for build-to-order and mass customization. Those who want to comprehend strategy, principles, techniques, implementation, and the business case should read the entire book, starting with Chapter 1 for those who want an overview first. Throughout the book are a few hundred *sidebars* which summarize key points near the related discussions. Endnotes (superscript numbers) refer to comments and referenced books and articles at the end of each chapter.

Book Organization

Part I presents an overview of the book in Chapter 1 and a summary of the shortcomings of mass production (Chapter 2), especially as it relates to the topics in the book.

Part II focuses on supply chain *simplification*, specifically product line rationalization (Chapter 3), part standardization (Chapter 4), material variety reduction (Chapter 5), and a discussion of outsourcing and integration (Chapter 6).

Part III describes how BTO&MC actually work starting with setting up a spontaneous supply chain that can *pull* parts and materials into assembly without forecasted parts inventory or ordering and waiting for parts (Chapter 7). Chapter 8 shows how to extend *lean production* techniques to build products *on-demand* spontaneously. Chapter 9 shows how to mass-customize products based on the foundations of spontaneous supply chains and on-demand lean production presented in the rest of the book.

Part IV focuses on implementation and shows how to concurrently engineer product families and flexible processes for lean, BTO, and mass customization (Chapter 10). Two chapters address the important issue of *total cost:* Chapter 11 shows how to *minimize* total cost and Chapter 12 shows some easy ways to *measure* total cost. Chapter 13 shows how to implement build-to-order and mass customization with implementation strategies and road maps based on several self-supporting coordinated steps.

Part V, the bottom line (Chapter 14), presents the business case for build-to-order and mass customization and summarizes competitive advantages for cost, responsiveness, and customer satisfaction. It shows how to achieve substantial growth of sales and profits.

Companies That Can Benefit the Most

Build-to-order and mass customization methodologies will be extremely valuable for manufacturing companies that have any of the following challenges:

Product Variety. Mass producers making a variety of products will always wrestle with the classic dilemma: too many different products to sell from inventory, but trying to build-to-order in a mass-production environment is too slow and costly while raising resource demands and lowering equipment capacity.

Unreliable Forecasts. Forecast accuracy is drastically reduced by increased market volatility and product variety, which are both rising. Forecast accuracy drops drastically as product variety increases, thus providing justification for implementing systems that do not depend on forecasts.

Inventory Problems. Selling products from inventory creates many problems: out-of-stocks, obsolescence, write-offs, inventory carrying costs, warehouse space, and discounting or write-offs of obsolete inventory. The inherent inventory dilemma is that trying to reduce out-of-stocks increases all the other inventory problems and vice versa.

Customization. Traditional attempts to customize products either use craft production, which is slow and expensive, or try to make custom products in a mass production environment, which is slow, has high resource demands, and disrupts the production and compromises the quality of standard products.

Response time. Manufacturers of products or industrial parts may try to build products "to-order," but will not be able to deliver them quickly if they have to wait for parts, setup changes, and equipment availability.

Build-to-order and mass customization solve all these problems by building a wide variety of mass-customized or standard products on-demand without forecasts or inventory.

The above criteria should be used for deciding (a) if a company can benefit, (b) how fast and thorough the implementation should be, (c) which product lines can benefit, and (d) which product lines to do first.

Preface to Students

For those readers that are students in Universities or other public classes, this book will teach you new principles and techniques that the business world needs to implement for competitive advantage or even survival.

This is leading-edge material, not conventional practice. Most companies may not be practicing this yet, but could benefit highly from these techniques. You can greatly benefit your employers – and your own career – by proposing and implementing these techniques.

You will probably see conventional practices in action when you arrive at a new company and, after reading this book, you will probably be able to realize their shortcomings and propose a better way. People operating in conventional ways may be set in those ways, so "turning around the ocean liner" may not be easy.

If there is significant inertia, the best approach would be to start with the prerequisites such as product line rationalization, standardization, material variety reduction, basic lean production, total cost measurement, and product development for flexible environments. Many of these programs can be justified and paid off as free standing programs. In addition, these prerequisite programs may take a while to implement; in the mean time, you can be working with like-minded people to gain support for more ambitious efforts.

About the Publisher

CIM Press works closely with Build-to-Order Consulting, so cooperation between author and publisher enhances the flow of reading resulting in abundant headings and sidebars, optimal page breaks, sensible figure placement, and the placement of related material on facing pages whenever possible.

The principles of the book were derived from years of research, plant tours, experience with a wide range of clients, and experience personally implementing these techniques as a design engineer, product development team leader, and, at Intel's Systems Group, Manager of Flexible Manufacturing.

The book is thoroughly researched with over 350 references, conveniently place at the end of each chapter. The book will be easy to read, search, or use as a reference with 300 sidebars, 450 Table of Contents headings, and 1600 index entries.

Much of the original material presented in this book was created for the author's seminars and first published in copyrighted class handouts. Some of that was then published in *Agile Product Development for Mass Customization* (1997, McGraw-Hill).

About the Author

Dr. Anderson is a management consultant who specializes in corporate training seminars, workshops, and consulting on implementing Build-to-Order, Mass Customization, and Design for Manufacturability.

His first book on BTO&MC was *Agile Product Development for Mass Customization* (1997, McGraw-Hill). Dr. Anderson was appointed guest editor for the Mass Customization issue of the Agility Forum's journal, *Agility & Global Competition* (Vol. 2, No. 2, Spring 1998).

Starting in 1989, he has facilitated the Suzaki/SME video program, *Techniques for Continuous Improvement,* on eliminating waste, reducing lead times, implementing flow, improving productivity, and reducing setup, batch size, and inventory.

He taught two courses on BTO for the Association for Manufacturing Excellence (AME), after which he lead two AME plant tours at Hoffman Engineering, which, based on Dr. Anderson's principles, build a $30,000,000 plant to build on-demand a wide variety of mass-customized and standard sheet-metal enclosures.

He "wrote the book on DFM," literally and figuratively: *Design for Manufacturability & Concurrent Engineering; How to Design for Low Cost, Design in High Quality, Design for Lean Manufacture, and Design Quickly for Fast Production* (2008, CIM Press). At the Management of Technology Program at the University of California at Berkeley, he created and taught the course "New Product Development." Relevant web-sites are *www.design4manufacturability.com* and his book-length site: *www.HalfCostProducts.com* (summarized on page 468).

Dr. Anderson has over 35 years of industrial experience and has trained many leading companies who have implemented principles from this book. As Manager of Flexible Manufacturing at Intel's Systems Group, he developed manufacturing strategies, designed flexible tooling, and initiated their successful programs for standardization and DFM.

For seven years, his own company, Anderson Automation, Inc., built special production equipment for companies such as IBM and Optical Coating Labs and generated production equipment design studies for SRI International, Clorox Manufacturing, and two divisions of FMC. He has been issued four patents and is working on more.

Dr. Anderson is a *Fellow* of ASME (American Society of Mechanical Engineers) and has been certified for 10 years as a Certified Management Consultant (CMC) by the Institute of Management Consultants. He holds professional engineering (P.E.) registrations in Mechanical, Industrial, and Manufacturing Engineering and a Doctorate in Mechanical Engineering from the University of California at Berkeley.

Dr. Anderson's web-site, *www.build-to-order-consulting.com* has information on in-house seminars, workshops, and consulting plus book descriptions, credentials, and a client list. His seminars and consulting services are described on page 469 - 470. Dr. Anderson can be reached at (805) 924-0100 or *anderson@build-to-order-consulting.com.*

INTRODUCTION

Most manufacturers are being forced to offer more and more variety to satisfy customers and compete in niche markets. Many also need to customize products for increasingly selective customers. However, if products are not designed well for this and manufacturing is not flexible enough, then trying to respond quickly to changing market demands, offering more variety, or customizing products will be a slow and costly ordeal.

Build-to-Order and Mass Customization (BTO&MC) solve all these problems by making operations *agile* or *flexible* enough to actually build products after the receipt of *spontaneous* orders and ship them directly to customers, without the forecasts, MRP, purchasing, expediting, inventory, or the tremendous waste of valuable resources that are consumed when operations run in the "fire-drill" mode. Customers receive what they want right away, including mass-customized products. No sales are lost because of forecasting inaccuracies, products being out of stock, part shortages, or lead-time delays.

BTO&MC is the practice of building customized or standard products as they are ordered and shipping them directly to customers, instead of building-to-forecast and shipping from inventory. Build-to-order is the next paradigm to follow the current build-to-forecast paradigm, which is getting bogged down in current times of turbulent markets, unreliable forecasts, and increasing customer demands for variety and customization.

CURRENT ATTEMPTS AT BUILD-TO-ORDER

There are four basic ways to build products. *Mass Production* builds products in batches based on forecasts and puts them in inventory in hopes that they will be the products that customers will order; this is also called *build-to-forecast* or *build-to-stock*. The shortcomings of mass production will be discussed in Chapter 2.

Pseudo "build-to-order" waits for an order and then orders all the materials to arrive at which time the product can be build (see below). Craft production is a variation of this.

Assemble-to-Order assembles products from forecasted parts inventory. This is the model used by Dell Computer (see below).

Spontaneous Build-to-Order, the subject of this book, builds mass-customized and standard products on-demand without forecasts or inventory.

Virtual BTO; Searching for Inventory

Some digital strategies accept the deficiencies of mass production and recommend "the illusion of build-to-order" by *searching the web for inventory!*[1] On the transportation side, overnight shippers offer the same pitch when they recommend that the way to reduce inventory levels is to know where the inventory is and how long it will take to ship it.[2]

Some have proposed *virtual build-to-order:* searching web for previously built inventory!

BTO from Forecasted Parts Inventory

One way used to build products to-order is to draw parts from forecasted inventory and *assemble-to-order* modular products. For this to work, the factory, supplier, or distributer must carry large (and expensive) inventories and be good at forecasting the assembler's demand. Ultimately the customer must pay for *carrying cost* of the inventory which over twenty-five years has averaged about 25% of inventory value per year (Ch. 2). Not only does inventory raise the product's cost and make the product less competitive, but it also makes the product's availability subject to outages and delays.

Dell Computer

Dell Computer should take credit for incredible growth and adequate profits in an industry where its competitors find their products are reduced to commodity status with eroding profit margins. Dell has been very successful because it has a better business model than the competition. Dell's competitors have had difficulty emulating Dell's model either because they could not free themselves from their dealer-based distribution system or because their production is heavily outsourced. Outsourcing is an inherent batch operation and outsourcers are usually not flexible or committed enough to do spontaneous build-to-order, as will be shown in Chapter 6.

Dell has been very successful because it has a better business model than the competition

The elements of the Dell model that are noteworthy (and correspond to the principles of this book) are the *direct sales* model that avoids finished goods inventory and makes Dell the first to market with new technology since it does not have to face the dilemma of either selling of the older products first or obsoleting the older products in the "pipeline." Dell also has a good web-site and phone sales organization. Further, it has very good relationships with corporate customers, in essence, performing many of the configuration tasks formerly done by customers' IT departments.

However, Dell should not be considered a benchmark example to emulate because it is a special case and its supply chain is not *spontaneous* as this book will show how to do. Dell's parts come from forecasted parts inventory, which can only supply the parts about three-fourths of the time. Fifteen percent comes from expediting and, if that is not successful, the remaining 10 percent comes from "moving demand," which means if Dell is out of 80 gigabyte hard disks, it offers the customer a "special deal" on the 120 GB version. More precisely, the Dell model could be defined as high-volume assemble-to-order of an inherently modular product from forecasted parts inventory.

The Dell model will be a difficult model to emulate because: (1) Dell has enough volume to force suppliers to warehouse the parts nearby and pay for the warehousing and the parts inventory carrying costs. (2) Dell's enormous sales result in very rapid turnover of parts, which makes the supply of parts less vulnerable to forecasts being off than would be the case at a lower volume producer. (3) Personal computers are inherently modular products that are easy to assemble-to-order.

> The Dell model will be hard to emulate because of its volume, turnover, and the inherent modularity of its products

This book has several chapters on simplifying the supply chain and arranging spontaneous supply chains that do not depend on supplying parts through forecasted parts inventory. Part of the author's consulting practice generates industry-specific scenarios for clients that show how to build-to-order (Ch. 8) and mass-customize (Ch. 9) many types of products based on spontaneous supply chains (Chapters 4 through 7).

Order Parts, Wait, and Assemble-to-Order

Many companies including those that build capital equipment (like machine tools, semiconductor fab equipment, production machinery, heavy vehicles, and airplanes) typically do not build products ahead based on forecast because of high cost inventory carrying cost and challenges with the amount of variety and customization. So they build their products "to order" – slowly.

After receiving an order, reactive assemble-to-order companies place purchase orders for all the materials, parts, and subsystem, wait for all of them to arrive, and then assemble them into a product. The delays will be even greater for every additional supply chain "tier" that has to go through the same process of ordering and waiting for *its* parts. Part manufacture in a "batch and queue" environment (mass production) also has delays caused by queues (parts waiting in batches) at each work station, lengthy setup changes,

> Many "BTO" operations order the parts, wait for their arrival, and then assemble them to-order

and waiting for equipment to become available. Instead of calling this build-to-order, it should be called *order parts, wait, and assemble-to-order.*

Traditionally, this slow process has been acceptable for "big ticket" capital equipment. However, capital equipment assemblers and their suppliers should take note of several compelling reasons to implement some of the techniques presented herein:

- The total delivery time may no longer be acceptable to customers after they realize that you – or your competitors – can reduce the delivery time. Customers may be lured away by the fastest manufacturer, especially when they can realize a competitive advantage from utilizing new equipment quicker.

- Fast moving technological developments and volatile markets offer opportunities for the fastest to deliver new technologies first.

- Similarly, the same technological changes and volatile markets also increase the vulnerability to obsolescence or canceled orders.

- Suppliers of subsystems to these markets *and their suppliers* will have to use BTO&MC techniques to decrease their delivery times as capital equipment assemblers realize the need to deliver faster.

- The mass customization techniques of Chapter 9 can help capital equipment assemblers and their suppliers more quickly and efficiently customize parts, subsystems, and products instead of the slow and expensive of "reinventing the wheel" in a fire drill mode.

- Production equipment manufacturers should understand BTO&MC principles well enough to offer production equipment that is flexible enough to enable *their* customers to build-to-order and mass-customize their products. Specifically, such customers may appreciate versatile equipment that can change setups and process variables quickly to support smaller batch sizes, ultimately approach batch-size-of-one capability.

PRECEDENTS

This book will cite dozens of examples where companies have implemented specific techniques discussed.

However, BTO&MC is so new that implementation cannot be based simply on emulating some "best practice" company. The usual definition of "best" practices is really not the best *possible,* only the best *reported* in the business press. In many cases, they are noteworthy simply because they are better than the competition, not because they are the ultimate practices.

> "Best practices" are not the best *possible* – just the best *reported*

Even "success" or lack of it can be misinterpreted; success could be attributed a combination of all business practices (which is really the case) or only the business practice of interest (which can be misleading). Some benchmark companies may be doing a very good job at the specific practices under study, but the business may be underperforming for other reasons. In its cover story on "25 Ideas for a Changing World," *Business Week* posed a new problem with benchmarking: "If one company seems to be faring exceptionally well, everyone else tried to figure out why and copy it. . . That process doesn't work right if one company is lying about its performance."[3]

Even if the benchmark company *does* has something worth emulating, by the time any of this can be implemented, the copying company will clearly be the follower, not the leader. One of the key points of Jim Collin's *Good to Great* is that the great companies took a decade of building "flywheel" momentum before their transitions were even noticed in the business press. On top of that, the great companies did not publicly proclaim ambitious goals at the outset.[4] Thus, companies that try to initiate change by emulating publicized benchmarks will be a decade behind. On the other hand, companies who want to enjoy the benefits of build-to-order and mass customization should start applying the principles of this book now

> Great companies quietly build up "flywheel" momentum for a decade before the press notices – Jim Collins, *Good to Great*

instead of waiting years to read about someone else who is quietly starting now.

STRATEGY FOR BTO&MC

The basic strategies for implementing BTO&MC are supply chain and operational simplification, the development of a spontaneous supply chain, concurrent design of versatile products, on-demand lean production, and the mass customization of variety. BTO&MC can actually build products on-demand *at less cost* than mass-produced batches, if "cost" is computed as *total cost*. Therefore, total cost measurements should also be part of this process, as discussed in Chapters 11 and 12.

The premise of *spontaneous* Build-to-Order is that products are "pulled" from customers. In other words, when a customer places an order for a product, the production process *begins at final assembly* instead of with parts ordering, as is typically done in Mass Production. The product is assembled with parts that are always available at final assembly using the following techniques.

> BTO&MC products are built on-demand with parts that are already available or built on-demand

Common parts, such as fasteners, are made available through *breadtruck* (free stock) deliveries where a supplier simply keeps the bins full and bills the company based on consumption at the end of the month.

Small and inexpensive parts are made available by *kanban* resupply techniques where parts are drawn from a bin until the bin empties (see Figure 7-1). When that happens, a full bin slides into place and the empty bin returns to its source where it is filled and returned to the assembly cell before the other bin runs out. In a similar fashion, raw materials are automatically reordered when the quantity in stock falls below the "reorder point." Bin size and reorder scenarios are all calculated so that a constant resupply of parts is assured for any spontaneous sequence of orders.

The prerequisite for automatic part resupply is intensive standardization and consolidation of parts and raw materials. One of the author's clients, who made sheet metal products, was accustomed to ordering 600 different sheet metal blanks. For their new BTO plant, the author recommended input sheet metal standardized to *one* sheet metal size: the longest, widest, and thickest, which was then cut by programmable laser cutters into a variety of sheet metal parts. Thus, sheet metal "ordering" would be as simple as arranging steady flows that match sheet metal input to the output of the plant.

Parts made by inflexible operations, like castings, moldings, extrusions, stampings, and bare circuit boards need to be *consolidated* so that many different parts can be made from a single "raw" part. Thus, a group of castings would be shaped so that all variations, in a size range, could be machined from a single casting. Plastic parts would have all the features molded in for a wide range of applications. Bare printed circuit boards would have enough traces and pads so that several different circuit board assemblies could be made from a single bare board. The consolidation of parts (Figure 4-3) requires total cost thinking backed up by total cost accounting to overcome resistance based on the common perception that consolidating parts may increase the cost of a single part. However, purchasing leverage, overhead cost savings, and the value of the flexibility should compensate for the part consolidation costs several times over.

Some parts can be made by programmable CNC (Computer Numerical Control) machine tools that can quickly fabricate unique parts from standard blanks, assemble unique circuit boards from standard bare boards, or load unique firmware into standard chips.

Parts that come from suppliers are "pulled" from the supplier quickly, using the same techniques. Similarly, a BTO&MC company's products may be part of *its* customer's pull system.

ENDNOTES/REFERENCES

1. Larry Downes and Chunka Mui, *Unleashing the Killer App, Digital Strategies for Market Dominance,* (1998, Harvard Business School Press), p. 146.

2. Don Tapscott, Ed. *Blueprint for the Digital Economy,* Chapter 11, "The New Logistics"

3. Michael J. Mandel, "Crimes Against the Information Age," *Business Week,* August 26, 2002, pp. 80-81.

4. Jim Collins, *Good to Great; Why Some Companies Make the Leap . . . and Others Don't,* (2001, Harper Business), Chapter 8, "The Flywheel and the Doom Loop."

EXECUTIVE OVERVIEW*

Build-to-order and mass customization offer the ultimate in supply chain management and lean production strategy that allows companies to build low-cost standard or mass-customized products on-demand without forecasts, inventory, or purchasing delays.

BTO&MC is a business model that offers an unbeatable combination of responsiveness, cost, and what customers want when they want it. BTO&MC companies can achieve substantial cost advantages from eliminating inventory, forecasting, expediting, kitting, setup, and inefficient fire-drill efforts to customize products.

BTO&MC companies can substantially *simplify* supply chains – not just "manage" them – to the point where parts and materials can be *pulled* into production without forecasts, MRP, purchasing, waiting, or warehousing. Stores and dealers can be rapidly replenished at high order-fulfillment rates without the cost and risk of inventory.

BTO&MC enables manufacturers to be the first to market with new technologies and efficiently mass-customize products for niche markets, countries, regions, industries, and individual customers.

Sales and profits will grow from expanded sales for standard, customized, derivative, and niche market products, while enhancing sales to existing customers.

---

*This chapter is an executive summary of the book. The organization corresponds to chapter numbers to facilitate more in depth reading on specific topics.

Shortcomings of Mass Production (Chapter 2)

Mass production thrived in the bygone era of stable demand and little product variety ("any color as long as it is black!"). In the mass production paradigm, the marketing department forecasted demand and these forecasts drove MRP (Material Requirements Planning) software to order the required parts and materials far enough ahead of time so that they would, hopefully, all arrive at the factory before the production run was scheduled. Assuming that was successful, equipment would be set up and a batch of products would be built to satisfy the forecast. These products were then put in a warehouse or series of distribution centers where they awaited the anticipated orders.

Even if this scenario could avoid the many pitfalls and defy the odds to succeed, it was not very agile or effective. Product availability was based on forecasts, which were really only estimates of customer demand. Many companies say their forecasts are only 50% accurate. Even if the forecasts were accurate *when generated,* market demand would probably change by the time the products have been manufactured. And the wider the variety of customer choices, the greater the chance that product availability from warehouses would not match current customer needs, resulting in inventory obsolescence or new product introduction delays until the manufactured products could be sold first. The inventory level dilemma gets worse whenever excessive equipment setup forces companies to build products less frequently in larger batches.

Since 1961, the average inventory carrying cost has been slightly over 25% of inventory value per year, as plotted in Figure 2-1.[1] Given this cost and a 10% profit margin, inventory carrying costs will exceed profit in just 5 months, after which finished goods are really being sold at a loss (Figure 2-2).

Market variety has an exponential effect on variety costs (Figure 2-3) while market volatility has an exponential effect on response difficulty (see Figure 2-4). Figure 2-5 compares the build-to-forecast to build-to-order for several criteria in production and fulfillment.

PART II: SUPPLY CHAIN SIMPLIFICATION

Although supply chain management is a much discussed topic today, most implementations concentrate on *management* of an overly complex system and fail to apply the basic lessons of Industrial Engineering 101: *"Simplify before automating or computerizing."* The simplification steps for supply chain management are part standardization (Ch. 4), material variety reduction (Ch. 5), shortening long and far-flung supply chains (Ch. 6), automatic resupply techniques (Ch. 7), and rationalization of the product line (Ch. 3) to eliminate or outsource the unusual, low-volume products that contribute to part variety that is way out of proportion to their profit generation ability.

The goal of supply chain simplification is to drastically reduce the variety of parts and raw materials to the point where these materials can be procured spontaneously by automatic and "pull-based" resupply techniques. Reducing the part and material variety will also shrink the vendor base, further simplifying the supply chain.

Product Line Rationalization (Chapter 3)

Product line rationalization is a powerful technique to improve profits, free valuable resources, and simplify operations and supply chains. It does this by *rationalizing* existing product lines to eliminate or outsource products and product variations that are problem-prone, have low sales, have excessive overhead demands, have limited future potential, may really be losing money, and are incompatible with new operational environments and corporate strategies, such as build-to-order and mass customization.

> Rationalization eliminates the losers and focuses on high profit products

Improving Profits. Rationalization can quickly improve profits by stopping the production of money-losing products and eliminating all of the excess overhead costs associated with "fire-drill" products. This, in turn, will allow precious resources to focus on the most profitable products instead of low-leverage products, which will increase sales and further lower costs. After rationalization, the

Fire-drill products consume too much overhead costs

remaining products will cost *less* because they will no longer have to subsidize the money losers or marginal products.

All these cost savings can be used to lower prices in price-sensitive markets or to increase profits. In fact, rationalization can be justified as a free standing program. A scenario in Chapter 3 will show that, by simply eliminating the lowest leverage products, profits can be tripled!

Product line rationalization may temporarily decrease revenue, but managers need to remember that the goal of a business is to make money, not blindly pursue growth (which one author,[2] described as "the road to hell") or market share, which is dead as a strategy, according to the authors of *The Profit Zone.*[3]

Focus. Product line rationalization encourages companies to focus on their best products by eliminating or outsourcing the marginal products. The resources that were being wasted on the low-leverage products can then be focused on growing the "cash cows." Revenue "lost" by eliminating the loser products can be quickly reclaimed with increased focus on the cash cows. In fact, one of the author's clients did this in two months!

Procedure. The rationalization procedure would be to: drop the least profitable, lowest-volume products (subject to certain considerations discussed in Chapter 3); outsource products that need to be in the catalog; keep the cash cows; and improve the remaining products.

Plot and segregate products into zones: Keep as is; Grow; Outsource; and Drop

The benefits. Product line rationalization can immediately increase profits, improve operational flexibility, free up valuable resources, improve quality, focus on most profitable products, protect most profitable products from competitive attacks, stop cross-subsidizes, and, most importantly for BTO&MC, simplify supply chain management and operations.

See Chapter 3 for complete discussions of all the principles, implementations, and benefits.

Part Standardization (Chapter 4)

Part variety can greatly be greatly reduced by designing new products around standard parts, which will simplify product development efforts, lower the cost of parts and materials, drastically reduce material overhead costs, improve availability and deliveries, raise quality, improve serviceability, simplify supply chain management, and therefore support lean production, build-to-order, and mass customization.

Part proliferation. Part proliferation happens because engineers don't understand the importance of standardization and because of other factors like arbitrary decisions, the "not invented here" syndrome, the minimum weight fallacy, qualifying part families, contract manufacturing, mergers, acquisitions, and the unknowing duplication of identical or functionally equivalent parts. Ironically, all of these are easily avoidable.

> Part proliferation is usually rampant, but can be drastically reduced with standardization

The Zero-Based Approach. Chapter 4 presents an effective "zero-based" approach that, rather than trying to whittle down an overwhelming list, starts at zero and asks the simple question, *"What is the minimum list of part types we need to design new products?"* Another way of looking at this is to ask, "What would be the minimum list of part types a *savvy competitor* would need to design new products, *especially a rationalized list of only the most profitable products*?"

The procedure starts with a Pareto sort of current part usage (for each category) and selects the most common parts to form a baseline list. Then new generation parts are added proactively rather than letting all the engineers pick many new parts without any coordination. Duplicate parts would be eliminated by consolidating similar and functionally equivalent parts. Similarly, multiple parts for various tolerances, grades, and strengths can be consolidated to the highest grade or strength. Any perceived cost "increase" would be easily paid for by the savings in material overhead; in addition, most products would benefit from higher quality parts.

This approach determines the minimum list of parts needed for *new* designs and is not intended to eliminate parts used on existing products, except when the standard parts are functionally equivalent in all respects. In this case the new standard part may be substituted as an equivalent part or a "better-than" substitution, where a standard part with a better tolerance can replace its lesser counterpart in existing products. Even if part standardization efforts only apply to new products, remember that in these days of rapid product obsolescence and short product life cycles, older products may be phased out in only a few years.

> Starting with zero, ask how many parts you really need to design new products

Standardizing Expensive Parts. Usually there will not be much resistance for inexpensive parts like fasteners and resistors. However, product designers may resist standardization on expensive parts if it is *perceived* to raise *one* product cost. This misperception happens whenever the cost system only quantifies parts and labor, while ignoring overhead costs that would be reduced substantially by standardization.

For instance, a client had 152 motors, but still resisted the author's recommendation to standardize during a seminar. When he asked their supplier, GE Consumer Motor Division, "What would be the savings if those 152 motors could be reduced to five or ten?" The one word answer was *"Massive!"* Why? Because each of those five or ten motors would be ordered in volumes that would be ten times their current order volumes, thus resulting in greater economies of scale. Further, the five or ten chosen would have been the most cost-effective in their line – the best designed motors that GE produced in high volume for other customers too. Chapter 4 has an illustration that graphically shows this effect (Figure 4-3)

> Savings from reducing 152 motors to 5 or 10 would be *"Massive"* – GE Consumer Motor Division

Encouraging Standardization. Standardization can be encouraged by varying the material rate – low for standard parts and high for oddball parts – to reflect the true overhead costs and encourage use. Standardization can also be encouraged by prequalifying standard parts, publishing spec books or web pages that include specifications of all standard parts, and making standard parts available to engineers on display boards or readily available floor stock.

The benefits of standardization are numerous: cost will be reduced in purchasing, inventory expenses, floor space charges, material overhead, and supply chain management. Quality will be improved with less chance of installing the wrong part (a surprisingly frequent mistake), improved continuous improvement that is possible with less inventory, and reductions in the supplier base. Flexibility and responsiveness will be improved since standardization greatly helps to eliminate setup, reduce inventory, enhance part availability, simplify internal logistics, and arrange steady flows of standard materials.

In general, standardization lists can be just a few % of the proliferated lists

In general, it should be possible to generate a preferred parts list that is just a few percent of the proliferated list. For very standard parts like fasteners or passive electronic components, it should be possible for the preferred parts list to be less than 1 percent of the current list.

See Chapter 4 for complete discussions of all the principles, implementations, and benefits.

Raw Material Variety Reduction (Chapter 5)

Raw material variety can pose serious supply chain problems for lean production, build-to-order, and mass customization. Most companies simply accept the variety and then implement complex information systems (like MRP/ERP) and "supply chain management" to try to handle the situation. But too many types of raw materials can thwart spontaneity and present manufacturers with the dilemma of having to choose between stocking all types and ordering them and waiting for delivery. Fortunately, it is not hard to significantly reduce incoming material variety.

Aggressive Standardization. The solution is to *aggressively* standardize incoming raw materials which are then cut into various shapes by programmable CNC equipment, such as laser cutters and machining centers. The ultimate scenario for spontaneous BTO is to reduce the number of raw material types within each category to *one,* in which case "ordering" material is as simple as matching the tonnage *in* to the tonnage *out*, so a *steady flow* of materials can be arranged. Such purchasing leverage and other material overhead savings would compensate for any cut-off waste and for some products getting "better" material than they need.

> If materials can be standardized on one type, then steady flows can be arranged without forecasts

As raw material variety is reduced, it becomes more feasible to use automatic resupply techniques, such as kanban, min/max or breadtruck (Ch. 7).

Material Reduction. Incoming material variety can be reduced tremendously by cutting the material on-demand to each required length or shape. This could apply to bar stock, tubing, sheet-metal, cloth, leather, wire, cable, and so forth. This can eliminate material order size variety by arranging a steady flow of, for instance, long bars, sheets, or ultimately reels, and cutting them by programmable CNC machine tools, such as laser cutters, cut-off machines, screw machines, CNC lathes, CNC machining centers, or less automated tools based on on-line instructions.

> Raw material could be cut-to-shape on-demand instead of forecasting and ordering hundreds of pre-cut pieces

Linear materials can be ordered in reels and cut-to-length without waste on-demand as needed or in batches to resupply kanban bins (Ch. 7), while avoiding the waste that is inevitable when linear materials are cut from discrete lengths. This applies to linear materials like wire, rope, plastic tubing, cable, chain, and so forth. Electrical wire can be cut and stripped programmably; further, color variety can be eliminated by printing labels on wire and tubing as it is dispensed.

Consolidation. Part variety can be further reduced by *consolidating* many inherently inflexible parts into a few very versatile standard shapes; this would apply to raw castings, molded parts, stampings, extrusions, custom silicon, power supplies, and bare printed circuit boards. Such a move might appear to raise part cost, but the combination of purchasing leverage and material overhead cost reductions will make this a net gain from a total cost basis.

Reduce inflexible raw part variety by consolidating them into versatile shapes

See Chapter 5 for complete discussions of all the principles, implementations, and benefits.

Outsourcing vs. Integration (Chapter 6)

Core Competencies. Lately, there has been a lot written about *core competencies,* usually presenting the seemly logical argument that companies should focus at what they are "good at" and outsource the rest to someone who is "better at it." However, this cart-before-the-horse strategy results in the corporate business model being determined by a collection of "whatever everyone happens to be good at the time." This often leads to the conclusion that manufacturing is not a core competency or that contract manufacturers could "do it better. "

The business model should determine core competencies; not the other way around

Rather, the business model should *decide* the set of core competencies that are needed and decide which core competencies need to be *developed* to support the business model. If the company wants to enjoy the benefits of the BTO&MC (Ch. 14), then it will have to make *conscious* decisions about core competencies that optimize the degree of outsourcing and internal integration .

Outsourcing. For the past several years, many companies have been outsourcing more and more of their operations in order to focus on "core competencies," such as research, development, sales, and only the final assembly aspect of manufacturing. Some companies have even outsourced all of their manufacturing.

In the last several years, much of part manufacture and material processing was outsourced to "specialists" who were assumed to be "better at it," thinking that this would save cost and increase profits. Some companies have outsourced production just to minimize the money tied up in capital equipment to try to maximize return on capital and stockholder value.

Outsourcing only *appears* to save money

The illusion of outsourcing saving "cost" will be shattered in Chapter 6, when companies measure total cost (Ch. 12) and consider the value of responsiveness, customization, and the "opportunity costs" not lost with an advanced business model.

Unfortunately, too much outsourcing has complicated the supply chain with more "links in the chain," more transfers, and longer lead-times, especially when there are sequential "tiers" through which materials and parts must pass on the way to the assembler. In the slow, stable days of mass production, this worked all right, since assembly could be safely scheduled in advance and there was plenty of time to order and deliver batches of parts from outsourcers. But now markets are changing so fast that the most competitive companies will be building products on-demand with their parts and materials delivered quickly on-demand. Thus, the long-lead times created by far-flung supply chains will no longer be competitive. Further, outsourcing is at odds with the inventory-less aspect of build-to-order, since outsourcing is usually a batch operation for manufacturing *and* shipping.

Far-flung supply chains slow down responsiveness

Outsourcing high-volatility parts can cause *product* shortages because contract manufacturers may not be flexible enough to rapidly increase production. Theoretically, an outsourcer could set up another general-purpose line to meet demand spikes, but only if the order was large enough to justify the conversion and other lines were not already committed to, and set up for, other jobs. In contrast, flexible factories could shift demand to other lines without such large step-function inhibitions.

In many cases, excessive outsourcing makes it hard for manufacturers to control costs and optimize profits which often drifts away from assemblers to suppliers.

Chapter 6 discusses the detrimental effect outsourcing has on total cost, true profitability, speed, quality, product development and leading edge processing. For these reasons, most outsourcing is not supportive of the BTO&MC business model.

> Separating Engineering from Manufacturing thwarts good *Concurrent Engineering*

Outsourcing manufacturing usually results in separating it geographically from product development engineers. This can thwart *design for manufacturability* and *concurrent engineering* and compromise product development success.

Great distances in supply chains slowed down the flow of parts and products. Womack and Jones, writing in *Lean Thinking*[4] summarize it succinctly: "Oceans and lean production are not compatible."

> Oceans and lean production are not compatible
> – Womack & Jones, *Lean Thinking*

However, there are appropriate uses for local outsourcing for the BTO&MC company: oddball products, standard parts, in-house customer-specific lines, or where special skills, talent, or costly equipment are required. Outsourced parts and subassemblies should be modular with clean interfaces and easy integration. Outsourced suppliers should have a cooperative working relationship, especially with respect to on-demand deliveries. Suppliers should be close enough to be able to quickly supply parts on-demand.

> BTO&MC companies outsource differently

Offshoring "To Save Cost." For this discussion, *offshoring* is defined as designing products and selling them in the home country but manufacturing them on another continent (or far away) in an effort to save cost.

Many companies are moving manufacturing operations overseas or to foreign countries because they *think* it will save cost. And their primitive cost systems make such a move appear to be justified. If all they quantify is parts and labor, then moving to a "low labor

> Offshoring to "save cost" rarely does based on total cost

rate" country labor will appear to lower labor cost. Since no other costs are quantified, it is a "case closed" decision.

However, when measured on a *total cost* basis, offshoring rarely results in a net cost savings. Further, greater distance between headquarters and engineers compromises control and concurrent engineering, while greater shipping distances shows responsiveness and makes it harder to implement build-to-order and mass customization.

> Overseas production poses major challenges for delivery, quality, innovation, control, and strategic focus

As with outsourced overseas production, company owned factories far from customers slows delivery to customers and makes BTO&MC much more difficult, maybe impossible, to implement. Quality may suffer if the new plant has not established an effective quality culture. Quality may also suffer if recurring defects are produced overseas and not detected until hundreds of defective products are discovered at the end of the long trans-oceanic "pipeline."

The author's book-length web-site, *www.HalfCostProducts.com* presents eight powerful cost reduction strategies, which are summarized in Chapter 6, along with the sobering conclusion that offshoring *thwarts, compromises, or inhibits six of the eight cost reduction strategies.*

Outsourcing and Quality. In order to produce high quality products, the quality of parts and materials must be assured *throughout the entire supply chain.* Internal production is under the direct control and influence of the manufacturer. And all the aspects of quality can be assured as part of a holistic approach. Under these conditions, Six Sigma programs have the best chance of achieving six sigma quality levels. On the other hand, it is often hard to consistently assure the quality of outsourced production.

> Many product quality problems are caused by outsourced parts

Outsourcing and Manufacturability. Many companies have drawn an arbitrary line between "us" and "them," simplistically defining their companies as assemblers and their suppliers as part manufacturers. However, such arbitrary demarcations can significantly raise cost by eliminating opportunities to optimize manufacturability. Outsourcing itself may require excessive modularization with more interface problems and preclude opportunities to fabricate monolithic structures on-site.

What *Not* to Outsource. Companies that selectively outsource must decide which products to outsource, but *most companies outsource the wrong products and parts.*

Many companies outsource their newest products and the cash cows – even the "crown jewels" – while retaining older legacy products, spare parts, and oddballs, which, ironically, are the ones that consume excessive resources and make the least profit on a total cost basis.

Most companies outsource the wrong products & parts

What *Should* Be Outsourced. Build-to-order and mass customization companies have different criteria for outsourcing. BTO&MC companies benefit from outsourcing older legacy products, spare parts, and oddballs.

Integration. Many articles and books have proclaimed vertical integration as "out" (as being so twentieth century!) and *virtual integration* as "in" for the twenty first century.[5] Books on about virtual integration define the *virtual organization* as "an opportunity-pulled and opportunity-defined integration of core competencies distributed among a number of real organizations."[6] But such a distribution can: (1) slow down the supply chain if suppliers are too distant or build outsourced products in batches, and (2) add unnecessary shipping, inventory, and other overhead costs, as discussed above.

The supply chain requirements for spontaneous BTO&MC are that *all* the parts and materials must be made available spontaneously. In order to accomplish this, the supply chain must be *tightened* by (a) arranging rapid on-demand delivery of parts and materials from suppliers or (b) bringing in-house any processes that have lead times or batch sizes incompatible with assembly responsiveness and inventory goals. Material cutting-to-shape, discussed in Chapter 5, may be one of these processes.

The supplier base needs to be narrowed to those who can make parts to the "pull signals" of the assemblers. In fact, product line consolidation (Ch. 3) will help shrink the supplier base by eliminating or outsourcing the most unusual products with the most unusual parts and suppliers. However, suppliers may not be suitable if (1) they are too far away for adequate delivery speed and frequency, (2) they are too entrenched in batch mass production, or (3) they are unwilling to make the necessary commitments and changes.

If it is not possible to arrange suitable suppliers, certain operations may need to be selectively reintegrated in-house to eliminate order/delivery lead-times and permit batch-size-of-one flexibility. This would greatly simplify or eliminate the MRP-based purchasing normally needed to procure these parts and materials from a diverse supply base. Further, in-house production of parts on-demand can immediately feed assembly lines.

Thinking about "core competencies" will have to be revised

Old decisions about "core competencies" will have to be revised to include supply chain responsiveness and operational flexibility as the basis of a *new* definition of core competencies. In some cases, long lead-time or large batch part production will have to be brought in (and streamlined) just to "complete the system," even if it is not considered a core competency.

Internal integration should be seriously considered when it is important for the assembler to understand, measure, and effectively control costs, quality, and delivery time. Integration would also be recommended where there is any significant variety, interface complexity, processing innovation, large hard-to-ship parts, or special/flexible tooling needs to be developed.

Integration give control over *all costs*

Globalization. The role of distance has a crippling effect on customer responsiveness at the distribution end of the supply chain. Even if the materials end of the supply chain is tightly coupled with close proximity and one-piece flow based on pull signals, the operation will probably not be able to provide the optimal customer responsiveness if finished products have to be shipped across oceans. Shipping delays will cause a significant competitive disadvantage for any company trying to take advantage of the strengths of BTO&MC. This is especially true for markets that

BTO&MC favors local production for local consumption

require fast responses: rapid replenishment of stores and dealers; individually mass-customized goods; internet sales; and on-demand part manufacturer for BTO assemblers.

As BTO&MC grows, exporting should diminish in both directions

Build-to-order and mass customization will favor local procurement, local production, and local distribution to ensure speed throughout the entire supply and distribution chains. And, as pointed out through this book, supply chain speed is what allows the BTO&MC company to build products on-demand and ship them to customers or stores as fast as the batch manufacturer can ship forecasted inventory, but without the costs of inventory or out-of-stock problems.

Thus, as build-to-order and mass customization grow, exporting should diminish in both directions. In developed countries, the mass marketers who mass-produced products overseas will gradually evolve to or be replaced by the BTO&MC companies who build mass-customized and standard products on-demand for quick shipment to domestic customers.

"Cheap labor" outsourcing countries will evolve to the dignified role of making their own goods on-demand

Developing countries will be able to raise themselves above the low status of being the "cheap labor" provider with all of its problems of workplace abuses, "sweat shop" allegations, and population dislocations. In its place, developing countries will to evolve to the dignified status of designing products suitable for their own cultures, building them in local plants by a proud, well-paid local workforce, and distributing customized or standard products quickly to local consumers. And in both developed and developing countries, companies, and ultimately consumers, will benefit from avoiding the rising costs of transoceanic shipping, inventory, obsolescence, tariffs, and duties plus minimizing macroeconomic problems of international currency fluctuations and balance of trade deficits.

PART III:SPONTANEOUS BUILD-TO-ORDER & MASS CUSTOMIZATION

Spontaneous Supply Chain (Chapter 7)

Spontaneous supply chains play a key role in BTO&MC to avoid the need to generate forecasts, count inventory on-hand, generate purchase order inputs through MRP (Material Requirements Planning) systems, place purchase orders, wait for parts to arrive, expedite those that are late, receive (and maybe inspect) materials, warehouse, group into kits for scheduled production, and distribute within the plant. BTO&MC can avoid all these costly and time-consuming steps with a *spontaneous supply chain,* which is able to *pull* in materials and parts on-demand. It is highly unlikely that this can be accomplished merely by asking existing supply chains to deliver all of your existing parts on-demand.

The concepts presented herein are labeled as the *resupply* of parts and materials as opposed to *procurement* or *purchasing.* This is to emphasize that most of supply chain management is a *resupply* function of parts and materials that have been procured before.

Automatic resupply of materials. If rationalization and standardization are successful in reducing material variety, then much, possibly all, of the procurement can be automatic, *based on demand,* without the need of forecasts, MRP, purchase orders, waiting, receiving, logging-in, warehousing, kitting, re-ordering, and so forth. Raw materials could be automatically resupplied using the following techniques (in order if increasing variety):

Steady Flow of Standard Raw Materials. The ultimate scenario for spontaneous BTO is to reduce the number of raw material types within each category to one, in which case steady flows can be arranged for each standard raw material. Ideally, there should only be only one type of *each* material. Then forecasting multiple types would be unnecessary and "ordering" would be as simple as matching the tonnage in to the tonnage out; in other words, the incoming flow of the standard raw materials would be equal to the monthly consumption of the plant. This steady flow can be arranged with confidence since the

> The ultimate resupply scenario is a steady flow of one standard material for each category

standard materials be used one way or another. Multiple types of materials in each category would allow the same spontaneity if material changeover is quick (and hopefully automatic) and the ratio is constant or predictable.

Material Cut-to-Length/Shape. Raw material can be cut to length or shape on-demand from the longest version or standard sizes by programmable CNC equipment, such as laser cutters and screw machines, by single axis programmable cut-off machines, or from less-automated tools based on on-line instructions (Ch. 5).

> Cut to shape on-demand, instead of ordering every possible shape

Min/Max Resupply. The "min/max" technique is an effective way to automatically resupply stacks of raw material like sheet metal, where material is consumed until the stack reaches the "min" level, usually marked on the rack or wall. This triggers a re-order of the material to bring it up the "max" level.

> When materials get to the "min" level, reorder up to the "max" level

Strategic Stockpiles. Until the above techniques can be implemented, it may be necessary to have strategic stockpiles of certain materials. The manufacturer could use *selective* stockpiles to temporarily compensate for any parts or materials that cannot be "pulled" or for temporary availability problems on standard materials. Stockpile ordering would have to be based on some kind of forecasts, but if the material was standardized, then the forecast would be easier to make for the *aggregated* demand for all consumption.

Order Material After Receipt of Product Order. Spontaneous resupply may not be feasible for unusual or seldom-used raw materials, especially on capital equipment with inherently high diversity of materials. If material order times are less than build times, these materials could be ordered after receipt of the product order.

Automatic resupply of parts

The typical response when suppliers are asked to deliver parts just-in-time to their customers' pull signals is to keep building the parts in large batches, try to stock enough in their finished goods inventory, and meter them out "on-demand." A special-case variation of this approach is the Dell model (see Introduction) where suppliers warehouse their parts next to the assembler's factories. Not many assemblers are big enough and powerful enough to force their suppliers into such an arrangement.

When asking to pull parts on-demand, most suppliers will try to dish them out from inventory despite all the problems

However, this is not really a pull-based supply chain and it is not conducive to spontaneous build-to-order. Parts availability depends on assemblers' forecasts, which are becoming increasingly less accurate, and the supplier's inventory, which is costly to carry and prone to obsolescence.

Parts could be automatically resupplied by the following techniques (listed in order, with the easiest first):

Steady Flow of Parts. As with standard materials, steady flows could be arranged for *very* standard parts, which would be used one way or another. The criteria for steady flow of parts would be standardization and widespread use.

Breadtruck. The easiest and "lowest hanging fruit" in resupply logistics is the *breadtruck* (sometimes called "free-stock") delivery system for small, inexpensive parts, like fasteners. Instead of counting on forecasts to trigger an MRP system to generate purchase orders, all the "jelly bean" parts can be made available in bins at *all* the points of use. A local supplier is contracted to simply keep the bins full and bill the company monthly for what has been used, much like the way bread is resupplied by the breadtruck to a market.

Breadtruck suppliers simply keep the bins full and bill for whateveris consumed. Thus, no overhead.

Kanban is a versatile technique that enables automatic resupply of parts that can be made in batches or have not-quite-spontaneous delivery times. In *kanban* resupply, parts with limited variety are made, maybe in batches, and resupplied automatically to replenish parts bins based on part consumption. This is one of the many *pull systems* used to "pull" parts into assembly operations. The resupply is automatic after "pull signals" are transmitted to suppliers. There are many simple ways to do this without complex information systems such as MRP or ERP. Thus, kanban resupply avoids the uncertainly of forecasting, the cost of purchasing, and the cost and risk of inventory.

> *Kanban* can automatically resupply standard parts, even parts made in batches

Kanban works best for semi-standard parts without too much variety, which would increase work-in-process (WIP) inventory and clutter assembly stations with too many part bins. Kanban parts can be made in mass produced batches, thus accommodating batch suppliers. Of course, the parts manufacturers may have to some implement setup and batch size reduction to be able to economically make batches small enough for kanban deliveries. Kanban is described in Figure 7-1 and the nearby discussion.

Parts Made On-Demand by Suppliers. Hopefully, it may be possible to find suppliers who can implement these techniques to make your parts on-demand in response to your pull signals. Pull signals need to be initiated early enough and response time needs to be quick enough so that parts arrive without causing assembly delays.

Spontaneous BTO of parts may require the development of supplier-partner relationships in which suppliers establish the ability to build parts in any quantity on-demand. The distance to the supplier must not be so great that part delivery is too slow for adequate on-demand assembly. Be wary of suppliers that are "pulling" parts from inventory because of the risk of not having enough (which they may blame on *your* inadequate forecasts) and higher than necessary cost for inventory carrying costs.

Parts Made On-Demand In-House. In order for spontaneous build-to-order to work, *all* parts must be available on-demand. If there are any key parts that are not suitable for kanban and no supplier can build them *and ship them quickly enough* to your pull signal, then you might have to bring those operations in-house, as discussed in Chapter 6 on selective vertical integration.

Parts may need to be built on-demand using BTO&MC techniques

Flexible Processing. Regardless of the sources of parts, spontaneous part manufacturing operations must be able to make parts on-demand efficiently in a batch-size-of-one mode without setup or inventory. CNC programmable machine tools and flexible assembly can produce a high variety of parts without setup from standard raw materials, as will be shown in the next chapter. Similarly, manual assembly can be made flexible as will also be discussed in Chapter 8. This may require concurrent engineering of product families, parts, and processing to eliminate all setup changes, as will be discussed in Chapter 10.

Strategic Stockpiles. Until the above techniques can be implemented, it may be necessary to have strategic stockpiles of certain parts. The assembler could use *selective* stockpiles to temporarily compensate for any aspect of the supply chain that cannot be pulled or for temporary availability problems on standard parts. Make sure that these parts are common enough to be readily available.

Order Parts After Receipt of Product Order. Spontaneous resupply can may not be feasible for unusual or seldom-used parts, especially on capital equipment with an inherently high diversity of parts. If parts order times are less than build times, these parts could be ordered after receipt of the product order.

Supplier Lead-time Reduction. As BTO&MC implementations continue to decrease the product build time, there will come a time when long-lead part delivery times will become the critical path and start to delay product shipments. *Before* this happens, the process of shortening supplier lead times needs to begin. Identify parts with long lead times. Explore

Shorten supplier lead times *before* they become the critical path ways to minimize the lead time. Investigate and understand reasons for long supplier lead times. Negotiate to improve lead-times. Look for faster suppliers. Work with suppliers to improve their lead times. Encourage suppliers to embark on lean production, build-to-order, and mass customization programs. Send a team to train suppliers or pay for outside experts to train them and help them implement BTO&MC. The resulting improvements to supply chain will be well worth the effort.

Dock to Line Deliveries. To be truly agile, incoming parts and materials must flow directly to the points of use, called *dock-to-line,* instead of going through the traditional maze, which includes the receiving department (where they are logged in), the incoming quality control (IQC) department (where they are inspected), the raw material warehouse (where they are inventoried), and the kitting department (where they are counted and grouped into batches), and kit distribution to all points of use.

> *Dock-to-Line* means deliveries go directly to all points of use

The prerequisites for dock-to-line are standardization and suppliers who can *assure quality at the source,* since dock-to-line parts go directly to production lines without going through the delays of incoming inspections.

Low-bidding for parts. Going for the low-bidding is something we management consultants have been trying to discourage for years, but now it has seen a resurgence just because it is easy to do on the internet. This started when many purchasing functions are under heavy pressure to "use the net" and "get into e-commerce." [7]

> Low-bidding finds the cheapest parts – *and that is the problem*

Bidding for the cheapest parts is not only an ineffective way to achieve *real* cost reduction, but it can substantially raise less-obvious costs and compromise other important goals like quality and delivery, which are especially important for BTO&MC. Suppliers who "win" bidding wars and do not

> Cheap parts usually come at the expense of quality, deliver, and flexibility

have cost-effective operations face the dilemma of cutting margins, "cutting corners," or beating up *their* suppliers, which, in turn, may also be similarly ineffective with respect to cost. Further, when an assembler focuses cost reduction efforts on part bidding, it distracts attention from *real* cost reduction opportunities, which are addressed throughout this book.

The Cost Reduction Illusion. Before manufacturers fall for a magic elixir, they should consider how part costs really would lowered under such pressure. One assumption is that either purchasing agents have naively offered to pay too much or that cavalier part makers have been gouging their customers. While this may have been true in sleepy industries of the past, it is rarely true in today's dynamic marketplaces.

Another assumption is that inefficiencies can be corrected by pressure after a supplier "wins" a contract at a lower-than-usual price. However, soon after a supplier wins a bid, it is expected to deliver the goods, and there will not be time to implement any meaningful cost reduction program. Thus, without a real means to lower costs, the supplier will either have to cut its margins (which will be resisted from the corner office all the way to Wall Street), cut corners , or do the same to thing to *its* suppliers, who will have the same difficulty achieving real cost reductions

> Without a real way to lower cost, bidders will either cut margins or cut corners

Cheap Parts – Save Now; Pay Later. Actually, this phrase should be, more precisely: *save a little now, pay a lot later.* Many times trying to save money on purchase cost has the unintended effect of driving up other costs many times the assumed savings, like the old English adage: *penny wise, pound foolish,* or the more colloquial *"you get what you pay for."*

Cheap parts usually have much higher quality costs and slower delivery, which raises material overhead cost and may thwart the spontaneity of build-to-order. Further, low-quality parts need incoming inspection, thus causing the assembler to miss out on the benefits of dock-to-line deliveries.

> Cheap parts can cause an explosion of other costs and problems

But, since part cost is a small percentage of *selling* price (Ch. 11), there is the compelling argument that other cost categories provide much greater opportunities for real cost reduction, as is emphasized throughout this book.

The Value of Relationships for Cost Reduction. In BTO&MC much of the real cost reduction opportunities are not just at the assembler or at the supplier, but rather *in their relationship*. The book that started the lean production movement, *The Machine That Changed the World,* notes that in lean production companies, suppliers "are not selected on the basis of bids, but rather on the basis of past relationships and a proven record of performance."[8] Honda's criteria for selecting suppliers is the *attitudes* of their management.[9]

> Suppliers should be selected on the basis of past relationships and a proven record of performance, not bids

Such inter-company cooperation offers significant cost reduction opportunities, for instance, if suppliers can build parts on-demand for BTO assemblers; then *both* avoid all the cost and risk of parts inventory in addition to minimizing many categories of overhead for procurement, material overhead, expediting, warehousing, internal distribution, and so forth.

Good relationships with suppliers are also very valuable for *concurrent engineering,* in which cooperative suppliers help manufacturers develop their products for the optimal manufacturability.

> Create relationships that work together to improve efficiency

However, switching suppliers every time a competitor drops its price is incompatible with these strategies and can jeopardize ongoing relationships and all their benefits.

On-Demand Lean Production (Chapter 8)

The ability to build mass-customized and standard products on-demand is the payoff for lean production programs. Lean production[10] has become a very popular program, but lean production efforts are usually confined to within Manufacturing Departments and existing MRP-based build-to-forecast scenarios.

> Building mass-customized and standard products on-demand is the payoff for lean production

On-demand lean production depends on a simplified, *spontaneous* supply chain (as discussed above) that can deliver parts and materials spontaneously without forecasts, purchase orders, part delivery delays, or warehoused inventory of parts and materials. When this supply chain is coupled to flexible batch-size-of-one lean production, the result is on-demand lean production.

In the phrase *batch-size-of-one*, the quantity "one" refers to the smallest order quantity anticipated, which could be one product or one case or one pallet. Batch-size-of-one capability means that setup delays have been eliminated or reduced to the point that production facilities can *efficiently* build any size order on-demand without having to batch orders together to spread out the setup delays.

Problems with Setup. Mass Production deals with setup by accepting it as a "necessary evil" and then spreads it over as many products as possible in *batches* or *lots* to minimize the set-up charge per part. For decades, industrial engineers have used formulas to try to calculate the "Economic Order Quantity" (EOQ). However, manufacturing in batches drastically raises costs and lead times because of its effect on space, throughput times, WIP inventory, recurring defects, flexibility, and disruptions caused by "rush" jobs interrupting scheduled production. For a treatise about the shortcomings of mass production, see Chapter 2.

> Mass production accepts setup as a "necessary evil" and wants to spread it over large batches

Setup & Batch Elimination. If successive products are to be unique and different, there cannot be *any* significant setup delays to get parts, change dies and fixtures, download programs, find instructions, or any kind of manual measurements, adjusting settings, or positioning of parts or fixtures. For a plant to mass-customize or spontaneously build products to-order, *all production setup must be eliminated,* not just the low-hanging fruit or "much as you can." Setup "elimination" is defined as reduced to the point where it is still *feasible* to operate efficiently in a batch-size-of-one mode.

One way to eliminate setup is to manufacture products on dedicated lines. High-volume mass production lines operate best when dedicated so no setup changes are required. BTO&MC factories also use this principle for medium-volume lines, where smaller, less-expensive machines are grouped into product family lines so they can operate without setup changes.

If parts are standardized, they can be distributed to all points of use, thus avoiding kitting

Part setup can be eliminated by eliminating kitting and distributing all parts at all points of use. Positioning setup can be eliminated by versatile jigs and fixtures. Tooling setup can be eliminated with universal tooling, rapid die changes, or quick-change inserts. Machining setup can be eliminated with optimal utilization of CNC machine tools. Programming setup can be eliminated by downloading machine tool programs on-demand or generating them on-the-fly. Process variable setup can be eliminated by standardizing on process variables. Manual setup to find and understand instructions can be eliminated by displaying instructions on on-line monitors.

Tooling setup. Design the product/processes to eliminate the need for tooling changes for cutting tools, dies, molds, tool plates, and fixtures. Design tool plates, jigs, and fixtures to be versatile enough to accept all parts in the family without having to change tool plates, jigs, and fixtures. The parts or standardized raw material must be able to be quickly positioned in the fixture without any need for manual positioning or measuring. Parts may need to be designed with common fixturing geometries.

Change tooling quickly or eliminate the need to have to change it

If that is not possible, develop ways to change tooling rapidly. Clever, universal die and mold mounting geometries can be developed to facilitate quick changeovers. Conveyors and carousels, which were first applied to moving parts and products, are now being applied to moving dies quickly in and out of presses and molding machines.

CNC to Eliminate Machining Setup. CNC machine tools are very versatile tools to eliminate setup since many operations can be done by multi-axis machine tools without having to reposition the workpiece. CNC machining centers can perform a wide range of operations such as machining, drilling, tapping, and so forth. The more operations that can be done in one operation, the fewer times the workpiece needs to be moved and set up. In fact, a key DFM and quality principle to ensure tight tolerances is to make sure all critical dimensions are cut in the same machine tool in the same clamping.

> Doing all operations in one versatile CNC eliminates a lot of moving and setups

Process Variable Setup. Develop processes, concurrently with product development as appropriate, to standardize on the variables that are the most time-consuming to change. For heat treating, oven brazing, or baking, standardize on furnace or oven temperature, since that would cause a long setup change to change; instead, vary the speed through a furnace or, better, design all products in the family to use the same processing. Standardizing process variables also eliminates the chance of using the wrong setting. This is an example of the Japanese quality assurance concept of *poke-yoke,* or mistake proofing.

> Eliminate process setup by standardizing on the hardest to change variables

Dedicated Lines. Medium-volume products can present a dilemma: the volume may not seem not high enough to justify their own dedicated mass production line, so they are run together with other medium-volume products or stuffed in to "fill" the high-volume line, with setup changes between every product change. The solution would be to eliminate setup by creating small enough lines that can be dedicated to a product family or group of parts, which would never have to be changed over.

> Dedicated lines of inexpensive machines have no setup

Since these smaller dedicated lines do not need the capacity of high-volume lines, they can be equipped with smaller *and less expensive* machines, either "low-end" new equipment or used machinery, which can be procured at a fraction of its original cost.

Flow Manufacturing. If setup can be eliminated or reduced enough to eliminate the need to manufacture in batches, then parts, sub-assemblies, and products can flow one piece at a time. One-piece flow assures the throughput and flexibility needed for BTO&MC operations. U-shaped lines can facilitate this feedback and enhance *visual control* while making more people available to help colleagues nearby.

> Without setups, waiting, and batches, production can *flow*

Assuring Quality with One-Piece Flow. One-piece flow has a distinct advantage for assuring quality. First, flow manufacturing eliminates the possibility that recurring defects may be built into several batches before being caught at a downstream inspection step (see "Defects by the Batch" in Chapter 2). Second, people working in flow manufacturing look for any visible deviation as each part is handed to "its customer" (the next station). Further, if the part doesn't fit or work in the next operation, the feedback will be immediate leading to quick resolution of the problem.

> Flow provides instant feedback and prevents "defects by the batch"

Cellular Manufacture. Flexible operations work best with dedicated cells or lines for every product family. Cells can be permanently configured so that within a product family, all setup has been eliminated. This strategy work best with many simpler dedicated machines instead of a single "mega-machine," unless the mega-machine can handle a very large family – enough to justify its expense.

> Cells are structured to make a product family without setup changes

Leveling production. Flow manufacturing works best with a fairly constant or *level* production output. Critics of one-piece flow cite demand fluctuations as justification for building lots of inventory. However most irregularities in factory workload are *artificially* induced by the following factors *that can be eliminated:* the "business cycle" (building up or working off inventories); production goals and quotas (for end of the month, quarter, or year); overstock promotions; capricious sales promotions not coordinated with production capability; quantity discounting; deal making; and lack of confidence in product availability (causing customers or dealers to "stock up").

> Most irregularities in factory workload are artificially induced
> – Womack & Jones, *Lean Thinking*

Seasonal Products. Spontaneous build-to-order techniques can accommodate seasonal demands better than stockpiling finished goods inventory, with all the associated inventory cost and risk of building too many or too few products. The first effort would be to try to level demand with off-season incentives based on actual differences in total cost. For the remaining seasonal variations, seasonal products can be manufactured on-demand with a minimum of part inventory and risk using BTO&MC principles. Several general techniques are presented in Chapter 8.

Capacity Challenges. Leveling production will, in itself, reduce factory demands for capacity and all the associated expenses in capital and resources. In addition, build-to-order and mass customization require less equipment for the same plant *output* capacity because of the elimination of setup and associated machine downtime. Peak capacity needs are also minimized if production can be shifted between flexible lines, thus eliminating the need to size *every* production line for *its* peak demand. This capability will also avoid the dilemma faced by many companies of having to close some plants but not being able to build enough hot models.

> Eliminating setup and delays raises utilization which, in turn, raises capacity

Long-term capacity challenges can be eased by permanently outsourcing the *least* efficient or *least* compatible products and operations.

When capacity does need to be expanded, it can be done cost-effectively by creating dedicated lines for medium-volume products by procuring low-end or used equipment, as discussed in Chapter 12. Removing these products and their setups from the expensive lines will increase their utilization and make it more likely that they will also be able to run with fewer setup changes.

Results of Setup Elimination/Batch-Size-of-One Flow. Eliminating setup itself can decrease throughput time and eliminate setup delays on expensive equipment. In addition, batch-size-of-one flexibility can allow *dock-to-line* part delivery, thus eliminating kitting, inventory carrying costs of raw parts inventory and the cost, and delays of incoming inspections.

In addition, one-piece flow can improve quality with rapid feedback to catch and rectify quality problems fast, eliminate fork lifts (including the labor, equipment, and floor space for the aisles), and foster psychological flow, improve job satisfaction, relieve boredom, and encourage continuous improvement.

In addition, eliminating WIP inventory can eliminate WIP inventory carrying cost, which is equal to 25% of value of inventory per year (Ch. 2), and cut floor space needs in half.[12, 13, 14] This is especially important in times of growth, but floor space savings should *always* be assigned a value to encourage more efficient utilization of space.

Build-to-Order Strategy. The overall mass customization strategy is to: *standardize all that you can and build or procure the rest on-demand.*

Standard materials arrive in steady flows (not forecast-based purchasing) and standard parts are available on-demand through kanban, min/max, and breadtruck automatic resupply techniques.

For everything else that *can't* be standardized, build the remaining parts on demand (from standard raw materials) and assemble with flexible automation or manual assembly stations feed by kanban bins and on-line instructions (Figure 8-3).

How BTO&MC Work. Chapter 8 concludes with three illustrated examples that show how BTO&MC factories operate for fabricated products (Figure 8-1), electronic products (Figure 8-2), and manual assembly utilizing *kanban* resupply and on-line instructions (Figure 8-3). These examples are very "do-able" in that they use readily available CNC machine tools so they do not depend on high volume or special automation. They even show how to use manual assembly to build mass customized or standard products on-demand.

Mass Customization (Chapter 9)

Mass Customization is the ability to quickly and efficiently build-to-order customized products. It uses all the techniques presented so far for the build-to-order of *standard* products and extends that to *custom* products. These products can be customized for individual customers or niche markets, such as versions optimized for certain market segments, industries, regions, or countries.

> Mass customization quickly and efficiently builds-to-order customized products

Mass customization has not yet reached a critical mass in industry because: (1) there are very few benchmark companies doing it well, (2) most of these companies have only *experimented* with mass customization and not extended these principles to their entire line of standard products, (3) writers of books and articles don't understand enough about manufacturing and supply chains to tell readers how to implement it.

Real mass customization is actually *manufacturing* custom products (not just assembling modules) quickly and efficiently in quantities as low as one.

The Buzz about Mass Customization. There has been a lot of hype lately about mass customization, but much of it is just *customization.* Mass customization is *not* internet sales of *craft* products that cost more than standard products and take weeks to build.

Many journalists and authors jump at the chance to say something about the latest management trends, but fail to comprehend the potential of mass customization because they don't understand the principles of on-demand production and spontaneous supply chains as uniquely presented in this book. Most writers still limit their view of mass customization as merely assembling *modules* to customer orders, although they are unable to explain how these modules can be always available for on-demand assembly and how materials can be always available for on-demand module production.

> Mass customization is much more than bolting modules together

The worst aspect about the buzz about mass customization are the misconceptions, ranging from PR to over-hyped implications that mass customization can build any odd-ball variation without any limits or constraints. In reality, mass customization can quickly and efficiently build any variation *within the capabilities that the mass customization system was established to build them.*

> Misconceptions range from buzz-word PR to over-hyping that MC can customize anything

How to mass customize. There are three ways to mass customize products: *modular, adjustable,* and *dimensional* customization. The most obvious way, modular customization, can customize a product by assembling various combinations of modules. Adjustable customization is a *reversible* way to customize a product, as in mechanical or electrical adjustments. Dimensional customization involves a *permanent* cutting-to-fit, mixing, or tailoring. The key to mass customization success is to optimize the combination of all these techniques, not just the obvious modularity. This optimization comes from a thorough

> There are 3 ways to mass customize: modular, adjustable, and dimensional

understanding of these techniques which are discussed at length in Chapter 9.

Postponement. Postponement is a mass customization technique that is applicable for certain products that can have their variety postponed until just before shipping. The factory builds basic "vanilla" platforms and adds "flavors" upon receipt of order. Postponement is most suitable for product architecture that has a major platform part can be built without variation and then customized by various adjustments, configurations, or bolt-on modules.

What to mass customize. Mass customization can efficiently and quickly build vast families of synergistic products, *but mass customization may not be able to build all your existing products.* In fact, it probably won't, especially if the product line has accumulated products for decades without ever having been rationalized (Ch. 3) to eliminate the most unusual products with the most unusual parts, materials, and processing.

MC can customize efficiently within limits, but not beyond the limits

The determination of which products to mass customize must be based on a combination of market needs *and* organizational capabilities. The common temptation is to only consider the market needs and offer to "customize anything," which may be so hard to do that operations revert to the inefficiencies and delays of craft production or frequent setup changes in a mass production environment. It is important to remember that mass customization operations can customize a certain range of products very efficiently, but just outside that range it may be difficult and beyond that it may not be feasible at all.

Extra Value-Added Opportunities. Mass customizers can use their versatile design and manufacturing capabilities to offer new, related, value-added products/services that expand the scope of their markets. These offerings may cost little extra to add, especially if they can be done in existing CNC operations. And yet, they may save customers so

Extra operations that are easy for you to do and valuable to customers = *high profit opportunities* much time and money that they would gladly pay well for the options, especially if these features are difficult for customers to make.

Order Entry. The main difference between build-to-order and mass customization is the order entry. Whereas build-to-order order entry is based on *standard* product model numbers or sizes, mass customized products are at least partially specified by some customization data or dimensions.

For simple customizations, these can be obtained from answers to salesperson queries, filling in forms on web-sites, or simply filling out paper forms. More complex customization can use software *configurators,* which keep track of all the options and features and all the rules that apply to their selection.

Marketing BTO and Mass Customization. The most important principle for the marketing of mass customization is that *all sales are not equal.* Product families that are structured according to the principles of this book will have the ability to satisfy customers with exactly what they want then they want it at the lowest cost on the market. By contrast, products outside these families would be nightmares for a BTO&MC company to produce and, further, would distract it from making the profitable product with the highest potential growth potential. Thus, as said many times, the sales force must resist all temptation to "take all orders." Mass customization offers great growth potential (Ch. 14), but only if done right.

> All sales are not equal.
> Some are dreams;
> Others nightmares.

When shifting to mass customization, Marketing will expand some markets (the BTO&MC products) while eliminating losers and outsourcing products that need to be in the catalog, but do not fit into the new operations (Ch. 3). BTO&MC products will have lower cost and much faster delivery, while outsourced products will have the original slow delivery and high cost.

> Expand the focus
> on MC products;
> Phase out the losers.

The biggest caution about mass customization is to make sure orders don't exceed the capability of the system. Some companies become victims of their own publicity if they oversell mass customization capabilities to the point that customers develop unrealistic expectation and salespeople think they can customize anything. It is important to keep emphasizing that a mass customization plant can only build products within its capabilities, which can be expanded to efficiently handle more variety, but it is imperative that current orders are *always* within the capabilities of the system. It may be necessary to find a way to handle oddball requests such as outsourcing to a versatile vendor (Chapters 3 and 6). Every mass customization plant must remember the "98% solution:" *Mass can quickly and efficiently make 98% of the products so don't dilute it with the other 2%.*

Learning relationships. Mass customization can yield great marketing benefits by developing *learning relationships*[15] where companies learn and adapt from each order, thus satisfying customers better on each order and progressively developing more committed customers.

Combined Mass Customization and BTO. There is a natural synergy between build-to-order and mass customization, hence the title of this book. They share the same batch-size-of-one operations and spontaneous supply chain.

Build-to-order and mass customization operations are equally efficient and very compatible, unlike situations where a mass customization experiment must be run separately from the "batch-and-queue" operations of mass production, or worse, trying to merge these inherehtly incopmatible paradigms.

> There is a natural synergy between build-to-order and mass customization

Building BTO and MC products on the same lines will often push the combined volume over the "critical mass" threshold necessary to justify these implementations. Higher total volumes makes it more likely to split production into more lines or cells, each of which would be more specialized with a smaller range and thus would be easier to establish. Thus, combined operations are more like to be approved and succeed than either attempted independently.

Once a viable mass customization operation is established, customization within the system will be free.

> Once MC is in place, *customization is free*

Reducing Cost & Weight with Mass Customization. In addition to cost-effectively building *custom* products, Mass Customization principles can also be applied to significantly lower the cost and weight of assemblies by enabling the efficient manufacture of "efficient structures" by mass customizing the structurally ideal shape for every part, instead of assembling mass produced standard parts, thus resulting in a suboptimal structure.

PART IV: IMPLEMENTATION

Concurrent Design of Versatile Products and Flexible Production (Chapter 10)

Challenges with existing products. Existing product designs may not be suitable for spontaneous BTO or mass customization. Products not concurrently engineered for flexible environments may impede implementations, diminish the payback, or even thwart success entirely. The product portfolio may have too many unrelated products that lack any synergy and, thus, too many different parts and processes. Even within a focused product portfolio, there may be a needless and crippling proliferation of parts and materials. The specified parts may be hard to get quickly. The products and processes may have too many setups designed in. Quality may not be designed into the product/process which results in disruptions when failures loop back for correction. The product/process design may not make optimal use of CNC. Ironically, potentially versatile CNC equipment is usually used in a batch mode, not flexibly.

> Existing designs may lack synergy and have too many parts and processing steps

Developing products for BTO&MC. To be successful at designing products for BTO&MC, product development teams must proactively plan product portfolios for compatibility with BTO&MC, design products in synergistic product platforms, design around *aggressively* standardized parts and raw materials, make sure specified parts are quickly available, consolidate inflexible parts into *very versatile* standardized parts, assure quality by design with concurrently designed process controls, and *concurrently engineer* product platforms *and* flexible flow-based processes.

> In BTO&MC, product development teams concurrently engineer families of products and flexible processes

Further, product development teams need to eliminate setup *by design* by specifying readily available standard parts and tools, designing versatile fixtures at each workstation that eliminate setup to locate parts or change fixtures, and making sure part count does not exceed available part bin capacity or space at each work station.

Finally, products must be designed to maximize the use of available programmable CNC fabrication and assembly tools, *without expensive and time-consuming setup delays.*

Designing for No Setup. Product design has a profound effect on setup. Excess proliferation of parts complicates internal part distribution and may make it impossible to have all parts available at all points of use. Even a moderate excess of part types will cause setup delays to distribute, find, and load parts into manual or machine bins. A serious excess of part types may make it necessary to kit parts for every batch, which is a significant setup. Designers can eliminate fixturing setup by designing parts for versatile fixturing (Ch. 8) which, if not already in place, may have to be *concurrently* designed with the parts.

> Product design has a profound effect on setup

Designers can eliminate tool change setup by designing parts around common tools (cutting tools, bending mandrels, punches, etc.), ideally one tool that never has to be changed. If multiple tools are required, designers must keep tool variety well within tool changing capacity for the whole line.

Design for Manufacturability (DFM) principles can greatly simplify assembly.[16] Designers need to work with manufacturing engineers to concurrently develop simple assembly procedures that can be understood in a few seconds either on a computer screen or on paper instructions that can be quickly located and understood.

Designing for CNC. Computer numerically controlled machine tools (hereafter referred to as "CNC") offer vast opportunities to eliminate machining setup, as discussed in the section, "CNC to Eliminate Machining Setup," in Chapter 8. CNC machine tools include metal cutting equipment (mills, lathes, etc.), laser cutters, punch presses, press brakes, printed circuit board assemblers, and basically any production machine controlled by a computer. Designers need to understand enough about CNC operation to use the versatility of CNC to eliminate setup.

Designers need to understand how CNC can make parts flexibly and eliminate setup

Designing around Standard Parts and Materials. Standardization of parts (Ch. 4) and material variety reduction (Ch. 5) are the most important design contributions to the feasibility of spontaneous supply chains. *Aggressive* standardization can enable the easiest technique of the spontaneous supply chain: steady flows of very standardized parts and materials (Ch. 7). If there are too many different parts and material types, steady flows can not be arranged because of the variety and unpredictability of demand. The total cost value of standardization and its contribution to the BTO&MC business model should motivate engineers and materials organizations to implement such aggressive standardization.

Product designers must design around *aggressively* standardized parts

Designing Around Readily Available Parts/Materials. A spontaneous supply chain depends on readily available parts and materials. Therefore, it is an important aspect of engineers' jobs to specify parts and materials that are readily available. Usually, design engineers choose parts based on functionality and maybe quality. But for BTO&MC, availability is equally important.

> An important aspect of engineers' jobs is to specify parts and materials that are readily available

Design for Manufacturability. Design for Manufacturability (DFM) guidelines can support BTO&MC by minimizing incoming variety of parts and materials, for instance, by eliminating right/left-hand parts, combining parts and functions into a single part, specifying prefinished material, and avoiding arbitrary decisions.

Arbitrary Decisions. Successful product development requires that designers proactively design products for a BTO&MC environment using the principles presented herein. But this may be difficult or impossible if designers make *arbitrary decisions* that preclude implementation of these principles. *All* design considerations – functionality, cost, quality, *and* flexibility – must be taken into account *early.* If they are not, then designers will probably make many *arbitrary decisions* that will make it much harder to satisfy omitted design considerations later.

> Design motto:
> *No arbitrary decisions!*

Modularity. Modularity can lower cost, lead time, time-to-market, lead to broader product lines, and make service and upgrades easier. On the other hand, modularity may not be feasible for highly integrated products or cover a wide enough range of customer needs. Module interfaces may increase product development expenses, add fabrication cost, and may compromise functionality by adding weight and seams, weakening structures, or slowing down electrical signals or even causing signal interruptions at low voltage.

Off-the-shelf parts. Off-the-shelf-parts can save a lot of money compared to reinventing the wheel with respect to: design, documentation, prototyping, testing, debugging, purchasing, manufacture, overhead, and administrative expenses. Off-the-shelf parts will also save a lot of time to design, document, prototype, build, test, debug, redesign, and manufacture.

Outsourcing Engineering. As with the outsourcing manufacturing (Ch. 6), many managers are intrigued with the prospect of outsourcing engineering as a way to "save cost" on product development with foreign engineers whose wages are a fraction of domestic engineers. Further there is the allure of speeding up product development by keeping engineering working in three shifts a day with work passed off around the world every day.

Engineering expenses are best minimized by making the *whole process* more efficient, not going abroad for cheap engineers

This might work with independent activities such as call centers or insurance or loan processing, but product development is a highly interactive and integrated team activity. Product development is most effective and efficient when products are designed by complete multifunctional teams that are "co-located" with manufacturing operations and close to customers and vendors.[17] Any proposals to save cost must be based on minimizing *total cost.*

The best way to lower product development expenses is to *maximize the efficiency of the whole process* through concurrent engineering (with co-located teams) and design for manufacturability (with continuous interactions with manufacturing people).[18] Optimal product development practices not only minimize *engineering* costs but also can substantially reduce *product* costs, which is one of engineering's primary goals.

Minimizing Total Cost (Chapter 11)

The key to achieving the lowest cost product is to base all thinking and decisions on a *total cost* perspective. Unfortunately, the typical company cost system reports only material and labor costs. All other costs are called *overhead,* which is spread over corporate activities according to some arbitrary allocation algorithm, for instance, proportional to material, labor, or processing cost. And yet, all products do *not* have the same overhead demands. In fact, much can be done to lower overhead costs, which will be discussed in Chapter 11.

Many of the recommendations of this book, when viewed narrowly, may *appear* to be more expensive, and therefore may be resisted. Rationalization may appear to lower sales (Ch. 3); Standardization and consolidation may appear to require most products to use better materials than they need (Ch. 4, 5); Offshoring may appear to save cost (Ch. 6); The lack of competitive bidding (inherent in vendor-partnerships) may appear to raise procurement costs (Ch. 7); Small batches, approaching one, and dedicated flow lines using simple equipment may appear to be inefficient from a mass production standpoint (Ch. 8, 9). However, these programs will really lower the *total cost.*

How *Not* to Achieve Low Cost. First, it is important to dispel some myths about cost. Low-cost products do *not* come from volume alone; that is the Mass Production paradigm. With build-to-order and mass customization, products can be made at low cost virtually *independent* of volume.

> Low cost does not come from cheap parts, low-bidding, cutting corners, or retroactive "cost reduction"

Low-cost products do not come from cheap parts, cutting corners, or "cost reduction" efforts, none of which are effective from a total cost perspective. Mercer Management Consulting analyzed 800 companies over a five year period. They identified 120 of these companies as "cost cutters." Of those cost-cutting companies, "68% did not go on to achieve profitable revenue during the next five years." [19]

> For 120 companies identified as "cost cutters," *2/3 failed to achieve profitable revenue* over the next 5 years – Mercer study

Moving production to "low labor rate" countries is another cost reduction fallacy (Ch. 6). Labor efficiency alone might cancel out the labor rate savings, for instance if labor cost is one third but labor productivity is also one third. Many decisions to move to production to low-labor-rate are based on labor-intensive designs. However, Design for Manufacturability can reduce labor content to the point where moving to low-labor-rate areas can no longer be justified.[20] Finally, cheap labor rarely stays that way and chasing it around the world is an expensive way to save money.

Low cost does not come from chasing cheap labor.

Cost is very difficult to remove after the product is designed. Eighty percent of a product's lifetime cumulative cost is designed into the product and is very difficult to *remove* later.[21] In addition, the *total cost* of doing the change may not be paid back within the expected life of the product.

There are also intangible impacts of an excessive focus on cost reduction: it absorbs effort and talent that should be applied to more productive activities, like developing better *new* products and implementing BTO&MC.

Low-cost products do not result from "saving" cost with layoffs or across the board budget reductions. Low-cost products do not result from manufacturing misconceptions like purchasing policies based on low-bidders or offshore manufacturing to lower labor costs, which too often raise many other overhead costs and lengthen delivery times.

Total cost accounting can quantify all costs (Ch. 12), but until that arrives, product designers must rely on total cost *thinking.*

Perceptions of Cost. Traditional cost systems typically only measure parts cost and labor cost, thus focusing all cost reduction thinking on only these two costs. The rest of the costs are lumped together in several categories, collectively called *overhead,* which is then averaged (allocated) over all products. The result of averaging overhead is *cross-subsidies* where the worst products (with the highest overhead) are subsidized by the best products. This, in turn, leads to distorted product pricing which overprices good products (making them less competitive) and underprices bad products (which may increase the sales of money-losing products). Further, considering only parts and labor is the root cause of bad cost decisions and cost reduction efforts that may raise other costs, such as quality. Distorted costs lead to a distorted view of profitability, which leads to poor product planning.

An important truism to remember is the following:

You don't compete on cost; You compete on price.

Customers don't care about *your* cost;
They only care about *their total cost* which is *your price*

> Customers don't care about *your* cost. They only care about *their* cost which is *your price.*

Total cost measurements provide a relevant cost breakdown (shown in Figure 11-2) and make it possible for decisions to be based on total cost considerations. Decision makers must acknowledge that overhead costs are significant (often more than labor and materials) and that companies really *do* have influence over overhead costs, which are summarized below and will be discussed in detail in Chapter 11.

Product Development. Minimizing product cost starts with design which determined 80% of cumulative lifetime cost.[22] In fact, 60% of cost is determined by the architecture phase, which is rarely optimized in the rush to show early "progress."

Product development costs can be minimized by using multi-functional teams in a concurrent engineering environment that designs products for manufacturability.[23]

The wasted cost of "reinventing the wheel" can be avoided in product development with maximum use of reusable engineering, versatile modules, and designing products around off-the-shelf hardware. A related cost saving is less debugging cost from modules that have already been debugged and proven. These approaches plus extending product life spans with upgradable designs save the cost of designing entirely new products.

Using concurrent engineering can cut the total product development cycle in half, which not only saves product development expenses, but also minimizes the chance of obsolescence, change orders, and redesigns.

Quality Costs. The "cost of quality" is really the cost of *poor* quality. Companies without strong Total Quality Management programs can have quality costs equal to 15% to 40% of revenue.[24] Advanced product development can *design in* quality. Process controls can *build in* quality. This dual approach to quality can substantially reduce the cost of quality.

Quality costs can also be minimized by selecting suppliers on a total cost basis that includes quality, which is Edward Deming's 4th point.

Operations. Many categories of operations cost can be minimized by BTO&MC techniques: eliminating setup, in turn, eliminates setup costs and improves machinery utilization. Eliminating inventory frees up a lot of valuable floor space. Designing products for manufacturability results in less assembly expense and less product-related fire-fighting. Developing products and processes concurrently minimizes tooling costs and results in more efficient production and less process-related fire-fighting. More efficient production minimizes overtime costs as does the ability to shift production to other flexible lines.

Material Overhead. Purchasing costs can be reduced if there are fewer purchasing actions for fewer part types. Standardized parts will cost less because of the greater purchasing leverage of higher volume parts. Furthermore, the "bread-truck" concept can be used where a supplier is responsible for keeping the bins full for common inexpensive parts, much like a bread truck keeps the shelves full in a grocery store.

Further, dock-to-line deliveries save the costs of incoming inspection, kitting, and internal distribution. Automatic resupply techniques, such as *kanban,* eliminate high overhead ordering and purchasing actions.

Working Capital. There is a lot of *working capital* tied up in various forms of inventory: raw materials and parts inventory; Work-in-Process (WIP) inventory; and finished goods inventory in factory warehouses, at distributers, and at the dealers. Fortune Magazine estimates that, for Fortune 500 companies, working capital averages an amount equal to 20% of sales.[25] Inventory carrying costs since 1961 averaged 25% of inventory value per year (Ch. 2).

Customization/Configuration. Mass customization is, by definition, an efficient way to customize or configure products, in contrast to the usual "fire drill" approach which adds unnecessary cost in order entry, engineering, and operation, and distracts companies from more profitable endeavors in product development, quality, and operations.

Measuring Total Cost (Chapter 12)

In order to realize all of the cost savings and revenue enhancements cited in this book, it would be highly advantageous to be able to *quantify all costs*. If costs were tracked on a *total cost* basis, then the cost saving potential of well-designed products could be known, and products could be given an appropriate overhead charge and thus a competitive price. However, if total cost is not tracked, then well-designed BTO&MC products (with have much less overhead demands) will have to subsidize the high-overhead products; this would be an unfair burden and may ultimately compromise the success of the BTO&MC products.

Most companies have such inadequate cost systems that it actually hinders good product development and distorts product development decisions. Merely reporting labor and material costs encourages companies to move operations to "low labor rate" countries. Trying to minimize return on capital encourages companies to outsource production. Both of these slow down the supply chain and actually increases total cost.

Allocating Overhead Rationally. Products with too much setup, inventory, "firefighting," engineering change orders, excessive part variety, low equipment utilization, and high quality costs should have a higher overhead rate. Products that are designed for quick and easy manufacture should have a much lower overhead rate. Overhead rates should be proportional to overhead demands, which vary by product.

Cooper and Kaplan[26] pointed out in an article with the profound title, *How Cost Accounting Distorts Product Costs,* that overhead costs "vary with the diversity and complexity of the product line." And, product diversity and complexity are increasing because of market pressures and perceived opportunities. Thus, it is important to quantify overhead costs, since they can be much greater than the typically reported costs of labor and materials.

Quantifying Overhead Costs. The first step is to acknowledge deficiencies in current product costing practices, which include distortions in product costing, cross subsidies, and decision making based on irrelevant or misleading cost data. The second step is to estimate the degree of cost distortions, which could be estimated subjectively or quantified for a few pilot investigations on parts or products identified by surveys are costing more than perceived.

The third step is to understand the value of total cost measurements for knowing true profitability, product line rationalization, product development prioritization, aggressive standardization, various trade-offs, justifications, quantifying quality costs, and identifying products that are losing money. The final step is to implement total cost measurements.

> Cost drivers identify the root causes of cost that should be quantified instead of averaged

Cost Drivers. The low-hanging fruit approach is the identification and implementation of simple *cost drivers* that make cost accounting more relevant and encourage behavior to lower these costs. A cost drivers is defined as the *root cause* of a cost – the things that "drive" cost. Identifying cost drivers make the root causes visible so that total cost can be measured and behavior that actually lowers total cost can be encouraged.

The cost driver approach identifies key drivers of cost that should be quantified instead of lumped in with all other overhead. The cost driver approach is easy to implement and starts with the most important overhead costs that need to be quantified. New data collection efforts are focused on only a few key cost drivers. Cost drivers can be based on estimates, as long as there is universal consensus.

For instance, material overhead is a cost driver. Conventional cost accounting would apply a fixed material overhead rate to all purchased parts and materials, which is unrealistic due to the vast range of procurement difficulty required. The cost driver approach would assign a low overhead rate for standard parts (as determined by the procedure in Chapter 4) and a high overhead rate for oddball parts. This not only reflects the true costs of procurement, but it also encourages engineers to use standard parts.

> It is better to be *approximately correct* than to be *precisely wrong* – Doug Hicks, *ABC for Small and Mid-Size Businesses*

There are two ways to determine the two rates. The full-blown Activity Based Costing approach would try to measure the time that material overhead personnel spent on standard and non-standard parts and then compute the rates accordingly. But there is an easier approach advocated by

Douglas Hicks' excellent "how-to" book, *Activity-Based costing for Small and Mid-Sized Businesses.*[27] It is based on the valid premise that *it is better to be approximately correct than to be precisely wrong; accuracy is preferable to precision.* Said another way, it would be better to be *approximately right* than *precisely wrong!* So using Hicks' approach, for instance, the director of all material overhead functions (purchasing, warehousing, internal distribution, etc.) would simply estimate what percentage of his expenditures were spent on standard parts. In most companies, the purchasing manager would say something like: "Hey, we only spend 10% of our effort on standard parts – most of them are routine reorders anyway. The other 90% goes to all those darn oddball parts – they're hard to find, cost more to ship, and often have higher charges for special setups and expediting, not to mention higher quality costs." And so material overhead rates would then be split 90/10 instead of the unrealistic 50/50 of conventional cost systems.

Implementing BTO&MC (Chapter 13)

Identify Goals and Drivers. Identify goals and challenges that can be improved by BTO&MC. Start with bottom line goals like profit improvement, growth, market share, stock performance, and so forth. Then identify the *drivers* for those goals, in other words, the activities and programs that would *drive* the achievement of those goals, for instance cost reduction, delivery time reduction, and better customer satisfaction through better cost, delivery, quality, and mass customizing specific needs.

Obtain Customer Inputs. Obtain customer input on the relative importance of their *preferences* and *competitive rankings,* which are two of the main inputs for the QFD process discussed in Chapter 10. Even if major new product developments are not forthcoming soon, obtaining this information will be helpful for BTO&MC planning now and can eventually be used in the full QFD process.

> Start with products that have high variety, labor, overhead, delays, inventory, and customization difficulty

Identify Where to Start. First, rationalize the product lines (Ch. 3) to remove, at least tentatively for the future: products, options, and variations that will not be in the

BTO&MC system. Then look for product families with the following characteristics:

- high *output* variety and potentially low *input* variety; trends indicating increasing product variety over time;
- high labor cost: increasing labor shortages; significant or erratic overtime; excessive firefighting;
- high setup costs and delays;
- unreliable forecasts;
- inventory problems: excessive carrying costs, obsolescence, write-offs, out-of-stocks, etc.;
- delivery response pressures or opportunities;
- the most difficult customizations;
- high visibility, the success of which will help spread implementations to other product lines.

For various potential activities, estimate relative benefits and cost in terms of human resources and capital. Plot benefit vs. effort and identify and group activities that are: "low hanging fruit" opportunities with a high payback for a low effort; related activities already supporting other programs and goals; and activities that may be needed to complete the system, such as bringing in-house long-lead-time parts production (Ch. 6). Extend mass customization capabilities to expand the ranges of customization or branch out to adjacencies, niche markets, derivatives, and so forth.

Decide which product line(s) to implement first and which activities with those product line(s) to address first. Tentatively map out a progression of subsequent implementations.

Free Up Resources. First, free up time for key implementers by: rationalizing product lines to eliminate high-overhead/low-profit products; delegating firefighting and routine tasks to the lowest levels; shifting parts and materials from MRP to automatic resupply techniques; and minimizing competing demands, especially those that may not be supportive of BTO&MC or be obsoleted, such as warehouses and many aspects of ERP systems.

Minimize Fears and Inhibitions. When implementing any change, it is very important to overcome any potential resistance to change and provide motivation to enthusiastically implement the changes. Chapter 14 presents the business case for BTO&MC and this should be clearly communicated to everyone, including how these changes will improve the company's performance and everyone's career prospects. All inhibitions, resistance, and fears should be addressed and nullified proactively. Part of overcoming inertia is to emphasize the risks of no change at all, which could be considerable in fast changing industries. The President of Olympus, Tsuyoshi Kikukawa, said *"If you don't take a risk on a new idea, that, in itself, becomes a risk."*

One of the greatest fears surrounding any changes that improve "efficiencies" is the fear that improving efficiencies may eliminate jobs. Unfortunately, there are enough horror stories circulating to support these fears. The most effective way to allay these fears is to issue a *no layoff pledge* that assures all employees that no one's career will be at risk in any way from implementing the improvements. In *Lean Thinking,* Womack and Jones advise that "you must guarantee that no one will lose their job in the future due to the introduction of lean techniques. And you must keep your promise."[28]

> Offer a *no layoff pledge* that no job will be lost to efficiency improvements

Training. For implementing something as broad as build-to-order and mass customization, training is an important early step. Some companies may be working on some elements, but they may be "islands of excellence" that may not be coordinated well with others or may not gel into coherent business model. Existing programs may not be going far enough: Lean programs may only be tackling the low-hanging-fruit and may only be *reducing* setup and batches, instead *eliminating* them whenever necessary for spontaneous build-to-order. Current standardization efforts may not have been *aggressive* enough for spontaneous resupply. In addition to all the elements, such as rationalization, standardization, integration, spontaneous supply chains, on-demand lean production,

> Training can show how to integrate many initiatives into a coherent, viable business model

training should show how these elements fit into an overall business strategy to achieve a viable business model.[29]

Create Implementation Road Map. Road maps for implementing BTO&MC should emphasize several parallel, coordinated tasks that can be implemented simultaneously to support the business model vision. Each category of tasks should be structured so that it can generate some *early useful results* and *a progression of useful deliverables* throughout the time line. This is important to deliver early paybacks, keep generating support, and support other activities.

Implementation activities summarized here are presented in Chapter 13 in time-line format:

Revolutionize your business model with *evolutionary* self-supporting steps

Overall Planning, including establishing the vision and establishing the overall implementation plan.

Product Line Rationalization, including data gathering on sales histories for products and variations, profiles to scrutinize new orders; criteria to reject; Pareto plots and analyses; and decisions to eliminate, outsource, or improve; and refocusing resources on remaining products.

Standardization, including data gathering of historical usage for each category, Pareto plots, analysis, consolidation, and standardization lists generated and implemented.

Supply Chain, including converting to automatic resupply, supplier lead time reduction, arranging to pull from suppliers on-demand, developing partnerships as necessary, and bringing in-house or acquiring as necessary.

Lean Production, including setup/batch/WP reduction, CNC utilization optimized, and manual integration followed by computer integration.

Product Development, including: developing products for lean, BTO, and mass customization; pre-engineering of anticipated options, variations, and customizations; parametric templates; and on-demand CNC program generation.

Configuration/Information Technology, including compiling knowledge used to evaluate orders, paper/spreadsheet "configurator," configuration software, and integrated information flow.

Total Cost Measurement, including profiles to identify low-profit products, cost drivers identified, cost drivers quantified, costing model developed, and compensation based on profit.

Focus and Staffing for BTO&MC. A BTO&MC site should focus on quick and efficient manufacture of a planned range of variation. The following activities should be focused and staffed as follows:

BTO&MC Focus & Staffing:

Sales/Marketing Focus	**Not:**
Take profitable orders within the BTO&MC system	Take all orders, regardless of profit or throughput implications
Compensated on profit	Compensated on units or revenue
Easy customizations	Difficult customization
Configurator quickly established validity and "what ifs"	Slow back-and-forth process to determine order validity
Orders complete and accurate	Order entry errors; surprises; changes
Few customer-induced changes	Many customer-induced changes

Engineering Focus	**Not:**
Parametric templates accept customization inputs	Manually modify or create unique drawings for every order
Automatic CNC program	CNC programs from separate efforts
Smooth predictable operations	Fire-drills
Most efforts improving system	Most efforts on each order

Manufacturing Focus	**Not:**
BTO and mass customization	Craft production or mass production
Automatic CNC fabrication	Skilled workers make parts manually
Many features in one setup	Many steps through many machines
Smooth operations	Fire-drills
Smooth pre-planned procedures	Rough *ad hoc* procedures
Easy integration	Integration problems
Few change orders	Many change orders

Supply Chain Focus	**Not:**
Simplify complexity	Try to *"manage"* complexity
Pulling parts and materials	Ordering parts and materials
Arranging automatic resupply once for each category	Purchasing parts and materials for each order

Information Technology Focus	**Not:**
Order entry, databases, & CNC	MRP & purchasing functions

Acquisitions Focus	**Not:**
Suppliers, distributors Help MC model at both ends	Competitors More parts, suppliers, & processes

Business Model Focus	**Not:**
Easily *earn* the numbers	Do what it takes to *make the numbers*
Grow sales with better cost, quality, time, satisfaction	Grow sales by mergers, acquisitions, or taking all orders
Grow profits with better focus, lower costs, & higher sales	Sacrifice profits for market share

Product Families. A key step in implementing BTO&MC is segregating products into *families,* which are groups of products that can be manufactured on-demand efficiently in a batch-size-of-one mode. BTO&MC parts and products are built in *families* of products that (1) can all be built efficiently on-demand without setup delays with (2) have all the parts and materials always available without the cost, delays, risk, and uncertainties of forecasts and inventory. This can be accomplished by building product families together on the same line.

BTO&MC Production Strategies:

- ***High-Volume/No-Variation Products.*** High-volume products that have no variations can be mass-produced *on dedicated lines that will never have setup changes.*

- ***Medium-Volume/No-Variety Products.*** Medium-volume products that have no variations can also be mass-produced on smaller dedicated lines each of which produces one product without any setup changes.

- ***Low-Volume/High-Variety and Custom Products.*** Low-volume products and high-variety or custom products can be built on flexible lines that are flexible enough to produce any product in a family without setup changes.

Range of Choices for Implementation. The following is a range of implementation choices for build-to-order and mass customization.

- ***Build Standard Products To-Order.*** A logical starting point for implementation would be to develop the capability to build-to-order *standard products* without forecasts, batches, or inventory.

- ***Build Custom Products On-Demand.*** The same spontaneous supply chain and on-demand lean production developed for the build-to-order of standard products can be extended to include *mass customized* products. The main difference would be the order entry system..

- ***Regional/Industry-Specific Plants.*** The BTO&MC expertise developed in a manufacturer's existing plants could be leveraged to regional or specialized plants to: (a) build certain retail products on-demand near market concentrations, distribution hubs, or individual customers; or (b) supply parts to specific industries or products to stores on-demand, especially if there are clusters of customers concentrated in certain geographical regions. These "mini-plants" would be able to build any mass-customized or standard products on-demand with delivery measured in minutes for customers in close proximity. For individual retail customers, mini-plants can interact directly with customers and then build their customized products on-demand.

- ***Customer-Specific Lines.*** A more focused version of industry-specific lines would be *customer-specific* lines inside, or next to, certain customer plants. Each customer-specific line would build on-demand any parts needed by each customer, for mass-customized or standard products. Customers would get near-instantaneous on-demand delivery in a seamless integration with their operations. The single-customer focus would benefit both customers and the manufacturer with learning relationships to continuously improve customer satisfaction.

- ***Expand Downstream***. If the BTO&MC supplier is convinced that the customer-specific line concept presents significant opportunities and all industrial customers decline such an opportunity, then the supplier could capitalize on these opportunities by *expanding downstream*, by setting up or buying existing plants to manufacture intermediate or end-products on-demand and offer them built to-order or mass-customized. In such a situation the supplier-turned-assembler would have unique competitive advantages and be able to grow and profit significantly in these new businesses.

- ***Expand Upstream.*** Chapter 6 addressed the issue of internal selective integration to manufacture parts in-house that cannot be pulled quickly enough from suppliers into BTO&MC operations. Such an analysis may reveal supply chain opportunities if your company and other potential customers would benefit from rapid part delivery at lower cost. If this looked promising, your company might consider acquiring (at least a controlling interest in) key suppliers and convert them to BTO&MC. Revitalizing these suppliers could increase profit and growth while ensuring fast delivery and low cost for your BTO&MC operations.

Downturn Strategies. Companies that have implemented BTO&MC will be much less affected by downturns, since they will be generating new growth, both in their cash cows and in new products. Wise companies can use any "calm between storms" to make a strategic investment, such as investing in improvement programs such as BTO&MC, expanding into related products, improving productivity, and immediately eliminating money-losing products, using total cost measurements (Ch. 11) to ascertain real profitability and the product line rationalization techniques of Chapter 3. The strategy here would be that if revenue is going to take a hit anyway, get rid of the money-losers to boost profits.

Companies should avoid "knee jerk" reactions to lay off people, halt training, curtail improvement programs, compromise product development, cut prices, or keep plants busy building inventory.

PART V: THE BOTTOM LINE

The Business Case for BTO&MC (Chapter 14)

Build-to-order and mass customization represent a business model that offers an unbeatable combination of responsiveness, cost, and what customers want when they want it. It enables companies to build any product on demand without forecasts, batches, inventory, or working capital.

> BTO&MC offers unbeatable responsiveness, cost, and customer satisfaction

BTO&MC companies can grow sales *and* profits by expanding sales of standard, customized, derivative, and niche market products, while avoiding the commodity trap. BTO&MC companies are the first to market with new technologies since distribution "pipelines" do not have to be emptied first. Specifically, BTO&MC companies have the following advantages:

Cost. BTO&MC companies enjoy substantial cost advantages from eliminating inventory, forecasting, expediting, kitting, setup, and inefficient fire-drill efforts to customize products. BTO&MC results in more efficient utilization of people, machinery, and floor space. Chapter 14 discusses 20 ways that BTO&MC reduces cost.

Responsiveness. BTO&MC companies build products on-demand, instead of having to forecast, order, wait for parts, build, and stock. For phone order and web-sales, 100% of orders can promptly be shipped directly from the BTO&MC factory.

BTO&MC delivers the goods fast to stores, OEM's, or directly to the ultimate customers

BTO&MC substantially simplifies supply chains – not just "managing" them – to the point where parts and materials can be *pulled* into production without forecasts, MRP, purchasing, waiting, or warehousing.

Customer Satisfaction. BTO&MC can provide unmatched customer satisfaction for industrial clients or the ultimate consumers. Consumers will be satisfied by products always available at the best prices. OEM and industrial clients will be satisfied by receiving parts on-demand to support *their* build-to-order efforts. Mass customization will enable even higher levels of customer satisfaction for customers who can quickly receive high-quality/low-cost products specifically customized to their individual needs. For customized products customers can make better choices and consider more "what ifs" with a good configurator.

BTO&MC delivers
what customers want when
the want it
at the price they want

Competitive Advantages. In general, competition now is *between business models* – the company with the best business model will be the best competitor. Ironically, the subject of several best-selling business books, *leadership* and *execution,* will only drive a company faster down the wrong path if they have the wrong business model. Dartmouth Business School Professor Sidney Finkelstein, writing in *Why Smart Executives Fail, and What You Can Learn from Their Mistakes,*[30] concluded that "The real causes of nearly every major business breakdown are the things that put a company on the wrong course and keep it there."

Competition is now between business models

As a business model, build-to-order and mass customization will compete well against competitors both large and small because of a superior combination of speed, cost, and customization.

The real cause of business breakdowns is being on the wrong course – Professor Sidney Finkelstein, author of book on business failures

Mass customization can efficiently customize products for niche markets, countries, regions, industries, and individual customers. Without a superior business model, companies might have to compromise profits to enhance market share. In the worst case competitive positions, products revert to commodity status with purchasing decisions made solely on price.

BTO&MC companies have the agility to expand business into adjacencies, niche markets, derivatives, and so forth. Finally, all the above advantages can create a reputation as a leader, which further improves sales, impresses investors, and attracts the best talent.

Growth. If companies don't use an effective growth strategy, like BTO&MC, then growth pressures may push them to "take all orders." Expanding sales by "dipping lower into the barrel" for low-volume, unusual, or marginal products may appear to satisfy growth goals, at least for a while, but in reality, it will have several unpleasant effects discussed in Chapter 14.

Trying to grow without a viable strategy means "take all orders" & "dip lower into the barrel"

Mergers are not an effective way to achieve growth because "pricing power" or "economies of scale" would only provide a benefit if the merged companies used the same parts and same processing equipment, which is highly unlikely. Further, merging two dissimilar product lines into the same factory would probably double the variety of parts and raw materials, which would thwart any progress made toward standardization, build-to-order, and mass customization.

> Merging dissimilar products into the same plant thwarts standardization & ultimately BTO&MC

Many companies desire growth, but a thorough study, based on two hundred case studies and a database on almost 2000 companies, concluded that 90% of companies failed to achieve sustained profitable growth over the past decade! The 10% that did owed their success to a strategy that is the title of the book, *Profit from the Core,* meaning that they focused on their core products, capabilities, customers, and channels.[31]

Build-to-order *enhances and grows* "the core" by building standard products on-demand at less cost without tying up working capital or incurring all the risks of forecasts and inventory. Further BTO adapts instantaneously to changing market conditions and enables the fastest introduction of new products. Mass customization *expands* the core to "adjacencies" (opportunities near the core) by offering variations of standard products and a vast range of customized products synergistic with core assets.

> Build-to-order *enhances and grows* the core. Mass customization *expands* the core.

Build-to-order and mass customization have unique abilities to generate significant growth in four ways: (a) expansion of standard product sales; (b) new market expansion into niche market derivatives of standard products; (c) new market expansion into customized products; and (d) maintaining sales to existing customers by avoiding the commodity trap.

Synergy between BTO and MC. There is a natural synergy between build-to-order and mass customization, hence the title of this book. They share the same batch-size-of-one operations and spontaneous supply chain. Build-to-order and mass customization operations are equally efficient and very compatible, unlike situations where a mass customization experiment must be run separately from the "batch-and-queue" operations of mass production. Building build-to-order and mass-customized products on the same lines will often push the combined volume over the "critical mass" threshold necessary to justify these implementations.

ENDNOTES/REFERENCES

1. Data from the "12th Annual State of Logistics Report," Published by Cass Information Systems and Prologis, June 4, 2001.

2. Richard Koch, *The 80/20 Principle; The Secret of Achieving More With Less*, (1990, Currency/Doubleday), p. 53.

3. Slywotzky & Morrison, *The Profit Zone,* Ch. 1, "Market Share is Dead."

4. James P. Womack and Daniel T. Jones, *Lean Thinking, Banish Waste and Create Wealth in Your Corporation* (1996, Simon & Schuster), p. 224

5. John A. Byrne, "What a Difference a Century Can Make," *Business Week*, August 28, 2000; Table, p. 87.

6. Steven L. Goldman, Roger N. Nagel, and Kenneth Preiss, *Agile Competitors and Virtual Organizations,* (1995, Van Nostrand Reinhold), page 205, "Virtual Organization Characteristics."

7. Philip L. Carter, professor of purchasing at Arizona State University, Tempe, and executive director of the Center for Advanced Purchasing Studies, was cited in *Industry Week* (February 12, 2001, p. 43) as observing that "Many purchasing organization are under heavy pressure form the corporate brass to implement some form of e-commerce."

8. James P. Womack, Daniel T. Jones, and Daniel Roos, *The Machine that Changed the World, The Story of Lean Production,* (1990, Harper Perennial), p. 146.

9. Jeffrey Pfeffer and Robert I. Sutton, *The Knowing-Doing Gap; How Smart Companies Turn Knowledge into Action,*(2000, Harvard

Business School Press), p. 23.

10. James P. Womack and Daniel T. Jones, *Lean Thinking; Banish Waste and Create Wealth in Your Corporation,* (1996, Simon & Schuster).

11. David M. Anderson, *Design for Manufacturability & Concurrent Engineering; How to Design for Low Cost, Design in High Quality, Design for Lean Manufacture, and Design Quickly for Fast Production*, (2008, CIM Press, 805-924-0200); Chapter 9; Guideline P14: "Design machined parts to be made in one setup (chucking)."

12. Jones, Daniel J., "JIT & the EOQ Model: Odd Couples No More!," *Management Accounting* v72, n8, Feb 1991, pp. 54 - 57.

13. Richard J. Schonberger, *World Class Manufacturing, The Lessons of Simplicity Applied,* (1986, Free Press), p. 83.

14. Ibid., pp. 229-236.

15. B. Joseph Pine, II, Don Peppers, and Martha Rogers, "Do You Want to Keep Your Customers Forever," *Harvard Business Review,* March-April, 1995, p. 103.

16. Anderson, *Design for Manufacturability & Concurrent Engineering.*

17. Ibid,. Chapter 2, Concurrent Engineering.

18. Ibid.

19. Robert G. Atkins and Adrian J. Slywotzky, "You Can Profit From a Recession," *Wall Street Journal,* February 5, 2001, p. A22.

20. Anderson, *Design for Manufacturability & Concurrent Engineering.*

21. Ibid., Ch. 1.

22. Ibid., Figure 1-1, "Product Cost vs. Time."

23. Ibid., Chapter 3, "Concurrent Engineering"

24. Phillip Crosby, *Quality is Free* (Mentor Books, 1979).

25. Shawn Tully, "Raiding a Company's Hidden Cash," *Fortune Magazine,* August 22, 1994, page 82.

26. Robin Cooper and Robert S. Kaplan, “How Cost Accounting Distorts Product Costs,” *Management Accounting* (April 1988).

27. Douglas T. Hicks, *Activity -Based Costing; Making it work at Small and Mid-Sized Companies,* Second Edition (1998, John Wiley & Sons).

28. James P. Womack and Daniel T. Jones, *Lean Thinking; Banish Waste and Create Wealth in Your Corporation,* (1996, Simon & Schuster), Ch. 11, “An Action Plan.”

29. For a outline of the author’s in-house BTO&MC seminar, see page 469.

30. Sydney Finkelstein, *Why Smart Executives Fail and What You Can Learn from Their Mistakes,* (2003, Portfolio/Penguin), p. 138.

31. Chris Zook, *Profit from the Core; Growth Strategy in an Era of Turbulence,* (2001, Harvard Business School Press).

2

THE SHORTCOMINGS OF MASS PRODUCTION

1923 – THE HEYDAY OF MASS PRODUCTION

Mass production thrived in the bygone era of stable demand and little product variety. At its peak, the "any-color-as-long-as-it's-black" Model T Ford had a 57% market share.[1] Despite its very low price of $245, however, it was pulled off its pedestal when General Motors offered variety such as color paint and other options. Ford was slow to respond, mostly because mass production's keys to its success – hard tooling, labor specialization, and economies-of-scale – *prevented* it from offering variety or adapting quickly to emerging trends.

IF YOU BUILD IT, THEY *MIGHT* COME

In the mass production paradigm, the marketing department forecasted demand and these forecasts drove complex MRP (Material Requirements Planning) software to order the required parts and materials far enough ahead of time so that they would, hopefully, all arrive at the factory before the production run was scheduled. Assuming this was successful, equipment would be set up and a batch of products was built to satisfy the forecast. In order to make a batch

Mass production should be called *batch and queue* manufacturing

of products, batches of parts would wait in *queues* before each operation – hence mass production is more aptly described by the less glamorous phrase, *batches and queues,* an apt phrase that was made popular by Womack and Jones, first in *The Machine That Changed the World*[2] and then in *Lean Thinking.*[3]

If all the parts did not arrive on time or if market projections were off, then more parts would have to be procured by expensive expediting. Finished products were then put in a warehouse or series of distribution centers where they awaited the anticipated orders, relying on the industrial version of the "Field of Dreams" philosophy: *If you build it, they will come.* More precisely stated for manufacturing, the phrase should be, *if you build it, they* might *come.*

The manufacturing "Field of Dreams:" *If you build it, they might come*

INVENTORY CARRYING COST- The Big Hidden Cost

Not only is it hard to satisfy customer demand from inventory, it is also very expensive. Since 1961, the average inventory carrying cost has been slightly over 25% of inventory value per year, as plotted in Figure 2-1.[4] In other words, $4 million in inventory will *cost* the company $1 million per year to pay for the interest, space, insurance, and administration. Refrigerated products cost more and frozen products cost more yet. Products that need upkeep, like charging batteries, also cost more.

From a competitive standpoint, it doesn't matter if the inventory is at the factory, distributers, or stores – *it will add to the selling price and customers will have to pay for it,* thus making it less competitive. If your company, or a competitor, eliminates inventory costs, it can either keep it as increased profit or lower the selling price for a competitive advantage.

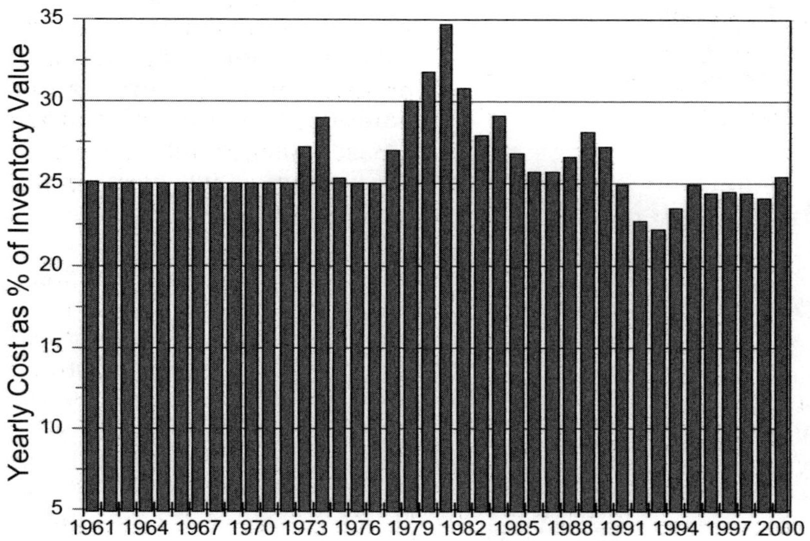

Figure 2-1: Inventory Carrying Cost Since1961

As pointed out in Chapter 11, the value of inventory cost is huge. For all manufacturing industries, inventory value *exceeds sales by one third.* The ratios of inventory to sales tracked by the US Department of Commerce indicates that the value of inventory equals about 1.3 times sales.[5] For example, a company with $3 million dollars of sales would, on average, have inventory worth $4 million dollars. The inventory carrying cost would be one-quarter of the inventory value or $1 million dollars, which would be a third of the sales value of $3 million dollars. *Thus, manufacturers, distributers, and stores together pay a third of sales dollars to pay for the cost of carrying inventory!*

On average, inventory carrying *cost* is 1/3 of sales dollars

> Inventory carrying cost can be three times as much as profit!

For the easy to compute case of a manufacturer who distributes and sells its own products, if profits are 10% of sales dollars, as is typical, then inventory carrying cost would be three times as much as profits! *Eliminating inventory would quadruple profits or allow prices to drop by one third!*

How Inventory Erodes Profits Over Time

Figure 2-2 shows how selling from finished goods erodes profits over time. The assumptions are fairly typical: profits at 10% and inventory carrying costs at 25% per year from the above discussion.

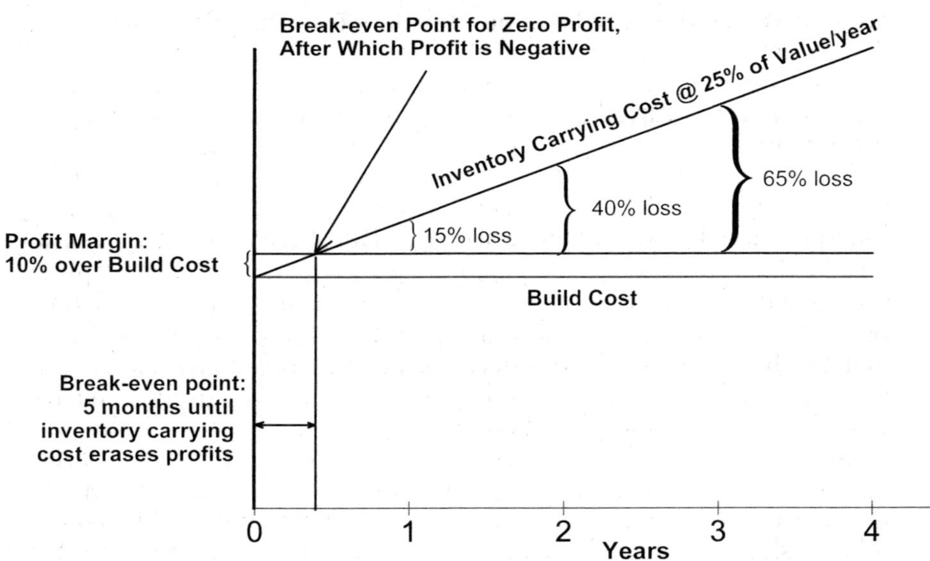

Figure 2-2: How Inventory Erodes Profit When Selling Finished Goods from Inventory

Under these typical conditions, inventory carrying costs exceed profit after 5 months, after which finished goods are really being sold at a loss, even though a primitive cost system would not indicate it. Losses are shown for 1 year (15%), 2 years (40%), and 3 years (65%). This analysis ignores the cost of selling (commissions, etc.) but this could be considered minor compared to the heavy losses due to inventory carrying costs.

The obvious message here is to build products on-demand and avoid the erosion of profits caused by inventory carrying costs.

FORECASTING – A HIGH STAKES GAMBLE

Mass production plants are usually too slow and inflexible to make products on-demand after receipt of orders. So, if customers can't or won't wait for the mass producer to order materials and wait for them to arrive, then the mass producer compensates by trying to *forecast* demand and order enough materials ahead of time to satisfy that future demand.

But forecast are never very accurate, especially in volatile markets and for varied product lines. In 2001, Cisco Systems paid for the biggest inventory write-down in history for parts ordered but not used: $2.2 billion! Much of the problem was that Cisco did not spot the coming plunge in technology spending early enough – an inherent vulnerability for forecasting. But there were other causes too. When routers were in short supply, customers would place orders with Cisco *and* its competitors, knowing they would only order from one. This artificially inflated forecasts *and* industry demand for parts, so Cisco ordered large quantities in advance. And to make matters worse, competing contract manufacturers also tried to lock up supplies to give them a competitive edge to win the next contract with Cisco.[6] Analysts blamed Cisco's vaulted information systems, but the latter effect would not have been a problem for internal manufacture (Ch.6) or if Cisco established vendor/partner relationships instead of having multiple contractors compete for the business (Ch. 7).

The general consequences of forecast inaccuracy are that high forecasts cause excessive inventory and low forecasts cause expediting, lost sales, and disappointed customers. However, this simplistic view only applied to mass producers who make one product.

Manufacturing companies that make any variety of products even in moderately volatile markets face much more complex problems and consequences from inadequate forecasts. Any manufacturer building more than one product will need to obtain forecasts for *each* product variation that uses different raw materials, parts, processing equipment, or labor skills. These specific forecasts would be needed to tell Purchasing to order the anticipated amount of *each* type of raw materials and parts. Inflexible equipment and narrow labor skills may have to be expanded or consolidated in anticipation of specific forecasts.

As product variety increases, forecasting problems escalate exponentially. If forecasts fail to predict the proper mix of products that customers will eventually buy, then the manufacture is in the sorry situation of having spent a lot of money buying the wrong materials and still *not* be able to build what customers are ordering. The results are: (a) missed customer sales and, for innovative products, missed windows of opportunity, and (b) inventory costs to carry the unused inventory until needed or write-off the inventory if it goes obsolete.

Forecasting problems escalate exponentially with product variety and market volatility

If all product variations could be made from the same standard materials, then forecasting would only have to guess the total demand and any excess materials would be consumed quickly (this shows the value of standardization, which is covered in Chapter 4 for parts and Chapter 5 for raw materials). And if forecasts are low, then there would only be a few types of standard materials to be expedited.

However, if different product variations are built from different materials – as is usually the case – then the forecasts will have to guess demand for *all* product variations. And if the forecasts are wrong, then there will be either: (a) many types of materials in excess that may not be consumed very quickly, if ever or (b) many types of materials that need to be expedited, which may raise cost and lead to availability problems.

Forecasting has to guess product demand for *all* variations

In mass production, each inflexible line has its own set capacity, so the associated product output can only be increased by overtime, which is expensive and may not be enough. However, in flexible BTO&MC plants, peak demand products can also be built on other lines, so demand peaks can be handled somewhere in the plant.

Trying to expand the capacity of inflexible lines ahead of time based on long-range forecasts is a capital-intensive gamble in which operations either expand prematurely, resulting in underutilized resources, or expand too late, resulting in missed sales, hasty outsourcing arrangements, or rapid, expensive, and distracting expedited expansion.

Matti Alahuhta, president of Nokia's mobile-phone division, says that "it is impossible to forecast the exact volumes of each product when the number of products is so high. The new challenge in logistics is to manage changes in the mix."[7]

DEFECTS BY THE BATCH

A recurring error can make the whole batch defective

The flow manufacturing movement is pointing out the inherent quality problems of batch production.

In mass production, batches of parts go from one machine or work station to another until they reach some testing function, which may discover that a recurring error at any one of the steps ruined the whole batch. Then the entire batch would have to be reworked or scrapped, thus raising costs, delaying delivery, *and* reducing production capacity.

In addition to recurring defects, large WIP (work in process) inventory levels cover up many other quality and efficiency problems when buffer inventory abounds, "just in case."

In *one-piece flow,* as each part is completed, it is passed on to the next worker, who looks for any visible deviations. If the part doesn't fit or work in the next operation, the feedback will be immediate leading to quick correction of the problem at the source (Ch.8). This is part of the continuous improvement, or *kaizen,* movement in which the production system is continuously learning and improving itself.

ITS ONLY OVERHEAD

Based on its early success with long runs of identical products, mass production has taken on the image of being the lowest cost way to build any product. Industrial empires were built on the principle of *economies-of-scale* and Frederick Taylor's specialization of labor. And even today, many people think that the *only* way to get cost down is to get volume up.

However, as product variety goes up and batch size is forced to go down, opportunities for economies-of-scale fade, except for standardized parts, which, ironically, are not used enough even by most mass producers.

Today, labor and raw materials are only a small proportion of the only cost that matters – the *customer's cost* which is the manufacturer's *price* (Ch. 11). But, since they are the only costs *measured* in most companies, they become the exclusive focus of "cost reduction" efforts, which often produce counterproductive results by buying cheap parts (Ch. 7), awarding the lowest bidder (Ch. 7), excessive outsourcing (Ch. 6), or moving production to countries with low labor costs (Ch. 6).

The majority of *total cost* is "overhead," which is rarely quantified, instead being allocated (averaged) over all products. And when overhead is not quantified, no one understands it or tries very hard to do anything about it. One of the author's clients had a controller who told engineers to "not worry about overhead – it's a *fixed* cost." But, in reality, these so-called fixed costs, like floor space, are not fixed at all. Flow manufacturing can cut floor space in half, eliminate WIP inventories, and eliminate recurring defects. Quality can be designed into the product[8] and built in with process controls. The costs of R&D and change orders can be minimized by good product development practices.[9] The "cost of variety" (see below) can be minimized by build-to-order and mass customization. Setup costs, finished goods inventory, and most of the distribution costs and material overhead can be eliminated by *spontaneous build-to-order*.

The majority of *total cost* is overhead, which is rarely quantified

TAKE ALL ORDERS

> "Taking all orders," grows revenue *a little* but grows costs *a lot*

Companies that have revenue growth as their cornerstone encourage the mantra, "take all orders," even if this brings in the *worst* variety and makes their mass production operations less efficient. This increase in variety shrinks batch size to the point where the increasing number of setups becomes a major cost and eats away at plant capacity.

When it comes to customization, most companies take an *ad hoc,* case-by-case approach to product customization where the mantra expands to "take all *customized* orders," followed by heroic efforts in Engineering, Purchasing, and Manufacturing where employees jump through hoops to get every customized product out the door. Often, such "fire-drill" customizations dilute other important programs when they borrow people from new product development and factory improvement programs.

Custom products and unusual standard products produced at low-volume are usually subsidized by the high-volume "cash-cows" (Ch.12). This has two detrimental effects: (1) the customized products and unusual standard products are not making nearly as much *real* profit as assumed and many are really losing money, and (2) the cash-cows *must be priced higher than they should be* to pay for the subsidization, thus lowering their competitiveness.

The quick and efficient approach to product customization is *mass customization,* which is accomplished by proactively developing the architecture of product families around dimensional (custom machining), adjustable, or modular customization (Chapters 9 and 10), implementing flow manufacture to achieve batch-size-on-one capability (Ch. 8), establishing a spontaneous supply chain around standard materials (Ch. 7), creating agile order entry systems based on configurators (Chapters 8 and 9), and building parametric CAD templates with automatic CAD/CAM links to CNC equipment (Chapters 9 and 10).

THE EFFECT OF VARIETY ON MASS PRODUCTION

Today, most companies offer hundreds or thousands of products or product variations, sometimes called SKUs (Stock Keeping Units). Even if most of the SKU variety is caused by "minor" labeling and packaging variations for sale in various regions of the globe, it still slows down the mass production and warehousing operations and makes them less efficient. And today, many companies are forced to customize products, to some degree, for increasingly selective customers or to compete in niche markets.

Unfortunately, most of these companies use a reactive approach to product variety, where streams of new products are introduced on a case-by-case basis without the benefit of systematic *product portfolio planning* or the standardization of materials, parts, modules, and processes. And, in the vast majority of companies, the old products are not discontinued (Ch. 3) – they just accumulate and dilute company resources away from higher leverage newer products.

Increasing variety is hard to handle in mass production

Relying on forecasts to order parts is becoming more problematic as product lines grow, build quantities shrink, and markets become more turbulent and unpredictable. As selling from inventory becomes more costly and less effective, some companies try to become agile and build products "to order," but do this reactively. Their inflexible factories still need to order an enormous variety of parts and schedule their production in small batches on production equipment that is efficient only in large batches – the legacy from the mass production era.

THE COST OF VARIETY

Chapter 3 of the author's book, *Agile Product Development for Mass Customization,*[10] introduced the concept of the *cost of variety,* which is the sum of all the costs of attempting to offer customers variety with inflexible products that are produced in inflexible factories and sold through inflexible channels. One way of estimating the cost of variety is to compare a company's current operation budget to the idealistic case of producing a single product with no variety manufactured in the same volumes as current operations. The difference between current operations costs and the single product scenario would be the *cost of variety.*

Figure 2-3 shows the relation between market variety and the cost of variety.[11] As market variety increases, the variety cost for mass production increases exponentially because of the compounding effect of all the inefficiencies discussed above. The "mass production" and "mass customization" lines in both Figures 2-3 and 2-4 could also be labeled "build-to-forecast" and "build-to-order," respectively.

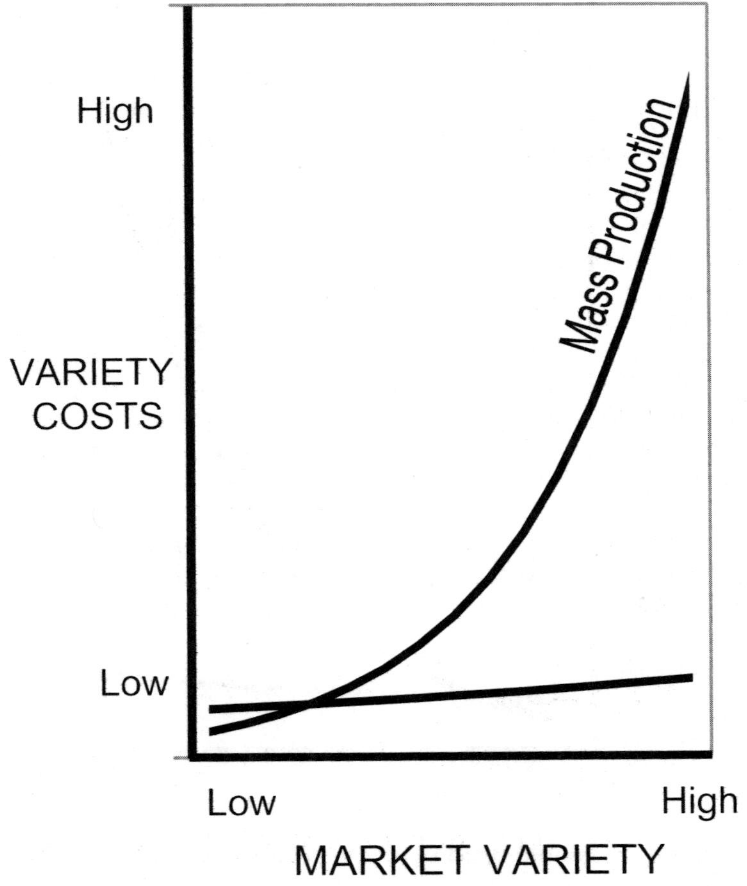

Figure 2-3: Variety Cost as a Function of Market Variety

Similarly, response difficulty rises exponentially with market volatility in Mass Production, whereas Mass Customization (and BTO) can respond easily to market volatility, as shown in Figure 2-4. This is because market volatility decreases forecast accuracy and therefore makes it harder to sell forecasted production from inventory.

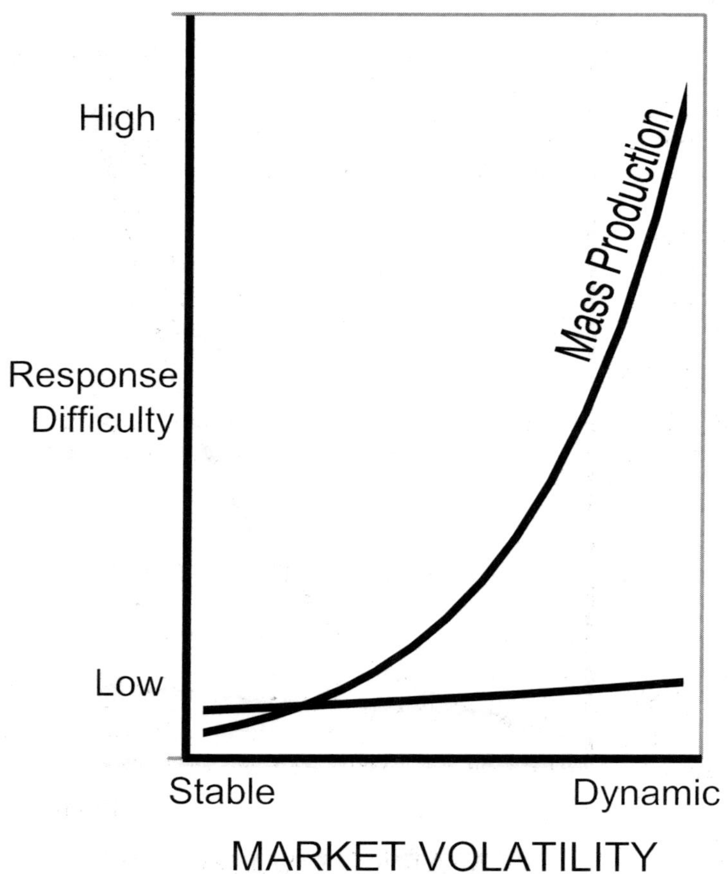

Figure 2-4: Response Difficulty as a Function of Market Volatility

Another way of computing the cost of variety is to add up all the actual costs of all the elements of cost of variety:

Inventory costs includes the carrying cost of the inventory itself (raw materials, work-in-process, finished goods, and spare parts) plus inventory related costs for labor (administrative, warehousing, and data processing), facilities (building construction, utilities, maintenance, and leases or mortgages), obsolescence (including damage, deterioration, and discounting), and internal transportation (labor, equipment).

Setup/Tooling costs include setup labor, loss of production capacity, lower machine tool utilization, kitting labor, changeovers of equipment, changeover induced quality problems, difficulty applying *kaizen* continuous improvements to non-dedicated lines, tooling for all variations, and tooling administration, handling, and maintenance.

Material overhead costs to handle the proliferation that is caused by product variety include MRP/BOM activities, materials administration, part selection/qualification, vendor management, purchasing, expediting, parts warehousing, and internal parts distribution.

Customization and configuration incur extra costs in product line management, sales, order entry, custom engineering, documentation, operations, distribution.

TRYING TO MAKE MASS PRODUCTION FLEXIBLE

Many companies ignore the shortcomings of mass production and then spend millions of dollars and precious resources trying to refurbish an inherently inflexible process to try to make it more flexible and responsive. These activities include building expensive warehouses and distribution centers in an attempt to move finished goods inventory faster to customers – that is, if the inventory is available.

As mentioned in the Introduction, some companies try a "web-based" version of selling from inventory and recommend offering "the illusion of build-to-order" by searching the web for inventory. Similarly, on the transportation side, overnight shippers come to the same conclusions when they recommend that the way to reduce inventory levels is to know where the inventory is and how long it will take to ship it.

Some think any problem can be solved by "throwing software at it"

Many companies, encouraged by ambitious software companies, think that they can solve any problem by "throwing software at it" and nowhere has this been more prevalent than with all the attempts to improve mass production with software. A few decades ago, MRP systems were created to allow purchasing functions to cope with rising product variety, an unnecessary proliferation of parts, and an ever expanding vendor base. But everyone forgot the basic premise of Industrial Engineering 101: *simplify before automating or computerizing*. There are many effective ways to simplify the supply chain with standardization (Ch. 4), automatic resupply techniques like *kanban* (Ch. 7), and product line rationalization (Ch. 3) which can eliminate the unusual *parts* by eliminating or outsourcing the unusual *products*.

Industrial Engineering 101: *simplify before automating or computerizing*

One of the biggest shortcomings of mass production is the reliance on forecasting as a basis to order parts and schedule batch production. So some companies are focusing their efforts on programs to improve forecasting. VF Corporation, makers of Lee, Wangler, Britannia, and Rustler jeans, spent $100 million in information systems that include sophisticated software to aggregate information on demographics and point-of-sales trends to try to improve forecasting. However, a much better approach would be to adopt a production system that could build products spontaneously without forecasts (BTO&MC).

Some companies accept the built-in inefficiencies of mass production and then conclude that the only way to deal with rising labor costs and falling prices is to move manufacturing off-shore where labor costs are cheaper. But this makes operations less flexible and less adept at handling variety (Ch. 6). They also accept *labor intensive designs* instead of designing products for easy manufacturability.[12]

TIME FOR NEW PARADIGMS

Now it is time to shift the paradigm. The overall paradigm shift could be described as evolving from *mass production* to *mass customization.* Another way of labeling the shift is from *build-to-forecast*[13] to *build-to-order.* Figure 2-5 and the related discussion shows the difference between build-to-forecast and build-to-order. Part of this shift is the change from *batches and queues* to *flow manufacturing.*

> It's time to shift from *Build-to-Forecast* to *Build-to-Order*

Spontaneous build-to-order is the ultimate in supply chain management and lean production strategy that allows companies to build low-cost standard or mass-customized products on-demand without forecasts, inventory, or purchasing delays.

Mass customization can build customized products for individual customers or niche markets like specific countries or regions. Mass customization also enables a steady stream of "new" products that are really planned "variations on a theme."

COMPARISON: Build-to-Forecast and Build-to-Order

Figure 2-5 compares the build-to-forecast (BTF) to build-to-order (BTO) for several criteria in production and fulfillment. The concepts apply to the production of products or parts, even though the word "product" is used predominately for simplicity.

Production Initiation. In BTF, production initiation is by *pushing* products through the factory based on a forecasted schedule, whereas in BTO, production is *pulled* through by customer orders. Imagine parts going in one door of a unidirectional factory and products exiting on the opposite side. In BTF, the parts would be *pushed* in one door; in BTO, products would be *pulled* out the opposite door.

Build Quantity. The build quantity for BTF is based on the forecast, which most companies admit are only 50% accurate. Underproduction results in unfulfilled orders and extra setup changes; overproduction results in unnecessary inventory costs and risks from obsolescence and deterioration. BTO operations build whatever is ordered for a 100% fulfillment rate and no inventory costs.

	Build-to-Forecast	**Build-to-Order**
Production		
Initiation	Push (by forecasts)	Pull (from demand)
Build quantity	Forecast	Orders placed
Min. quantity	One batch	One product
Timing	Scheduling	On-demand
Flow	Batches & queues	One-piece flow
Sold from	Inventory	Production
WIP inventory	Batch between stations	One on each station
Recur. defects	Batches made wrong	One, caught the next work station
Fulfillment		
Fulfillment rate	Never 100%; costly to try	100%
Factory inv.	*f* (var, fore. acc, ful. rate)	None
Distribution inv.	Intermediate buffers	Only transfers, if at all
Inventory cost	1/4 of inv. value/year	None
Out-of-stocks	*f* (inv., variety, volatility)	None
Obsolescence	*f* (life, inventory, volatility)	None
Spoilage	*f* (life, inv., fore. acc.)	None
Discounting	*f* (inv., variety, volatility)	None
More variety	Exponentially worse	No difference
More volatility	Exponentially worse	No difference
Customer Satisfaction	One-size-fits-all	Products mass-customized for regions, niche markets, stores, or individuals

Figure 2-5: Comparison: Build-to-Forecast and Build-to-Order

Minimum Quantity. For BTF operations, the minimum order quantity would be the batch. Mass production efficiencies are best with large batch sizes – the bigger, the better. Lean production seeks to reduce batch size. BTO operations reduce the batch size to a "batch size of one" in which case the build quantity equal the orders placed, which could be as low as one. As discussed in Chapter 8, the quantity "one" refers to the smallest order quantity anticipated, which could be one product or one case or one pallet.

Production Timing. The production timing for BTF is based on scheduling, which is more complex with lengthy batch changeover times and high product variety. Complexity and variety cause more setup changes. The timing for BTO is based on product demand – flexible production lines can build any quantity of any product in the family in any order. (Ch. 8).

Flow. The "flow" in BTF is really the movement of *batches* waiting at each work station in *queues,* hence the label "batches & queues." In BTO "flow" means a literal flow of products as a one-piece flow, as presented in Chapter 3 in *Lean Thinking.*[14] Using one-piece flow to achieve *on-demand build-to-order* is presented in this book in Chapter 8.

From Where are Products Sold. BTF products are sold from inventory, if the forecasts were correct or more than necessary were built. If the products are not available from inventory, the manufacturer is in the dilemma of choosing between not filling the order until the next batch is scheduled (which might cause the loss of the sale) or interrupting scheduled production, running a less-than-efficient size batch, and paying for *two* extra setup changes (Ch. 8). BTO products are sold from the production line on-demand.

WIP Inventory. BTF operations usually have at least one batch of pieces between each work station, which could be a machine tool, robot, or manual assembly station. Sometimes, there is a batch (or more) on the incoming *and* outgoing side of each work station. BTO operations have one piece in each work station and possibly one in-between. The BTF operations may have hundreds or thousands of times more WIP inventory than the one-piece flow of BTO. Since inventory carrying cost is 25% of the value of the inventory per year, WIP inventory costs often exceed profits!

Recurring Defects. As pointed out early, recurring defects may be built into several batches before being caught at a downstream inspection step in BTF operations. In the one-piece flow of BTO, there is virtually no WIP inventory. So, if the next work station sees anything wrong or the part does not fit, the feedback, and corrective actions, will be immediate.

Fulfillment Rate. BTF products issued from inventory can only achieve a 100% fulfillment rate if the forecasts were perfect (which is rare) or there was overproduction (which is expensive), but this might be feasible for a single product factory – the proverbial Model-T plant. However, if there are even moderate degrees of product variations, two principles will doom the order fulfillment rate to unacceptably low levels: (1) forecasts get exponentially worse as variety increases, especially if markets are volatile, and (2) compensating for such inaccurate forecasts with extra inventory of all product variations would be an intolerable financial burden for any company. On the other hand, building products on-demand would have a consistent 100% order fulfillment rates if companies simplify supply chains (Chapters 3 through 6), arrange spontaneous supply chains (Ch. 7), and implement on-demand lean production (Ch. 8).

Factory Inventory. BTF factory finished goods inventory will be a function of (indicated by the math notation, *f*) variety, forecast accuracy, and fulfillment rate. Inventory will go up exponentially with increasing variety and decreasing forecast accuracy, especially as companies try to approach a 100% order fulfillment rate. BTO factory finished goods inventory would be zero.

Distribution Warehouse Inventory. BTF manufacturers often carry buffer inventory closer to stores and customers in an attempt to improve order fulfillment rates, but this would only be effective for standard products without variety for reasons discussed above. If BTO companies used distribution hubs, they would be only for transfers, not for buffer inventories. When applicable, BTO companies may air freight products directly to stores or customers, with the extra freight cost paid for by the cost savings from eliminating inventory and distribution centers. A compromise strategy for bulkier, heavier products would be to ship “vanilla” base products to configuration centers and then add the “flavors” there as discussed in the *postponement* section in Chapter 9.

Out-of-Stocks. If BTF companies do not have enough inventory to fill an order on-time (and if they can't build it on-demand), they will have to tell their customers that the products are *out-of-stock*. If the company has a monopoly or customers are patient, the sales might be deferred to the next scheduled production. Otherwise, the sale will probably be lost. An even worse consequence would be alienating stores by missing order fulfillment targets established by stores and dealers. BTO results in no out-of-stocks because everything is built on-demand and this can be achieved without the expense of holding inventory.

Obsolescence, Damage, and Spoilage. An important aspect of inventory carrying cost is the cost of obsolescence, damage, and spoilage. Products vulnerable to obsolescence would be those with short life-cycles, rapidly evolving technologies, quickly changing stylistic preferences, or volatile markets. When inventory goes obsolete, manufacturers face a difficult dilemma: they can either write it off or try to sell it at a discount, which would delay the launch of *new* products until the old products can be sold off.

Fragile products might be damaged just going into and out of the warehouse. The damage would be even greater if they have to be moved around to make room for other inventory.

Some products may deteriorate over time. Others may deteriorate more rapidly if the climate has too much heat or humidity; some plant relocations have discovered this the hard way. Food, drugs, and any products with limited "shelf life" (even ink-jet cartridges) will be vulnerable to spoilage if stored too long.

The amount of obsolescence for BTF would be a function of product life, inventory levels, and market volatility. The amount of spoilage would be a function of shelf life, forecast accuracy, inventory, and volatility. The amount of obsolescence, damage, deterioration, spoilage for BTO would be none.

Discounting. Obsolete inventory presents several dilemmas for the BTF producer. Simply doing nothing and hanging on to it costs 25% of its value per year (as graphically shown in Figure 2-1) and will become money losers after only a few months (as shown in Figure 2-2). Unfortunately this is common practice when companies do not understand the problems of inventory and no one keeps track of it. When companies realize that inventory is going obsolete, they usually try to sell it at discounts, but this will probably cancel out the intended profit. Discounting unsold products is endemic in the clothing industry with only 50% to 60% of products sold at full price.[15]

As products become more obsolete, the discounting may cause significant losses. If the inventory is too obsolete to be sold, it then would be written off. BTO avoids all the financial losses of discounting by build products on-demand and never letting products go obsolete or accumulate until profits disappear (as shown in Figure 2-2).

More Variety. As product variety increases, the cost and difficulty satisfying demand go up exponentially, as shown by the mass production line in Figure 2-3. In BTO there would be virtually no difference, as shown by the mass customization (BTO) line in the figure.

More Volatility. As market volatility increases, the response difficulty goes up exponentially, as shown by the mass production line in Figure 2-4. In BTO, there would be virtually no difference, as shown by the mass customization (BTO) line in the figure.

Customer Satisfaction. Since the BTF business model has difficulty manufacturing much variety, mass producers try to stick with the traditional mass production approach of "one size fits all." BTO can efficiently offer a greater variety of standard products in addition to mass-customized products for various regions, niche markets, stores, or individuals.

ENDNOTES/REFERENCES

1. Robert Lacey, *Ford; The Men and the Machine,* (1986, Little, Brown, and Co.), p. 285.

2. James P. Womack, Daniel T. Jones, and Daniel Roos, *The Machine That Changed the World; The Story of Lean Production,* (1991, HarperPerennial).

3. James P. Womack and Daniel T. Jones, *Lean Thinking; Banish Waste and Create Wealth in Your Corporation,* (1996, Simon & Schuster).

4. Data from the "12th Annual State of Logistics Report," Published by Cass Information Systems and Prologis, June 4, 2001.

5. Data was drawn from the US Department of Commerce web-site page "Real Inventory-Sales Rations for Manufacturing and Trade, Seasonally Adjusted," for late which is tabulated for various industries at www.bea.doc.gov/bea/an/0400niw/table3.htm for later 1999 and early 2000.

6. Paul Kaihla, "Inside Cisco's $2 Billion Blunder; How the world's most admired supply chain screwed up, and how CEO John Chambers plans to fix it," *Business 2.0,* March 2002, pages 88 - 90.

7. David Pringle, "How Nokia Thrives by Breaking the Rules," *The Wall Street Journal,* Jan. 3, 2003.

8. David M. Anderson, *Design for Manufacturability & Concurrent Engineering; How to Design for Low Cost, Design in High Quality, Design for Lean Manufacture, and Design Quickly for Fast Production,* (2008, CIM Press, 449 pages; 805-924-0200), Chapter 10, "Design for Quality."

9. Ibid., Chapter 2, "Concurrent Engineering," and Chapter 3, "Designing the Product."

10. Ibid., Chapter 3, "Designing the Product."

11. This figure was created for the author's seminars and first published in copyrighted class handouts. It was then published in *Agile Product Development for Mass Customization,* by David M. Anderson (1997, McGraw-Hill).

12. Anderson, *Design for Manufacturability & Concurrent Engineering.*

13. Some people call this "build-to-stock," which may sound deceptively less alarming because "stock" may sound more comforting (when inventory is "in stock") than "forecast," in which few people have much confidence.

14. James P. Womack and Daniel T. Jones, *Lean Thinking; Banish Waste and Create Wealth in Your Corporation,* (1996, Simon & Schuster).

15. Edward Feitzinger and Hau L. Lee, "Mass Customization at Hewlett-Packard; the Power of Postponement," *Harvard Business*

Review, January-February 1997, pages 116 - 121.

3

PRODUCT LINE RATIONALIZATION

Product line rationalization is a powerful technique to improve profits, simplify operations and supply chains, and free valuable resources to implement BTO&MC. It does this by *rationalizing* existing product lines to eliminate or outsource products and product variations that are problem prone, have low sales, have excessive overhead demands, have limited future potential, may really be losing money, and are incompatible with new operational environments and corporate strategies, such as build-to-order and mass customization.

> Rationalization eliminates the losers and focuses on high profit products

Rationalization can quickly improve profits by stopping the production of money-losing products and eliminating all of the excess overhead costs associated with "fire-drill" products. This, in turn, will allow precious resources to focus on the most profitable products instead of low-leverage products, which will increase sales and further lower costs. After rationalization, the remaining products will cost less because they will no longer have to subsidize the money losers or marginal products.

All these cost savings can be used to lower prices in price-sensitive markets or to increase profits. In fact, rationalization can raise profits enough to be justified as a free standing program. The following scenario will show that, by simply eliminating the lowest leverage products, profits can be tripled!

Pareto's Law for Product Lines

All companies experience some Pareto effect, typically with 80% of profits or sales coming from the best 20% of the products.

This happens because almost all companies keep *adding* products to the portfolio without ever *removing* any. Further, sales incentives and emphases on growth and market share encourage the sales mantra of "take all orders," thus overloading production operations and the supply chain with too many low-volume products that have unusual parts and manufacturing procedures. This causes many setup changes, incurs excessive overhead costs, lowers plant capacity, complicates supply chain management, and dilutes engineering and manufacturing resources.

> Usually, 80% of profits come from 20% of products – so why keep making the rest?

Few companies realize these problems because their cost systems allocate (average) overhead costs, which implies that all products have the same overhead costs, a very unlikely situation.

Focus

Product line rationalization encourages companies to focus on their *best* products by eliminating or outsourcing the *marginal* products. The resources that were being wasted on the low-leverage products can then be focused on growing the "cash cows."

Robert Atkins and Adrian Slywotzky, writing in a Wall Street Journal about profiting during a recession contend that "spreading resources evenly among all customers is bad even in good times; in bad times, it's disastrous."

HOW RATIONALIZATION CAN TRIPLE PROFITS!

The following scenario shows the power of this methodology using a simple example illustrated in Figure 3-1.

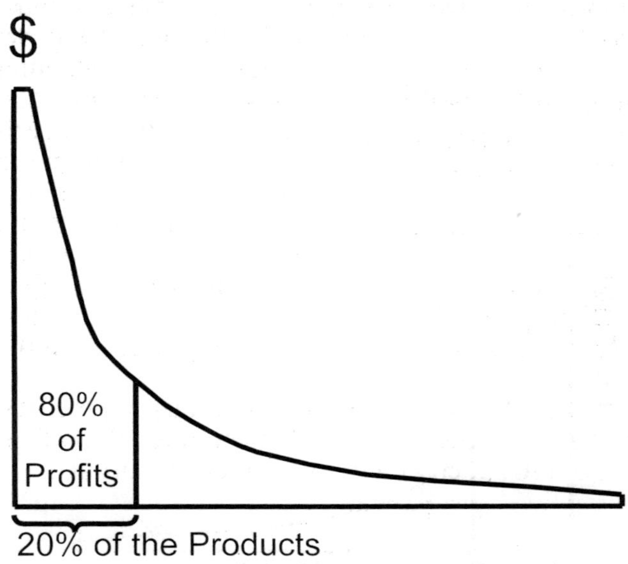

Figure 3-1: Pareto's Law for Products

The actual product line rationalization takes many more factors into account, but this example shows the profit increasing potential for rationalization.

If a company kept the 20% of the products that were making 80% of the profits and dropped the other 80% of the product line, it would result in only a 20% drop in revenue.

> Fire-drill products consume most of the overhead costs

The cost reduction can greatly exceed the revenue drop because of the way cost is distributed. Direct costs such as materials, parts, and labor would be proportional to revenue – in other words, the more products sold, the more materials, parts, and labor would be consumed. However, overhead has the *opposite* effect. Indirect cost (such as procurement, manufacturing engineering, and other support functions) would be low on the cash-cow products because they are better designed for manufacturability, parts are procured routinely, quality issues have been resolved, and processes have been stabilized due to the focus that is usually applied to higher-volume production. In contrast, overhead cost on the low-volume products is high, probably 80% of the total, because of all the inefficiencies inherent in building many low-volume, seldom-built products. Further, those products may be less well designed for manufacturability and have much higher quality costs. So the breakdown of direct and indirect costs would be as shown in Figure 3-2.

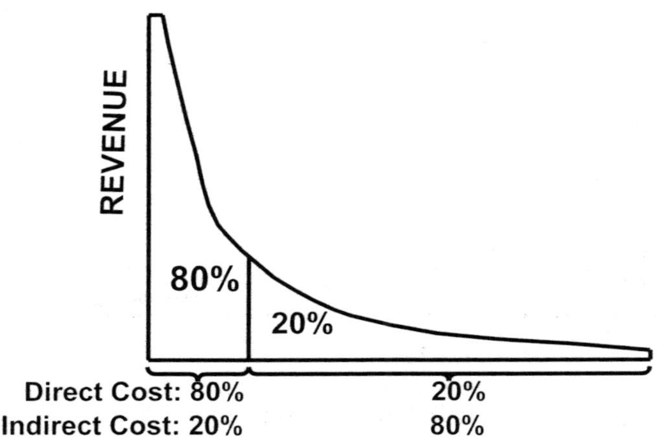

Figure 3-2: Cost Breakdown

In order to make this example relevant, Figure 3-3 converts these percentages into dollars for a $100,000,000 business, which, according to the common Pareto's law effect experienced by most companies, generates $80 million in revenue from cash-cow products and $20 million from all the others. The cost breakdown shows indirect (overhead) costs as half the total cost, which is not an unreasonable assumption. In fact, in many industries, overhead cost is greater than half of the total cost (as shown in Figure 11-2), thus resulting in an even stronger case.

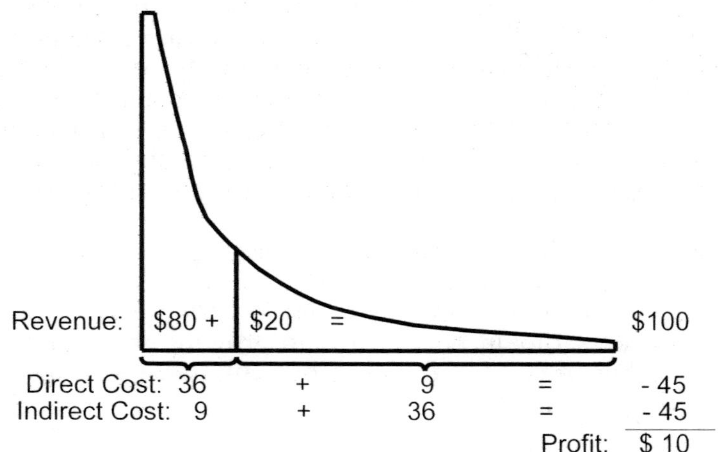

Figure 3-3: Cost Distribution in Dollars

Now to show the powerful effect of this procedure, simply eliminating the 80% of the low-volume products and keeping the 20% cash-cows will have the following effect, as shown in Figure 3-4.

The bottom line: revenue drops 20% (this will be discussed later) but eliminating the high-overhead products eliminated most of the indirect cost, so that profits are 3.5 times *that of the full product line!*

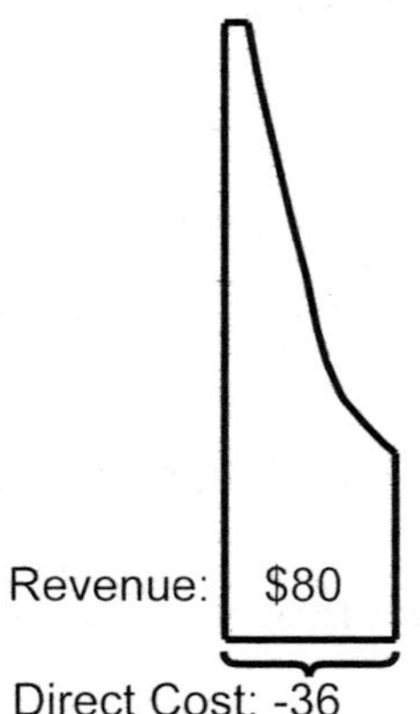

New Profit: $ 35 = 3.5X Previous Situation

Figure 3-4: Results after rationalization

> "Liberated" people should be invested in growing the most promising products

For this motivational scenario, overhead cost savings are assumed to be eliminated to show the effect on profits. This could be realized in a rapidly growing company where the focus shifts from current loser products to upcoming growth products.

However, it would be a shortsighted strategy to lay off the workers to realize a short-term overhead cost savings (see the last section in Chapter 13 on why not to lay off workers). Rather, "liberated" workers should be *invested* in growing the remaining aspects of the business. Breakdowns of actual cost savings and reinvestments appear next.

COST SAVINGS FROM RATIONALIZATION

The cost savings from rationalization comes in two forms: cash and human resources, which will be itemized below. Rationalization, like lean production, effectively *liberates* many types of resources and their disposition should be planned ahead of time. The lean production bible, *Lean Thinking,* has one section titled, "Deal with Excess People at the Outset." [1]

Product line rationalization results in the following short term cash savings and resource investment opportunities.

Short Term Cash Savings:

- Avoid the purchase of parts and materials for rationalized products, which may have less purchasing leverage, higher procurement costs to find, and higher than normal setup and expediting costs.
- Avoid quality costs of unusual products, which may be higher than normal.
- Limit, postpone, or cancel hiring for growth and attrition replacement
- Avoid overtime
- Phase out temps (temporary workers) as long as they do not have critical knowledge or skills.

- Bring in-house currently outsourced services, especially those that improve flexibility (Ch. 6).
- Delay facility expansion; unusual products generally require more space than those that fly through the factory; in addition, continuous improvement programs, like lean production (Ch. 8), also save space.

Investments:

- Focus on improving sales on the remaining products, which now can sell at higher profits (or lower prices) without having to cross-subsidize the "losers" (see the next section on redirecting resources).
- Improve quality and lower the cost of quality
- Continuously improve operations and productivity
- Expand into related services
- Get certificated (ISO 9000, QS 9000, etc.) or win awards (Best Plants, Baldridge, Shingo Prize, etc.) to improve stature with customers
- Upgrade CAD tools, information systems, and web presence.
- Upgrade skills with investments in training
- Implement new capabilities like build-to-order and mass-customization
- Improve product development. Without the daily fire-drills in operations and procurement, manufacturing people will be more available to participate in product development teams, which is a key element to successful product development (see Ch. 10).
- Invest in internal start-up ventures
- Use liberated cash and resources to enter new businesses or buy and transform related supportive businesses up and down the supply chain (*not competitors!).* The book *Lean Thinking* cited an example: "Each time Wiremold's vacuum sucks up a batch-and-queue producer, it spits out enough cash to buy the next batch-and queue producer!"[2]

SHIFTING FOCUS TO THE MOST PROFITABLE PRODUCTS

After product line rationalization, resources that were being wasted on the low-leverage products can now be focused on improving the remaining products, as shown in Figure 3-5.

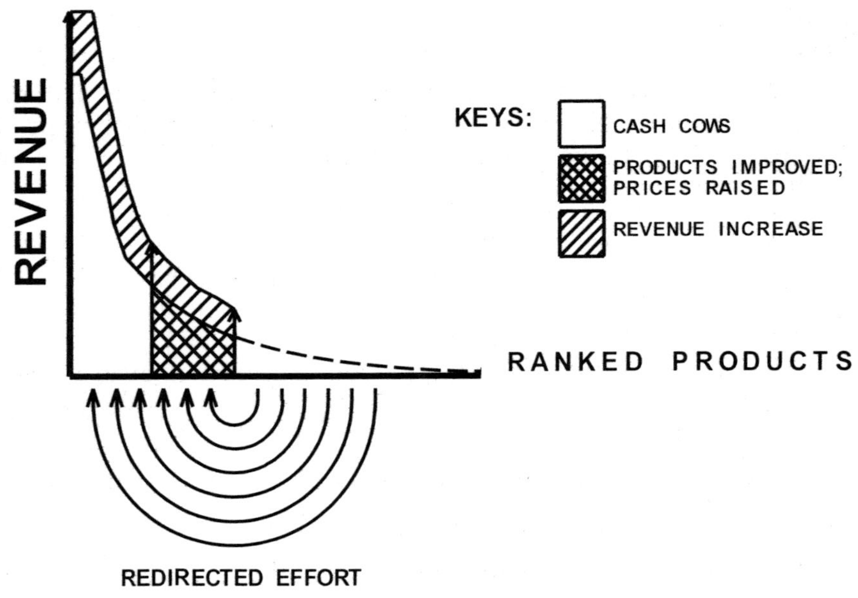

Figure 3-5: Redirecting Focus to Cash Cows

One of the author's clients, a telecom equipment company, reported that after dropping their marginal products, they recovered their original revenue within two months! This came from focusing liberated resources (both personnel and money) on the cash cow products with improved efforts in advertising, sales channels, product design, operations, and supply chain management. All of these efforts will be easier and more efficient with fewer products to focus on. Operations and supply chain management will benefit from greatly reduced number of parts and processes to deal with. Multifunctional

> After rationalization, one company got its revenue back in just 2 months!

product development teams will make faster progress and develop better products with more people available from operations and purchasing. In addition, the money saved not having to subsidize "the losers" can fund these improvement activities, be used to lower prices, or simply go to the bottom line to improve profitability.

In addition to improving the known cash cows, it may be possible to "raise worthy dogs," or focus some improvement on selected worthy products, especially if they are related to cash cow products. For instance, consider raising their prices, if the market accepts them; they may survive and simply make more money. Look for products that have *easy* opportunities to lower their total cost, but be wary of any "cost reduction" effort that does not pay off the cost of the effort within the expected life of the product. Make sure good products are not unfairly burdened by inappropriate overhead charges, like paying the "loser tax" to subsidize marginal products. In addition to these techniques, selected products may also benefit from all above mentioned improvements in advertising, sales channel, product design, operations, and supply chain management.

During the transition to BTO&MC, product variations that *look like* they will be part of the implemented system should not be rationalized away, but should continue to be offered, even if this means using current operations, until more efficient BTO&MC capabilities come on line.

Professor Kim Cameron of the University of Michigan Business School recommends that during downturns companies should "exit from weak businesses entirely."[3]

Sometimes companies can profit from shifting the focus from products that are only *good* to those that would have the potential to be *great.* Jim Collins, author of *Good to Great,* points out that "few executives have the guts to get rid of profitable businesses where their company can only be good, but never great."[4]

What Is the Goal of a Business?

Many managers still have trouble with the issue of dropping revenue 20% even if it triples profit. And this brings up the issue of what is the goal of a business.

> The real goal of a business is to maximize profits, not productivity, growth, or market share

As Eli Goldratt[5] and others have pointed out, the real goal of a for-profit enterprise is to make money, not optimize other common measures like productivity, growth, market share or "growth at any cost."

The opening chapter in Slywotzky & Morrison's book, *The Profit Zone; How Strategic Business Design Will Lead You to Tomorrow's Profits,*[6] is titled: "Market Share is Dead," in which are found the following quotes.

> "Market share is dead" – Slywotzky & Morrison, *The Profit Zone*

"The two most valuable ideas in the old economic order, market share and growth, have become the two most dangerous ideas in the new order."

"Paradoxically, the devout pursuit of market share may be the single greatest creator of no-profit zones in the economy."

Similarly, the book about the largest research project ever devoted to corporate failures, *Why Smart Executives Fail, and What You Can Learn from Their Mistakes,*[7] states that market share is the wrong *scorecard* because "market share does not translate into profitability, since significant investments are typically needed to build share in the first place."

One of the themes of Richard Koch's book *The 80/20 Principle,*[8] is:

> "Market share does not translate into profitability" – Professor Sidney Finkelstein, author of book on business failures

"Successful firms operate in markets where it is possible to generate the highest revenue with the least effort."

Volume Growth Strategies

Another quote from *The 80/20 Principle* is:

"The road to hell is paved with the pursuit of volume." [9]

Of course, "pursuit of volume" here means a volume growth strategy. Many companies, who have been stumbling lately, have volume growth as the cornerstone of their corporate strategy, with quarterly and yearly growth goals. But, the underlying theme of Pareto's law is that:

All opportunities are not equal and do not make an equal contribution to profitability.

But a volume growth strategy encourages the "take all orders" mantra. Thus, when there is pressure to grow the business, the company ends up taking any business it can, not just the most profitable. This behavior is built into the system if sales incentives are based on volume growth, instead of profit. This is the sales equivalent of piece-part incentives that have been abandon long ago because they favored one metric (volume) over another (quality).

All opportunities do not have equal profit potential

The key to ***profitable*** *growth is to focus on the products with the most potential, not to dilute resources on the most products.*

Rationalization Prerequisite - Eliminating Duplicate Products

Before doing the rationalization procedure, there are certain first steps that can simplify the process. Eliminate overlapping or duplicate products. Search out and eliminate or consolidate duplicate products, overlapping products, and superceded products, even if some customers are still using the older product. You may have to encourage or force customers to switch from the older, less-advanced products that they have been ordering because of arbitrary decisions, inertia, or lack of awareness about newer/better replacements.

THE RATIONALIZATION PROCEDURE

The rationalization procedure divides product line into four zones as shown in Figure 3-6.

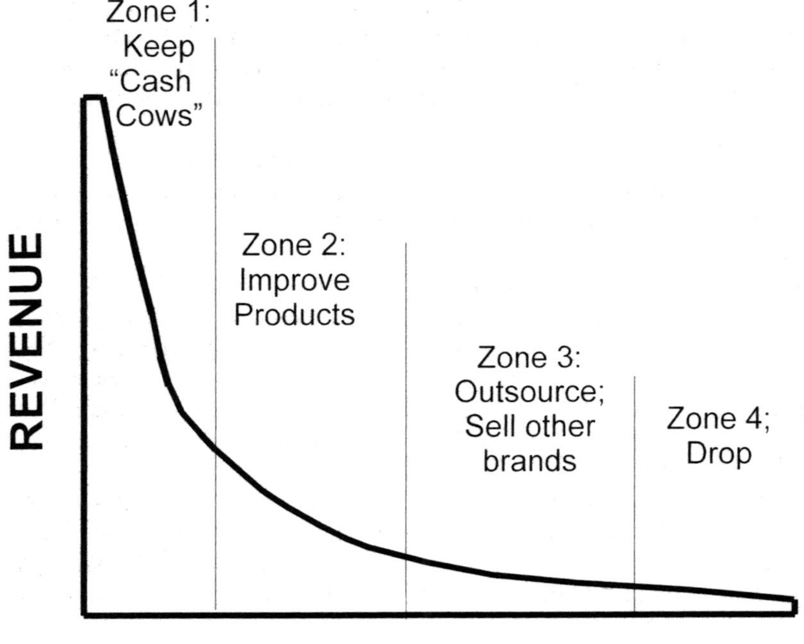

Figure 3-6: Rationalization Procedure

The least profitable, lowest-volume products would be mostly dropped (zone 4) subject to certain considerations discussed in the product family section below. Products that need to be in the catalog could be outsourced (zone 3), thus simplifying the in-house supply chain and manufacturing operations. Cash cows would be kept (zone 1). As mentioned in the focus discussion, the remaining products would be improved (zone 2).

> Plot and segregate products into zones: Keep as is, Grow, Outsource, and Drop

Zone 1) "Cash Cow" products should remain since they are probably making 80% of the revenue and profits. Some may be just fine in dedicated mass production lines. No change in production *per se* may be required for some products while others may be made together on flexible lines.

Zone 2) This zone includes products that could be improved and grow with redirected efforts. These products could receive the greatest benefit from build-to-order and mass customization.

Zone 3) This category consists of products that do not fit into either (1) or (2) but still need to be in the catalog for completeness, to satisfy loyal customers, or for service obligations. These products may remain in the catalog, *but they do not have to be designed and built in-house.* They could be outsource and manufactured by a supplier under the company name/label. Or the company catalog could simply carry another source's product to complete product line, assuring customers about appropriate equivalency.

Zone 4) This zone contains the products that should be dropped from product line. Sell off the rights to losing products if possible; someone else may be able to make money if they are a better fit with their products and operations. This certainly would not be a competitive threat, since those products would not be making as much money as new products designed with better focus and built by BTO&MC facilities.

TOTAL COST IMPLICATIONS

Chapter 11 will discuss total cost measurements and their impact on business decisions. The cost accounting system has significant implications for product line rationalization. Most companies average (allocate) overhead costs so much that the reported "costs" of individual products do not reflect reality. What happens is that good products usually subsidize bad products. And since profitability is based on cost, it will be distorted too. When total cost measurements are implemented, the profitability of products becomes more realistic, as shown in Figure 3-7, and it becomes apparent how many of the products are really losing money. These then become the prime candidates for elimination in zone 4 in Figure 3-6.

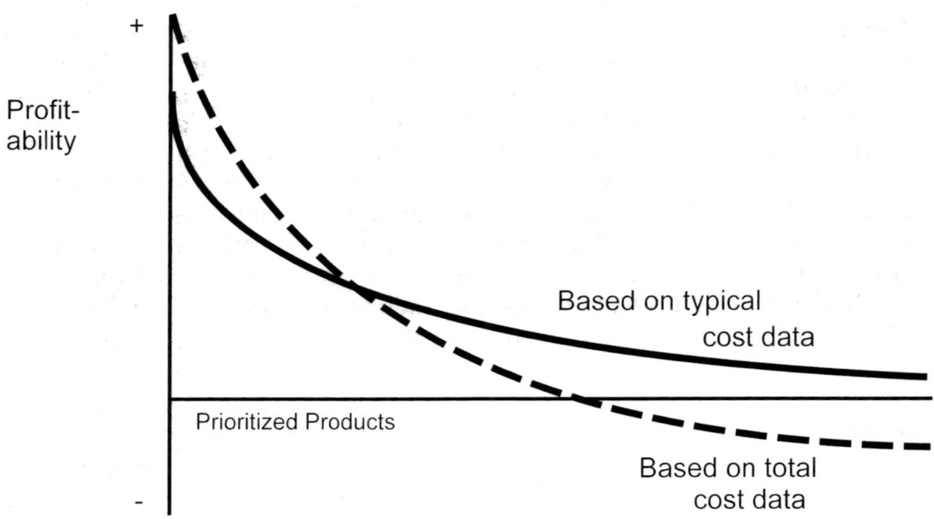

Figure 3-7: Prioritized Profitability; Typical or Total Cost

You cannot have a profitability strategy unless you can have total cost measurements and can break down "corporate profits" into profits for products, market segments, and so forth.

Sometimes implementing total cost measurements can resolve a rationalization impasse. A Harvard case study (9-186-272)[10] analyzed seven products made by Schrader Bellows. In a factory where most products had annual volumes of thousands per year, there was one product that they only build 53 per year. Of course, production management wanted to get rid of it because they understood all the inefficiencies, but the current cost system said, quite illogically, that it had the highest "profit margin" in the plant, so they had to keep building it. However, after the total cost analysis was performed, the product that originally was thought to have the highest profit margin of the group was shown, in reality, to have a negative 59% "margin."[11] This shows the dangers of using profit margins for reasons cited next.

The Margin Trap

Making decisions based on product profit margins can be a dangerous practice when costs are not based on total cost accounting practices and are not updated often.

Low-profit or money-losing products will keep selling if there is a *perception* that they have high "margins." Many misleading situations occur when products have margins computed after high-volume builds and these numbers stay in the system, even though the order volumes have since decreased. In this case the obsolete margin data mislead decision makers and may thwart rationalization efforts.

Seldom Built Products

Making decisions based on product profit margins can be a dangerous practice

Products "revived from the dead" have very high overhead costs, despite their reported "margins."

Another rationalization procedure is to investigate which products have not been built recently, for instance, in the last year, two years, three years, four years, and five years. These products that are "revived from the dead" have very high overhead demands because of the effort to remember how to build them, find all the documentation, procure unusual materials, and

Products "revived from the dead" have very high overhead costs, despite their reported "margins" find the tooling. In one company, tooling for a seldom built product had been sitting outside in the snow for months and required much rehabilitation. Some companies can simply implement a policy that all products that have not been built in, say three years, will be dropped immediately, or at least discouraged or given special scrutiny.

Obsolescence Costs

When mass producers build too many low-volume products, it increases the risk and cost of obsolescence due to market changes and engineering change orders. An APICS article on "product proliferation" stated:

> "Low-volume products are particularly prone to [obsolescence] since batch sizes are often increased to produce a six-month or more supply in order to reduce the number of changeovers required in manufacturing. With this amount of inventory on the shelf, the risk of becoming obsolete due to engineering changes or changing levels of demand increases dramatically."[12]

OVERCOMING INHIBITIONS, FEARS, AND RESISTANCE

Despite the fact that product line rationalization can easily raise profits, free valuable resources, and simplify operations and supply chain management, many managers have inhibitions, fears and resistance:

Growth Emphasis. When there is too much emphasis on revenue growth, it may be hard to get companies to do what may *appear* to reduce their revenue stream, even temporarily, in order to eliminate the low-leverage products, improve the cash cows, improve profitability, and ultimately grow revenue.

Cost System Deficiencies. When overhead is allocated (averaged), all products will appear to have close to equal profitability. Total cost measurements will then be necessary to flush out the low profit products.

Inertia. Many people resist change in general, especially if the cost system or personal communications do not show them the "big picture," – how product line rationalization can help supply chains, operations, and programs to implement build-to-order.

Departmentalism. Often resistance comes from those who have been optimizing their department's performance, for instance Sales, at the cost of the company as a whole. So metrics may need to be changed to direct behavior toward maximizing *company* goals.

Unhealthy Attachments. Every product at one time was the "baby" or someone or several people. But remember that the goal of a business is to produce the most profitable products, not a collection of favorites.

Fears Overstated. Usually fears about negative consequences are overstated, especially the fear of losing customers who, in reality, will most likely switch to better equivalents from the same catalog. When low-leverage products are eliminated:

1) The company can *increase* sales on the remaining products with better a focus in each of the following areas: research, product development, manufacturing, quality, and marketing.
2) Without having to subsidize low-leverage products, the remaining products can generate more profit or be priced lower.
3) Customers can switch to remaining products from older, less-advanced products that they have been ordering because of arbitrary decisions, inertia, or lack of awareness about newer/better replacements.

The following is a list of most common questions that come up in the author's workshops, followed by his answers.

What about the "complete product line" argument?

If the catalog *must* really be complete (zone 3 in Figure 3-6), that doesn't mean *you* have to manufacture all the products, options, or variations in your factories. Outsource unusual products/options/variations so that the resources that were formerly squandered on them can now be invested in development a *better portfolio* of new products and building them in BTO&MC operations.

Won't customer satisfaction drop?

Overall customer satisfaction will actually be *lower* if you waste your resources making low-leverage products instead of capitalizing on *opportunities* to give customers better innovation, lower cost, better quality, and efficiently manufactured variety. However, they will probably miss the good deals you were giving them on your money-losing products.

Won't we be limiting customer choice?

Often customers keep ordering older/unusual products because of arbitrary decisions, inertia, and lack of awareness about newer/better replacements. Point out how the remaining products have been improved; drop prices on them.

What about loss-leaders?

If low-leverage products are to be retained as "loss leaders," then management should *know* how much money is really being lost, including the *opportunity* losses of what those resources could have accomplished. If loss-leaders are still a valid strategy after implementing total cost accounting, then the loss-leader products should be outsourced to avoid distracting the factory from its most efficient tasks.

But what about the perception that we have to sell some oddball products to get the big sales?

First of all, customers may have been spoiled by years of unnecessary concessions made by salespeople who may not know or care about the trauma that low-volume products cause in operations and supply chain management. Customized products would fall into this category if not done efficiently through mass customization.

Second, ask objectively if the customer will really terminate a long and viable relationship over a few low-volume products.

Third, point out that if customers continue to buy older products, they will become increasingly vulnerable to product obsolescence and availability problems.

Finally, if customers still insist on package deals, the manufacture of oddball products could be done by outsourcing so those sales do not distract from new product development and improvement programs line BTO&MC.

If all of these arguments are not convincing, then consider "Competitive Scenario" (next)

Competitive Scenarios

A savvy competitor would not blindly compete against your entire product line. It would offer only the most profitable products. Not burdened by your low-leverage products, it would be able to make the *rationalized* product line quicker, better, and at lower cost.

Another competitive scenario is "cherry picking," when an existing or new competitor skims off the most profitable products.[13] Thus, competitors could steal your "cash cows" leaving you with the "dogs."

A savvy competitor could steal your "cash cows" and leave you with the "dogs"

Role Playing. It may be a valuable exercise to role-play competitor scenarios, pretending you, or your brainstorming group, are a well-financed new competitor. Ask the questions:

- Which products would you want to have in your product line?
- How would you group products and structure production lines and cells for the greatest efficiencies?
- How could a concise product line benefit from standardization and vice versa?
- What is the minimum list of materials needed?
- How would you build them flexibly to be able to reach the broadest possible markets?
- What would you like to do that you cannot do because of existing product line limitations?

Start with your discrete product line first and then conduct additional sessions to group products for build-to-order and mass customization.

Here is what Downes and Mui reported in their book, *Unleashing the Killer App:*[14]

> "One organization had teams of executives play the role of well-funded outsiders, both new entrants and existing competitors, and asked them to devise business plans that attacked the organization's prime markets and stole away its most profitable customer segments. Knowing the blinders of the organization, and the exposed flanks of its offerings, these teams easily put together alliances and business propositions that realistically challenged the status quo."

Similarly, General Electric uses "Destroy-Your-Business.com" task forces to identify business threats.[15]

The Dartmouth Business School study that produced the book on corporate failures, cited earlier, recommended as a high-return activity the practice of "convening 'devil's advocate' groups that are assigned the task of spotting vulnerabilities in past and current policies."[16]

Three years after Quaker lost $1.4 *billion* acquiring Snapple, which was a complete mismatch with Quaker's operations, the CEO admitted, "We should have had a couple of people arguing the 'no side' of the evaluation."[17]

IMPLEMENTATION & CORPORATE STRATEGY

The rationalization approach depends on the business model.

Scenario for Mass Production

For mass production, which builds batches of products for inventory based on forecasts, individual products decisions are made independently. Be sure to quantify the total cost of all setup, inventory and all the other overhead costs to identify which low-volume products are making the least profit. Then eliminate them or, if they must be in the catalog *at any cost*, outsource them to free in-house people for new product development. Note that the "cost" will go up if those products were subsidized before, because outsourcers will charge the true total cost.

Scenario for Mass Customization/Build-to-Order

Flexible cells can build a range of any product in a "family" (or platform) with minimal setup costs and delays. In such an environment some low-volume products can be retained if they can be grouped into a *product family* and quickly and cost-effectively build on either dedicated mini-lines or flexible product lines.

Products that do not fit into any family group should be removed from all *flexible* operations, either eliminated or outsourced.

Thus, an inflexible mass production plant will eliminate more low-volume products that a flexible (lean, build-to-order) plant.

Implementation Steps

Data gathering. Create Pareto plots by plotting revenue (or sales units) against ranked products in the format of Figure 3-6. First, plot all products; this is for motivation and buy-in, so it does not have to be rigorous. Then plot each product family, market segment, or other logical grouping with product identifying numbers displayed.

Polls and surveys. Polls and surveys can quickly identify difficult-to-build products for scrutiny. Just ask the following question to everyone involved in building products, procuring their parts, and performing custom engineering or configurations: "What products or variations cost us more, and delay us more, then we think?" Then plot out the results and start scrutinizing from the top of the list.

Seminar/workshop. Arrange for training on Product Line Rationalization for all people that will be involved. This could be one subject in overall BTO&MC training discussed in Chapter 13. The training should be interactive enough to discuss principles, address concerns, and get buy-in to process. The workshop phase of this event should start the rationalization process based on preliminary data gathered in the previous steps.

Profiles. Profiles can be quickly created to "red flag" certain products for special scrutiny. The profile could be based on any criteria that should raise red flags: low-volume, infrequent manufacture, special materials, hard-to-get parts, unusual processing, difficult customizations, or any other unusual demands. Profiles are valuable because they can be created *immediately,* before the rationalization process is complete, based on anecdotal criteria in addition to available data. If neither of these are available, conduct polls and ask everyone in operations to vote on which products they think are making less money than assumed and which are distracting them from their jobs.

At first, red-flagged products would receive special scrutiny and, before the order could be accepted, would require signatures from Manufacturing, Purchasing, Engineering, and so forth. A senior manager may need to be appointed to quickly arbitrate disputes.

For seasonal products, the profile could block or delay orders that do *not* satisfy the current peak seasonal demand. As profiles mature, they could automatically block unacceptable orders.

Configurators. Profiles can be built into a *configurator,* which is order-entry software that has the added ability to:

- Contain all the rules, profiles, data, and formulas to certify a valid order and provide customers with instant cost quotes and delivery schedules.

- Quickly provide customers with many "what if" scenarios showing the cost and delivery time for standard or custom orders.

- Transmit the data needed for processing the order, doing engineering work, procuring the materials, setting up production, and launching the product.

Configurators are discussed more in Chapters 8 and 9.

Data analysis. Segregate Pareto plots as shown in Figure 3-6. Scrutinize the low selling products to see which should be dropped or, if necessary to be in the catalog, which should be outsourced. Look for opportunities to shift resources to improve worthy products as shown in Figure 3-5.

Recommendations. Make recommendations on which products to drop, outsource, or improve and how these actions should be executed.

Total cost. Improve the costing system (as discussed in Chapter 12) to the point where all costs are quantified for all product variations, so that:

- All product variations can be plotted by true profitability (as in Figure 3-7), which will greatly improve corporate strategy, product portfolio planning, and product line rationalization.

- Pricing can be objectively based on the total cost for all product variations, which will result in an automatic and enduring rationalizing effect. The result of this will be that previously subsidized products, with high overhead costs, will have their prices *raised,* so the market will rationalize away products that do not provide a good value to customers. On the other hand, efficient products will have their prices *lowered* (or profit raised) because (a) they will no longer have to subsidize the "losers" and (b) their price will reflect increased efficiencies in manufacturing and supply chain management, which will become even more effective as inefficient products are removed from the system.

***Immediately* adjust responsibilities, incentives, and compensation** for people and groups associated with discontinued products to minimize disruptions and resistance now and in the future. This will be easier if total cost measurements can quantify the poor financial performance of the eliminated products.

Implement recommendations. Get necessary approvals and implement the recommendations. Adjust sales compensation policy as necessary to focus on the most profitable products and products that are compatible with manufacturing and supply chain strategies.

The author's rationalization workshop is described on page 470.

HOW RATIONALIZATION IMPROVES QUALITY

Quality metrics are a summation of the quality of all products. Rationalization will *raise corporate quality* by eliminating the unusual, low-volume products which usually have the lowest quality because:

- Unusual, lower volume products get less *kaizen* focus (continuous improvement) and have less sophisticated tooling and procedures.

- Infrequently built products may have: missing or vague instructions, procedures, and "know-how;" rusty or damaged tooling; or missing or discarded tooling, build fixtures, test fixtures, or repair tools, resulting in costly and error-prone manual or "plan B" procedures. More difficult setups generate more scrap before the first good units can be successfully built.

- Older products may have: worn tooling; less sophisticated diagnostics, tests, and repair tools; less effective design for manufacturability and quality; and old materials that may have deteriorated, which is especially likely after "end-of-life buys" (buying a lifetime supply of parts before they go out of production). Older generation *parts* have lower quality because all the above effects apply to older/low-volume parts too.

Not only will rationalization raise the quality of *existing* products, it will also make quality improvement programs, like six-sigma, more effective and easier to implement because (a) quality improvement efforts can be better focused better on the remaining products, (b) these efforts will not have to deal with products that have inherently lower quality for reasons cited above, and (c) program results will not be pulled down by those products with inherently low-quality and little prospects for improvement.

THE VALUE OF RATIONALIZATION

Eliminating or outsourcing *low-leverage* products will ***immediately:***

- **Increase profits** by avoiding the manufacture of products that have low profit or are really losing money because of their (unreported) high overhead demands and inefficient manufacture/procurement

- **Improve operational flexibility** and make BTO&MC production implementations quicker and more successful because, typically, low-leverage products are inherently different with unusual parts, materials, set-ups, and processing. Often, these are older products that are built infrequently with less common parts on older equipment using sketchy documentation by a workforce with little experience on those products. Rationalization had the following effect on a drill bit manufacturer:

  > "Efforts to convert to cellular manufacturing using small batch flow became immeasurably easier. By eliminating the very low volume product line, the company was able to set up a simple kanban system between finished goods and the manufacturing cells, which eliminated the need to operate a complicated, computer-based work order system."[18]

- **Simplify Supply Chain Management.** Eliminating the products with the most unusual parts and materials will greatly simplify supply-chain management. Rationalization enabled the same drill bit manufacturer to reduce bar stock from 24 different types to only six.[19]

- **Free up valuable resources** to improve operations and quality, implement better product development practices, and introduce new capabilities like build-to-order and mass customization. One of the author's clients summarized the resource gains as follows:

  "Product line rationalization freed up a lot of people!"

  - Jon Milliken, Vice President of Engineering,
  Fisher Controls Division, Emerson Electric

- **Improve quality** from eliminating older, infrequently-built products, which inherently have more quality problems than current, high-volume products that have benefitted from continuous improvement and current quality programs and techniques.

- **Focus on the most profitable products** in product development, manufacturing, quality improvement, and sales emphases. Focusing on the most profitable products can increase their growth and the growth of similarly profitable products. According to Richard Koch, writing in *The 80/20 Principle,*[20]

  "If you focus on the most profitable segments, you can grow them surprisingly fast -- nearly always at 20 percent a year and sometimes even faster. Remember that the initial position and customer franchise are strong, so it's a lot easier than growing the business overall."

- **Better quality.** Similarly, getting rid of the worst products raises existing quality and enables quality improvements to focus efforts better, as discussed earlier.

- **Protect the most profitable products** from "cherry picking" (launching a competitive attack on the most profitable products), which can be a threat when agile competitors can skim off the most profitable products.[21]

- **Stop cross-subsidizes.** Remaining products will no longer have to subsidize the "dogs" and so they can generate more profit or offer a more competitive selling price.

- **Ensure resource availability** to implement BTO&MC and ensure that multifunctional product development teams have all the specializations available early to design products well for BTO&MC.

ENDNOTES/REFERENCES

1. James P. Womack and Daniel T. Jones, *Lean Thinking, Banish Waste and Create Wealth in your Corporation,* (1996, Simon & Schuster), p. 257.

2. Ibid., p. 147.

3. Jon E. Hilsenrath, "Many Say Layoffs Hurt Companies More Than They Help," *Wall Street Journal,* Feb. 21, 2001.

4. Jim Collins, "Beware of the Self-Promoting CEO," *Wall Street Journal,* November 26, 2001.

5. Eliyahu M. Goldratt, *The Goal,* Second Revised Edition, (1992, North River Press).

6. Adrian J. Slywotzky and David J. Morrison, *The Profit Zone; How Strategic Business Design Will Lead You to Tomorrow's Profits,* (1997, Times Business/Random House), Ch. 1, "Market Share is Dead."

7. Sydney Finkelstein, *Why Smart Executives Fail and What You Can Learn from Their Mistakes,* (2003, Portfolio/Penguin), p. 142.

8. Richard Koch, *The 80/20 Principle; The Secret of Achieving More With Less*, (1998, Currency/Doubleday), p. 53.

9. Ibid., p. 93.

10. The Schrader-Bellows case study is described in Harvard Business School Case Series 9-186-272; A summary of the findings appears in "How Cost Accounting Distorts Product Costs," by Robin Cooper and Robert S. Kaplan, *Management Accounting,* (April, 1988).

11. Robin Cooper and Robert Kaplan, "How Cost Accounting Distorts Product Costs," *World-Class Accounting for World-Class Manufacturing,* Edited by Lamont F. Steedle, (Institute of Management Accountants, 1990), p. 122.

12. C. Karry Kouvelas, "Getting a Grip on Product Proliferation," *APICS - The Performance Advantage,* April, 2002, pp. 26 - 31.

13. Downes and Mui, *Unleashing the Killer App*, p. 140.

14. Larry Downes and Chunka Mui, *Unleashing the Killer App, Digital Strategies for Market Dominance,* (1998, Harvard Business School Press), p. 171.

15. Thomas H. Davenport and Laurence Prusak with H. James Wilson, *What's the Big Idea? Creating and Capitalizing on the Best Management Thinking,* (2003, Harvard Business School Press), p. 37.

16. Sydney Finkelstein, *Why Smart Executives Fail and What You Can Learn from Their Mistakes,* (2003, Portfolio/Penguin), p. 185.

17. Ibid., pages 79 and 98.

18. C. Karry Kouvelas, "Product Proliferation," *APICS - The Performance Advantage,* April 2002, pp. 26-30.

19. Ibid.

20. Koch, *The 80/20 Principle*, p. 90.

21. Downes and Mui, *Unleashing the Killer App*, p. 140.

4

PART STANDARDIZATION

This chapter presents a powerful, yet easy to implement, standardization procedure, which can reap enough benefits to be well worth the effort as a stand-alone program. As a prerequisite to build-to-order and mass customization, its benefits are even greater.

Standardization of parts and materials is an important prerequisite of build-to-order and mass customization. Standardization can also simplify product development efforts, lower the cost of parts and materials, drastically reduce material overhead costs, simplify supply chain management, improve availability and deliveries, raise quality, improve serviceability, and support lean production, build-to-order, and mass customization.

Standardization supports the fundamental precept of lean production, build-to-order and mass customization: *All parts must be available at all points of use,* not just "somewhere in the plant." This eliminates the setup to find, kit, distribute, and load parts. Standardization makes it easier for parts to be *pulled* into assembly (instead of ordering and waiting) by reducing the number of part types to the point where the *remaining few standard parts can receive the focus to arrange just-in-time deliveries based on customer demand.*

PART PROLIFERATION

Competitive pressures are forcing manufacturers to look at all ways to lower total cost and improve flexibility. One of these thrusts is to investigate and reduce the cost of part and material variety. As a result of such an investigation, Nissan realized that, *in the one model lineup alone,* it used 110 different radiators, 300 different ashtrays, 437 dashboard meters, 1,200 floor carpet types, and 6,000 different fasteners![1] This type of variety does not add value to the customer. Of course, customers want their carpets to match interior fabrics and exterior paint, but this could be accomplished with far fewer than 1,200 carpet types.

An electronics company had 1,500 different types of resistors, including 120 different kinds, sizes, and tolerances of 1,000 ohm resistors.[2] That company was able to reduce the 1,500 resistor types to less than 200 actively used for manufacturing.

One of the author's clients, who made consumer products, discovered that it was using the following numbers of different part types: 1,248 wire assemblies, 152 motors, 151 screws, 74 switches, 67 relays, 65 capacitors, 37 valves, 16 transformers, 62 types of tape, and 1,399 different "standard" labels.

Every company has similar "horror stories" with regards to part proliferation. One might ask, *why?* – especially considering the facts that part proliferation raises part and overhead cost, impedes flexibility, and, as will be shown later, is really unnecessary.

The Cost of Part Proliferation

Part proliferation is expensive. A Tektronix study determined that half of all overhead costs related in some way to the number of different part numbers handled.[3] Most companies do not even know how much that cost is in dollars. A survey of several Fortune 500 manufacturing companies revealed that *not a single company or division had an accurate estimate of the cost of a part over its lifetime!* [4] According to Venkat Mohan of CADIS, Inc., who markets parts management software, "intuitive estimates range from $5,000 for a standard part to as high as $60,000 or even $100,000 per part for custom parts."[5] James Shepherd, director of research for Advanced Manufacturing Research (AMR), Boston, says that in electronics, the cost of just *entering* new purchased components is between $5,000 and $10,000 per component.[6] The Ernst & Young *Guide to Total Cost Management* states that "It is not surprising that manufacturers have estimated the annual administrative cost of each part number to be $10,000 or more." [7]

In addition to these official "materials" costs, excessive part proliferation adds cost to field service and manufacturing in important, but rarely measured, ways related to the setup, inventory, floor space, lower machinery utilization, and other flexibility issues.

Part proliferation also lowers assembly productivity. Writing about the automobile industry in a Wharton Business School report, Fisher, Jain, and MacDuffie write that "Part variety also appears to have the greatest negative impact on assembly plant productivity."[8]

Why Part Proliferation Happens

Part proliferation happens for the following reasons, which are all easily avoidable:

1. **Engineers don't understand the importance.** Most product designers do not understand the importance of part standardization, and therefore, do not attempt to design around standard parts. An example of this attitude was discovered when the author was soliciting feedback from engineers on a proposed standardization list (which was generated by techniques described below) for resistors. One electrical engineer commented *"Why are we standardizing on resistors; Aren't they cheap and aren't they in the computer?"* What this engineer did not realize is that, regardless of a part's cost and the company's ordering/tracking sophistication, *every* part must be physically delivered to the plant, possibly inspected and warehoused, and then distributed to each point of use. One solution to eliminate this erroneous perception would be training and education that stresses the importance of part standardization and how it helps to satisfy corporate goals.

2. **"Not invented here."** Sometimes standardization is resisted because of the "not-invented-here" syndrome, but that can be countered by teamwork and training.

3. **Arbitrary decisions.** Product designers make many *arbitrary decisions* when specifying parts. They may arbitrarily specify a fine pitch 5/16" bolt with a button head that is 7/16" long when a more common course-pitch 3/8" hex bolt that is ½ inch long could have done the job just as well. In the DFM section of

Chapter 10, the general problem of arbitrary decisions in product development is discussed.

Electronic engineers at Intel's Systems Group said that for digital circuitry, they did not really need any resistor value between 1,000 ohms and 2,000 ohms. From this feedback, those values were immediately deleted from the approved part's list.

4. **Many versions of the same part to "save cost."** When product designers are pressured to lower cost, and all that is measured is part cost, they sometimes specify the cheapest version of a part for *each* application. This may appear to lower the part cost by a few pennies, but such a proliferation will generate much greater overhead costs. For example, one manufacturer of medical equipment had one printed circuit board had 11 different versions of 1K resistors on a single circuit board! The engineers thought they were saving cost by specifying the cheapest tolerance and wattage for each application. However, not only did this practice cause rampant part proliferation, it also forced the board to be run through the assembly machine *twice* because there were more parts than bins available on the assembly machine. The standardization principles presented herein encourage using the best single version for all applications.

5. **The minimum weight fallacy.** A phenomenon that may be contributing to part proliferation is the fallacy that all parts have to be sized "just right" to have the minimum weight and be made of the minimum amount of materials. Engineers may resist standardization because the standard part may be the next-larger-size to ensure adequate strength and functionality. The following rules-of-thumb may help guide engineers past this obstacle.

   If it doesn't fly or accelerate fast,
   use the next larger size standard part.

   If it isn't made of precious metals,
   use the next larger size standard part.

6. **Qualifying part families.** A related cause of part proliferation is the practice of *qualifying* (for entry on approved parts lists) entire *families* of parts, like fasteners, resistors, and capacitors. This process makes available many more parts than are needed and allows engineers to keep proliferating the active part base without any approval or authorization.

7. **Contract manufacturing.** Sometimes, a short-sighted business strategy undermines standardization efforts. Some companies, in response to downturns, respond to the pressure to "fill the factory" by, in essence, becoming a contract manufacturer. This "use-it-or-lose-it" approach may bring in some additional revenue, but, often loses money in the long run. This loss would be realized if computations included *total* cost issues such as the overhead expense to support such part diversity, and, of course, the substantial learning curve expense needed to gear up to build many new products.

8. **Mergers and acquisitions.** Another cause of excessive internal variety is the merger of dissimilar products through corporate mergers and the acquisitions of products and companies. Products that originated in different companies are likely to have very different parts and processes. Thus, manufacturing flexibility should be added to the list of primary factors that determine such decisions.

9. **Duplicate parts.** When product designers do not know what parts exist, they will often "add" a "new" part to the database, *even when the identical part already exists.* Even if they suspect that the needed part exists, they will probably specify a new *purchased* part if it takes less time than finding an existing one. Similarly, they will probably *design* a new part if it takes less time than finding an existing one. Many times engineers find it much harder to search through awkward databases than to design new parts or purchase them.

   Many companies have hundreds of incidents of duplicates, triplicates, or even several versions of the exact same part, existing under different company part numbers, and even more situations where a *close* existing part could have been specified instead of introducing a new part.

   Upon the author's suggestion, one aerospace company investigated this and found that it had 900 different types of spacers! Apparently, it was easier to design number 901 than to search through all 900 existing spacers.

   When a large machine tool company investigated this phenomenon, it discovered that it had 521 very similar gears. They eventually reclassified all those gears into 30 standard gears.[9] There were 17 times more gear types then necessary! Every gear had an average of 17 duplicates.

The Results of Parts Proliferation

The net result of part proliferation is that most companies have thousands or even hundreds of thousands of different part types (unique part numbers). Such internal variety is rarely necessary, and is usually the result of this careless – often a rampant – proliferation of parts. In the absence of awareness and standardization goals, designers keep choosing *new* parts for new designs, without any consideration of prior usage of similar parts.

> Part proliferation is usually rampant, but can be drastically reduced with standardization

Every company can investigate the extent of part proliferation by simply looking up the total number of active part numbers for all part categories. Plot out all parts usage per year in Pareto format (without any part number labels) just to show the degree of the proliferation. In almost all cases, these plots show a much steeper Pareto effect then the generic Pareto curve in Figure 4.2.

Another revealing investigation would be to summarize the "materials" budget for all the overhead expenses related to parts. Hopefully, these investigations will provide the motivation to eliminate existing duplicate parts and to substantially reduce part types for new designs using the very effective procedures presented below.

PART STANDARDIZATION STRATEGY

New Products

Part standardization has the most opportunities for *new* designs or redesigns. Usually it would cost too much to convert existing designs, part by part, and there may be issues with replacement part backwards compatibility. In addition, changing existing designs may be precluded by current qualifications of existing products.

Existing Products

The main opportunities for existing products are for "better-than" substitutions, in which wide tolerance parts would be replace by tighter tolerance parts/materials or better grades. In most cases the overhead savings and supply chain benefits resulting from standardization would far exceed any perceived cost "increase" for better materials.

For manufacturing families of parts, subassemblies, and products in flexible *cells,* it may be worth the effort to convert a group of existing designs to a standard set of parts and materials. If this is too difficult on existing designs, such a conversion may have to wait for the next generation design.

As new products with standard parts and materials are phased in, manufacturing operations and procurement will realize more benefits. When the products with the standard parts and materials reach a certain critical-mass threshold, the factory will benefit from upgrading the remaining unusual parts and materials to the standard parts. At this point, the old products can be redesigned around standard parts or outsourced.

EARLY STANDARDIZATION STEPS

Eliminate Approved but Unused Parts

Many parts are approved in families and entered into the approved parts list *en masse.* Many of these parts have never been used. If left on the list, any engineer could add a new part into manufacturing without any approval or authorization.

So the first pre-step would be to identify approved parts that have never been used and immediately remove them from the list for new designs. A variation of this procedure would be to identify parts not used in so many years and remove them.

At Intel's Systems Group, the author's standardization task force was surprised to find that, of the 20,000 parts approved for new designs, *7,000 parts had never been used!* In other words, more than one third of the approved parts were not used on any product. And, yet, any engineer *could have* arbitrarily chosen one of those unused parts and entered a new part into the system without any approval or authorization. In this case, those unused parts were immediately deleted from the approved parts list.

Eliminate Duplicate Parts

The problem with multiple part numbers for the same part goes beyond the obvious extra material overhead cost of carrying extra parts. Most likely, the similar parts would be ordered separately for each different application. This would prevent Purchasing from obtaining quantity discounts and just-in-time deliveries that would have been possible with a consolidated order. Further, the smaller order quantities increase the chances of shortages for a given part. Ironically, the missing part that delays production might be sitting in another bin under a different part number.

The problem becomes more severe in a lean production, build-to-order, or mass customization environment, if different products are using the same part under different part numbers. This may require automation assembly equipment, for instance, for printed circuit board assembly, to load the same part in multiple bins, since bins are assigned by the part number listed on the bill-of-materials. Even if the machine operator notices that some of the parts *seem* similar, the operator does not have the time nor the authorization to consolidate parts on the spot. This duplication alone may prevent flexible operations if there are more parts (including duplicates) in the product family than there are bins available. If this is the case, then parts would have to be reloaded *twice* for each product, which is the type of setup that must be eliminated for flexible operations.

Tektronix deactivated 32,000 duplicate parts out of 150,000

Using part management software, Tektronix was able to deactivate 32,000 part numbers from an active base of 150,000. Bob Vance, Tektronix VP and Chief Information Officer, summarized the return on eliminating excess parts: "There are few areas where a manufacturing company can make such a significant impact to its bottom line with so little effort. We want to invest our resources in product innovation and customer services, not carrying an overburdened parts inventory."[10]

An easy way to stop the introduction of duplicate parts is to make it easier to *find* existing parts than to release new ones. The following techniques will do that.

Part Listing

To *immediately* stop the proliferation, simply list all existing parts in some logical order (as shown in Figure 4-1), circulate the lists to the design community, and encourage engineers to use existing parts whenever possible. A procedure will be presented below for determining *preferred* part values from these lists.

> If designers don't know what parts are already in use, they will probably add new ones

Part Type	Listing Order
Threaded fasteners	Thread diameter, pitch, length, head type, material/coating, grade
Washers/spacers	O.D., I.D., thickness, material, finish
Gears	Pitch, number of teeth, face width, material
Gearboxes	Ratio, horsepower, shaft orientations, shaft diameters
Motors	Horsepower, voltage, phase, shaft diameter, mount
Pumps	Pressure, Flow rate
Power supplies	Output voltage, wattage
Resistors	Ohms
Capacitors	Microfarad
Integrated circuits	Generic numbering system (e.g., 74F00)

Figure 4-1 Examples of Part Type Listing Orders

STANDARDIZATION PROCEDURE

The Zero-based Approach

There is an easy-to-apply approach that is more effective than *part type reduction* measures, which require tremendous efforts for their return. Reducing active part numbers, say from 20,000 to 15,000 will, in fact, lower material overhead somewhat, but may not reach the threshold (eliminating part-related setup) that would enable the plant to build products flexibly without delays and setups to get the parts, kit the parts, or change the part bins.

Instead, the most effective technique to reduce the number of different parts (part types) would be to standardize on certain *preferred* parts. This usually applies to purchased parts and materials, but it could also apply to manufactured parts. The methodology is based on a *zero-based* principle that asks the simple question: *"What is the minimum list of part types we need to design new products?"*

Answering this question can be made easier by assuming that the company, *or a new competitor,* has just entered this market and is deciding which parts will be needed for a whole new product line. One of the advantages of new competitors is the ability to start *fresh* without the old "baggage": too many parts. Just image a competitor simultaneously designed the *entire* product line around standard parts. Now, imagine doing the same thing internally. This is called the *zero-based approach.*

> Starting with zero, ask how many parts you really need to design new products

The zero-based approach, literally, starts at zero and adds only what is needed, as opposed to reducing parts from an overwhelming list. An analogous situation would be cleaning out the most cluttered drawer in a desk, a purse, or a glove compartment; removing unwanted pieces would take much effort, and still would not be very effective. The more effective zero-based approach would be to empty everything into a box labeled "garbage" and then take out only the items that deserve to be retained. The difference in these approaches is where the "clutter" ends up: still in the drawer, purse or glove compartment or in the garbage can. Similarly, part reduction efforts have to work hard to remove the clutter (excess part variety) in the system, whereas the zero-based approach excludes the clutter from the beginning. The clutter is the unnecessary part proliferation that would not have been needed if products were designed around standard parts. Not only do these excess parts incur overhead costs to administer them, they also lower plant efficiency and machine utilization because of the setup caused by products that are designed to have more parts than can be distributed at every point of use.

This approach determines the minimum list of parts needed for *new* designs and is not intended to eliminate parts used on existing products, except when the standard parts are functionally equivalent in all respects. In this case the new standard part may be substituted as an equivalent part or a "better-than" substitution, where a standard part with a better tolerance can replace its lesser counterpart in existing products.

Even if part standardization efforts only apply to new products, remember that in these days of rapid product obsolescence and short product life cycles, all older products may be phased out in only a few years.

Standard Part List Generation

To determine a standard parts list, the company must achieve a consensus on the minimum number of parts necessary to design new products. It is important that all engineering groups, in particular, agree on the list because *they* will be the ones that will be designing products using the standard parts. Consensus can be reached by forming a team of key representatives from Engineering, Purchasing, Quality and Manufacturing departments. The steps are listed as follows:

1. **Determine usage histories.** Start with a baseline list based on usage history of existing products. It would be counterproductive to create a standard parts list by *adding* even more *new* parts. Usage history can be based on the total quantity of all parts purchased per year or the number of products that use the parts (which can be found by generating a "where used" report from MRP systems). Exclude parts used in products that are near the end of their lives and should be phased out soon anyway. Plot the data in the format of Figure 4-2 – in reality, most real life examples are actually steeper then this generic Pareto curve.

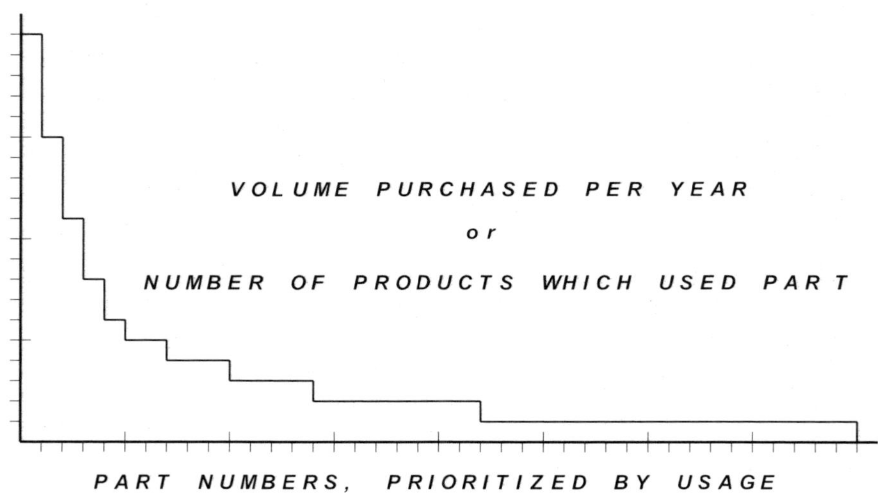

Figure 4-2: Pareto Chart of Existing Part Usage

2. **Establish Baseline List.** The next step is to establish a *baseline list* which will be based on the inherent commonality of existing parts plotted above. Obviously, the "high runners" on the left of the curves should be part of the baseline list and the low-usage parts on the right should not. Where to draw the line (that separates common parts from the rest) will require some judgement. This judgement should come from a consensus of all engineering groups, Manufacturing, Purchasing and those who qualify parts and suppliers (Quality or Materials Engineering Departments). Material qualification departments would be interested because minimizing the number of part types does have a positive impact on procurement and quality activities, since fewer part types will mean that more attention can be focused on procurement, supplier qualifications, and quality issues for the remaining standard parts.

If the usage profile looked like Figure 4-2, the left forth or fifth of the parts might qualify for the baseline list. When parts proliferation has been more rampant, the low-usage and single use parts would extend much farther to the right, in which case, the standard parts may be only the top 5 or 10 percent of the parts.

In some cases, there may be anomalies, such as parts used in low volume, but used in very many products. The widespread use, in itself, may qualify that part for the list, despite the low volume. In other cases, a part may be used on only one product, but in very high volume. This may require some investigation to determine if the design team has discovered a clever use for the part that may apply to future product designs.

These parts become the baseline list, which would then be arranged in some appropriate order, as discussed earlier.

3. **New Generation Parts.** Add new generation parts (that have no usage history) to the baseline list, by asking which new generation parts are envisioned by researchers and engineers from "advanced" Design and R&D groups, Materials Qualifying groups, Manufacturing, Purchasing and Quality.

4. **Consolidate Duplicates.** As discussed before, many duplicates get into the system and good parts management should eliminate the exact duplicates. When companies start investigating duplications, they often discover slight differences between the duplicate parts. This presents an opportunity to select the part that could most likely replace the others. This may involve choosing the "better" part, using the procedure discussed next.

5. **Consolidate Parallel Lines of Parts.** Next, consolidate parallel lines of parts. If whole families of parts are available in multiple tolerances, grades, quality levels, thread pitches or material finishes, the team should consolidate them into one set of parts, even if it has to standardize on the more expensive parts. Usually, any increase in part cost, due to using the "better" parts, will be dwarfed by the overhead savings from having fewer parts overall. If the company cost accounting system cannot quantify the value of this, then the company must recognize the *qualitative* value to manufacturing flexibility and lowering material overhead costs (see discussion below on "Standardization of Expensive Parts"). For example, various grades of bolt strength or resistor tolerance could be consolidated, as indicated by the following example.

> Consolidating on the better grade can cut variety in half for each category

When the author was initiating a part standardization program at Intel's Systems Group, there were two completely different families of resistors: 5% (tolerance) carbon resistors and 1%

metal film resistors. Using this approach, they were consolidated into one line of exclusively 1% resistors, thus eliminating hundreds of part numbers for new designs. These parts were also substituted in existing products since this was considered a "better-than" substitution. A few years later in a public DFM seminar, one company reported that the purchasing leverage resulting from making all resistor purchases the same tolerance (1%) canceled out the perceived "extra" cost of raising the 5% parts to the 1% level, thus resulting in overall cost savings.

Note that consolidation based on higher quality parts may, in reality, raise product quality, even if the lower quality parts were *theoretically* adequate.

6. **Optimize Availability.** The availability and sourcing of all parts and materials on the standard parts list should be investigated and optimized. Standard parts and materials should be selected to be readily available from multiple sources *and be available over the expected life of the products.* Investigations should include the number of sources, the technical and business strength of the sources, and if the parts are being used in *other products* that have similar lifespans.

7. **Structure List.** Structure the list into appropriate order, as shown in Figure 4-1, by values of diameter, pitch, ratio, power, flow rate, voltage, ohms, microfarads, etc.

8. **Review the Lists.** Review the tentative baseline lists for each type of part and feature by involving representatives of all relevant engineering departments for feedback and approval. This could be a formal process that is part of the procedures to generate standard parts lists. Or, it could be an informal process that would solicit informal feedback from some experienced engineers who could be assumed to be representative of their departments. Earlier participation of representatives of these departments on the task force should minimize surprises at this stage.

9. **Circulate the Lists.** Circulate the *tentative* lists to all engineers with an explanation of why the standardization is important to company goals to simplify product development efforts, lower part cost, reduce material overhead costs, simplify supply chain management, improve availability and deliveries, raise quality,

improve serviceability, and support lean production, build-to-order, and mass customization. Solicit feedback about whether the tentative baseline list has the right parts for new designs. Follow up is anyone says any part on the list is superfluous or if any important part was wrongly omitted.

10 **Finalize the Lists.** Review feedback from all those reviewing the list. Investigate promising suggestions and add or subtract appropriate items to or from the list. Finalize the lists and prepare it for implementation.

11 **Determine Scope of Implementation.** The scope of the standardization effort should be matched to implementation resources and the general company awareness of the importance and value of designing around standard parts. Some companies may choose to start with the "low hanging fruit" first, say fasteners or resistors. Success here may then be leveraged to other types of parts. As companies embrace operational flexibility, it becomes imperative to implement strong standardization efforts.

12 **Educate the Design Community.** Before standard parts lists are issued, the design community needs to be educated on the importance of using standard parts in new designs. Point out how important this is to manufacturing flexibility and lowering overhead costs. Another educational and motivational technique is the "embarrassing statistics" technique: reveal the scope of past part proliferations and discuss their causes, which are usually caused by designers who arbitrarily chose a low-usage part when a high-usage part would have worked.

An excellent example is the company mentioned earlier that had 900 different types of spacers. Why so many spacers, when the product line really only needed a small fraction of that number? Every time engineers needed a spacer, *they just designed a new one!* The fallacy in their thinking was that spacer design appeared to be "easy," but, in reality, the documentation, procurement, manufacture, storage and distribution of 900 spacers were not "easy."

Design engineers need to realize that *no matter how simple a part appears, every part number incurs a material overhead burden* to document, procure, store, distribute, re-supply, and, most significantly, to manufacture in low volume.

13 Determine the Strictness of Adherence. The strictness of adherence to the standard parts list should reflect the company's need for manufacturing flexibility, automation utilization, overhead cost reduction, and ease of service. High volume, dedicated lines with expensive automation might require 100% adherence because grippers are created for specific parts. Lean production, build-to-order, and mass customization environments require enough adherence to eliminate setup and assure dock-to-line deliveries.

Many companies just want to reduce the active parts base and encourage engineers to use parts that are already in use. For instance, General Electric Lighting has established a goal of 90% parts reuse in all new designs.[11]

14 Issue the Standardization Lists. Designate the parts on the standard lists officially as *standard, common,* or *preferred* parts and give them special emphasis, at least, with an asterisk or bold type on the larger approved part lists. A more effective method is to present the preferred standard parts on a separate list, perhaps in the front of the section containing that category of parts. Intel's Systems Group presented preferred parts on a gold page, followed by the existing approved parts on white paper. When standard parts are to be used exclusively, the standard parts list would be the only parts list issued to engineers.

PART STANDARDIZATION RESULTS

This part standardization approach was implemented by the author at Intel Corporation's Systems Group. Starting with 20,000 parts for printed circuit boards and computers, this standardization approach generated a preferred parts list of 500 parts, which is 2.5% of the original! For resistors, capacitors, and diodes, 2,000 values were reduced to 35 values (less than 2%).

Fasteners for computer systems were standardized on one screw! This is how the standardization process worked: Service wanted a Phillips head so they, and customers, could keep using the same tools. Quality wanted a captivated "crest cup" washer to protect surface finishes and yet still have a

Intel used this procedure to reduce 20,000 active parts to 500 for printed circuit boards and computers locking effect. Engineering wanted the 6-32 size screw to be only a quarter inch long. Manufacturing recommended that the screw be three-eights of an inch long so that it would not tumble as it was fed to *auto-feed screwdrivers.*[12]

Previous designs had so many different screws that Manufacturing could not use their auto-feed screwdriver at all. The next design used the standard screw in 40 locations. This standardization, and the correct screw geometry, made use of the more efficient auto-feed screwdriver practical. In order to feed the screw, it had to be one eighth of an inch longer, but this meant that the screw would protrude beyond the fastened material. This violated a workmanship standard prohibiting such protrusions; some people even thought the standardization was doomed. But the workmanship standard was modified to allow the protrusion as long as it did not pose a safety hazard or compromise product functionality in any way.

Intel's enforcement goal was not 100%, as might be required for a totally flexible operation, but we felt that even 95% usage would result in significant material overhead savings and yet not be stifling to design engineers.

In general, it should be possible to generate a preferred parts list that is just a few percent of the proliferated list. For very standard parts like fasteners or passive electronic components, it should be possible for the preferred parts list to be less than 1 percent of the current list. When the author lead a product development team to redesign a food processing machine (which had 150 *types* of fasteners), he standardized the fastener list to only two – large and small!

> In general, standardization lists can be just a few percent of the proliferated lists

TOOL STANDARDIZATION

A subject related to part standardization is *tool* standardization, which determines how many *different* tools are required for assembly, alignment, calibration, testing, repair, and service (fabrication tool standardization will be discussed next under "Feature Standardization").

Tool standardization enhances manufacturing flexibility by eliminating the setup to locate and change tools needed in the manufacturing process. If adjustments are to be provided by dealers or users, ideally no tools should be required. But if tools are required, the product should be designed around *standard* tools that are easy to use and would be available to dealers or users. A single tool that performs all repairs and adjustments may be supplied with the product.

Some designs may require several lengths of screws, but if they had the same head geometry, then one screwdriver could be used for all of them. Tool standardization becomes even more important if service people have to be mobile or have to perform service in awkward situations such as clean rooms, crawl spaces, catwalks, utility poles, diving under water, "space walks," and so forth. Tool standardization can also help minimize the expense of providing repair tool kits with products (as provided with some automobiles) and enable the user to perform more repairs.

Tool standardization should be based on *standard,* readily available tools. Often *special* tools are required because tool specification was not really part of the design process. Sometimes special tools are required simply because tool access was not designed into products.

Company-wide tool standardization can be determined as follows: First, analyze tools used for existing products. Prioritize usage histories to determine the most common of existing tools. Work with people in manufacturing and service, in addition to dealers and uses, if appropriate, to determine tool preferences. Coordinate standard *tool* selection with standard *part* selection. Issue standard tool lists with standard parts lists.

FEATURE STANDARDIZATION

Features, such as drilled holes, reamed holes, punched holes, and sheet metal bend radii, require special tools, such as drills, reams, hole punch dies, and bending mandrels. Unless there is a dedicated machine for each tool, the tools will have to be changed, and this will result in a setup change every time the tool needs to be changed. An exception would be machines with automatic tool changing capabilities, but they are limited in the number of tools they can store.

Ironically, most sheet metal bends do not need to be any specific value within a reasonable range. But designers must enter a bend radius value to complete the drawing. Unfortunately, most designers specify an arbitrary

> An arbitrary bend radius forces the shop to change bending mandrels

bend radius, which often requires the shop to locate and change mandrels to bend the sheet metal to an arbitrary radius. In the worse case, a special tool would have to be provided just for an arbitrary decision! Using feature standardization, designers would use the shop's most common bend radius as the best dimension. At an in-house DFM seminar at Hewlett-Packard, where key vendors had been invited, the sheet-metal vendor stood up and said he could generate only four bend radii for bending sheet metal. And then he identified the *one* mandrel that was usually mounted on the bend brake, so if designers didn't use that bend radius, there would be *two* additional setups: one to take off the dominant mandrel *and* another to replace it.

A more subtle, but still important, form of feature standardization is standardization around cutting tools used on machine tools such as lathes and milling machines. As with bend radii, many designers specify arbitrary fillet radii when a standard fillet radius might satisfy the needs of an entire product family. Multiple fillet radii may force a machinist to change cutting tools often or make it difficult to arrange automated machining in CNC machining centers.

When designing parts for milling, designers should specify the same standard fillet radius throughout the product family so that a single cutter or "end mill" may be used. This ensures that all parts in the family can be milled without setup changes and, thus, ensure flexibility and high machine tool utilization on expensive equipment.

To implement feature standardization, standardize features around standard production tools, making sure not to exceed the tool storage capacity on the machine. Investigate the tools used by the plant *and* by key outside vendors (whether currently used or not). A safe approach is to choose only features that can be easily built by *all* (or at least most) potential production facilities and vendors. Based on the production tool availability and capabilities, compile a feature list and issue it with the standard parts lists, hand tools, and raw materials.

PROCESSES STANDARDIZATION

Standardization of processes results from the *concurrent engineering* of products and processes to ensure that the processes are actually *specified* by the design team, rather than being left to chance or "to be determined later." Processes must be coordinated and common enough to ensure that all parts and products in the product families can be built without the setup changes that would undermine flexible manufacturing.

One of the processes affected by part standardization is mechanized screw fastening. The auto-feed screwdriver is a very cost-effective mechanized tool that orients and feeds a screw and then blows it down a tube to the screwdriver head, where it waits for the operator to activate the power drive by pushing down on the screwdriver handle. A preset torque-limited mechanism makes sure the screw is fastened consistently. This useful production tool can feed any style of screw (machine threads, self taping, etc.), but *only one size and type at a time.* Changing screw sizes is possible but would cause too much of a setup to be used in flexible operations. Thus, auto-feed screwdrivers can only be utilized effectively if the product and process are concurrently designed so that *only one* screw type is used at each work station.

Another concurrent engineering issue is that the screw specified must be longer than it is wide so that it will not tumble as it is blown down the feed tube. Each manufacturer has specific guidelines for specifying these dimensions.

Similar devices, based on the same principle, can fasten screws automatically when mounted on robots or on assembly mechanisms specifically designed to dispense screws, such as those used in Hewlett-Packard's DeskJet factory in Vancouver, Washington.

STANDARDIZATION OF EXPENSIVE PARTS

Usually, standardization programs for inexpensive parts, like fasteners, do not meet serious resistance, since the standard parts are perceived to cost no more than non-standard parts (in reality, they cost less). But, as the cost of the parts increases, standardization efforts confront more resistance because of the *perception* that specifying the next larger (and more expensive) standard component would cost more than one that "just satisfies" its requirements in the product. However, considering the *total* cost of standardization can encourage standardization, even for expensive components.

Savings from reducing 152 different motors to 5 or 10: *"Massive"* – GE Consumer Motor Division

The following experience illustrates the resistance and the opportunities involved in the standardization of expensive parts. While training a company that manufactured heating, ventilating, and air conditioning (HVAC) equipment, the author discovered the company used 152 different types of motors. When he challenged the designers, they insisted they needed every size to specify "just the right" motor for every application. Then we asked their supplier, GE Consumer Motor Division, "What would be the savings if those 152 motors could be reduced to five or ten?" The one word answer was *"Massive!"* Why? Because each of those five or ten motors would be ordered in volumes that would be ten times their current order volumes, thus resulting in greater economies of scale. Further, the five or ten chosen would have been the most cost-effective in their line – the best designed motors that they produce in high volume for other customers too.

The upper graph in Figure 4-3 shows the *apparent* implication that the standard parts would cost more than parts that have "just enough" performance for a given product. However, if *all* company products use company standard parts, the cost of those parts will be less, due to purchasing leverage and material overhead savings. Thus, there would be a net company savings for expensive parts, as shown in the lower graph in Figure 4-3. Some products may be forced to use a more expensive part than is required (as shown for one size by black shading), but most products would be able to use less-expensive standard parts (as shown by the crosshatch shading). The result is a net cost savings for the company plus the flexibility that is essential for lean production, build-to-order, and mass customization.

Standardization may *appear* to raise cost for one product, but will generate substantial cost savings in purchasing leverage and material overhead

Further, when a standardization task force standardizes on expensive parts, it can select the most cost-effective parts made in large enough quantities to take advantage of suppliers' economies of scale. These parts are usually the ones that have better availability and are more likely to have better quality and reliability. Design engineers, however, may not be aware of these nonlinear relationships between performance, price, availability, and quality.

The logic of this section also applies to *consolidation,* where versatile parts are designed for widespread use (see Chapter 5).

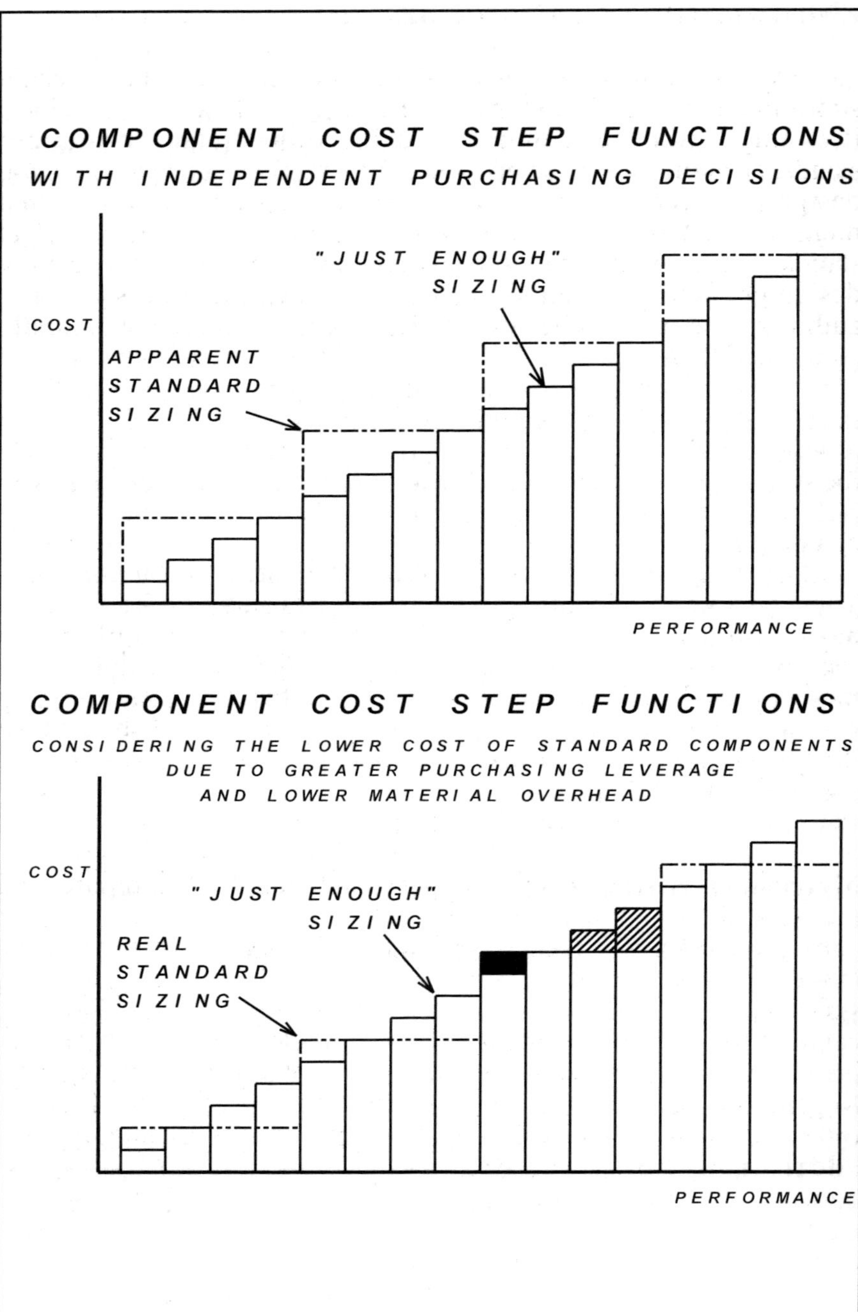

Figure 4-3: Standardization of Expensive Parts

ENCOURAGING STANDARDIZATION

Given the importance of standardization for cost, supply chain simplification, and optimizing current *or future* flexible operations, it is imperative that manufacturing companies implement standardization as soon as possible. This really means *encouraging* design engineers to design around standard parts (even ones that might appear too expensive), specify standard design features, select standard tools, base designs on standard materials, and concurrently design products to be built on standard or flexible processes. The author's experience indicates that design engineers are not naturally committed to these goals. In fact, engineers may actually be pushed the other way by poorly conceived metrics, such as emphasizing "part cost" and low bids for single products instead of *total* costs for product families and all part related expenses. Similarly, narrowly focused *value analysis* on only one product may steer designers to pick the "cheapest" part, regardless of total cost implications for the company.

The following steps can be taken to encourage standardization. They involve "discounting" material overhead rates for standard parts, prequalifying standard parts, making samples and specifications of standard parts readily available, and emphasizing total cost thinking, preferably incorporated into total cost accounting systems. In addition to these procedural steps, managers should take every opportunity to emphasize the importance of standardization in goals, policies, directives, "pep talks," training, and incorporating standardization into the corporate culture.

Material Overhead Rate. The procurement of standard parts and their distribution through the plant will incur less overhead burden compared to the usual excessive internal variety. Therefore, the material overhead rate for standard parts should be less, to reflect the lower overhead because standard parts are easy to procure and most purchase orders are just reorders. In addition to being a more accurate reflection of overhead costs, lower material overhead rates for standard parts should motivate engineers to specify standard parts.

> Standard parts need less overhead and should be charged less to reflect true costs and encourage use

The simplest method is to establish a two-tiered overhead rate, as done at Intel's Systems Group: a general material overhead rate and a lower rate for standard parts. This is a very logical approach since standard parts really do consume less overhead expense, because of the reasons pointed out throughout this chapter. In order to compensate for a lower material overhead rate for standard parts, the general material overhead rate may have to be raised from the previous single rate. It is also logical to assign a higher overhead rate to low-usage parts because of their higher overhead demands. Thus, if engineers choose standard parts, their design will be "rewarded" with the lower material overhead rate. Conversely, if they choose non-standard parts, the overhead rate will be higher than even the previous single rate.

In every DFM class, the author asks the top Purchasing Manager in the room what percent of all the purchasing agents' time would be devoted to re-ordering parts from a standard parts list that is just a few percent of the current proliferated list. The answers from all classes averages about 10%, which means that standard parts should have 1/10 of the material overhead as non-standard parts!

Another method of establishing overhead rates for standard parts would be a variable rate that would be inversely proportional to volume, so that a very high-usage part would have a very low material overhead rate and a very low-volume part would have a much higher overhead rate. This approach is used by the Portable Instruments Division of Tektronix as a "cost driver" to discourage engineers from using low-volume parts.[13]

Pre-Qualified Standard Parts. With fewer types of standard parts for new designs, these parts can be more thoroughly evaluated and their suppliers can be more thoroughly scrutinized than would be possible with many more parts from many more suppliers. The standard parts can be prequalified for immediate use. This can accelerate product development, since design teams do not have to wait for this qualification.

Floor Stock. Usually, the list of standard parts is small enough to allow a *floor stock* to be kept in the engineering area, so that engineers can always have samples of the standard parts available in the design area. Having floor stock samples can help the design team visualize concepts based on standard parts, and, thus, encourage standard part usage. Floor stocks can make standard parts readily available for engineers to evaluate parts, conduct experiments, and build breadboards. Floor stocks can also be mounted on display boards in prominent places near the design team.

Personal Display Boards. For small, inexpensive parts, like fasteners, every engineer could be issued a personal display board with the standard parts mounted with labels that list the generic value plus the company part number. Hewlett-Packard, at the author's recommendation, followed the above procedure and reduced the number of fasteners for large format plotters from 49 types to only seven. Samples of these seven fasteners were mounted on an aluminum plate with values and part numbers printed on paper that was affixed to the plate. These personal display boards were issued to all engineers to encourage them to use the standard fasteners.

Spec Books. Part specifications for standard parts could be reproduced and compiled into a single "spec book" or database. This would encourage engineers to use the standard parts when they can by providing all specifications in a single reference. Similarly, pre-drawn CAD *symbols* of standard parts should be readily available both to speed product development and encourage standardization.

Cost Metrics. If cost metrics are based on total cost for product families, they will encourage standardization. If they are based on part costs alone for a single product, they may *discourage* standardization. If accounting systems cannot quantify total cost, then engineers should be encouraged to balance the costs that are reported with the qualitative benefits of standardization as addressed in the next section.

WHY STANDARDIZATION IS SO IMPORTANT

Implementing standardization for parts, features, tools, and materials can benefit any company, even before it embarks on operational flexibility programs like lean production, build-to-order, or mass customization. Since standardization is a prerequisite for these flexible paradigms, standardization programs should be one of the first steps to implement. Since standardization generally applies only to new products, the programs should be implemented as soon as possible in order to reap the benefits as soon as possible. The following discussion presents the benefits of standardization under four categories: cost reduction, quality, flexibility, and responsiveness. The cost reduction aspects of these benefits will be discussed in more detail in Chapter 11.

Cost Reduction

- **Purchasing costs.** Standard parts will incur much less procurement cost because fewer parts are being purchased in larger quantities. This not only results in fewer purchasing actions, but also results in better *purchasing leverage* that entitles the company to take advantage of suppliers' economies-of-scale and get quantity discounts and better delivery. Ordering fewer types of standard parts in larger quantities is the key to delivering parts to all points of use, which supports flexible operations. The cost of expediting unusual or hard-to-get parts can also be eliminated.

- **Inventory cost reduction.** Part standardization can, directly or indirectly, help reduce all three categories of inventory (raw materials, WIP, finished goods) and their significant carrying costs. Use of standard parts and materials results in fewer types of parts in incoming parts inventory and fewer types of materials in raw materials inventory. Part and feature standardization help eliminate setup, thus supporting lean production which, in turn, reduces WIP inventory. Standardization makes parts available for build-to-order, which reduces finished-goods inventory.

- **Floor space reduction.** Reducing inventory and eliminating kitting can significantly reduce floor space requirements. Floor space can also be saved by eliminating the fork lift aisles needed to move large bins of parts, since flow manufacturing moves single parts between work stations. Reducing floor space needs can be a very attractive alternative to expanding buildings or moving to larger facilities. This issue may be a real "sleeper" until the company needs to expand facilities or move to where more floor space will be available. If floor space reduction was part of a continuous improvement program, the need to expand facilities or move may be postponed or averted entirely.

- **Overhead cost reduction.** In addition to purchasing, inventory reduction, and floor space reduction, standardization also reduces the other costs that constitute "materials" overhead, including the documentation, administration, qualification and distribution of parts. Feature standardization lowers tooling costs by minimizing the number of tools needed to fabricate features like punched holes and sheetmetal bends.

Quality

- **Product Quality.** Having fewer part types in the plant means that there will be less likelihood of using the wrong part. A manufacturer of semiconductor processing equipment had so many different screws that the wrong screw, one that was too long, was used to fasten the cover to a case that housed light-sensitive sensors. Since the screws were too long, they "bottomed out" in the blind tapped holes and, thus, did not adequately seal the case from extraneous light. This caused the equipment to malfunction when placed in service. The company and its customer wasted much time on the diagnostic effort, first checking the sensors and related circuitry, before discovering the light leak caused by the wrong screws.

- **Continuous Improvement.** In addition to the enormous cost savings of eliminating WIP inventory, discussed earlier, inventory reduction is also a key element to *continuous improvement* programs.[14] Inventory hides many problems; eliminating the inventory exposes the problems and, thus, forces solutions.[15]

- **Supplier Reduction.** Standardization programs reduce the number of suppliers because fewer parts usually come from fewer sources. Dealing with fewer suppliers can strengthen ties with those suppliers, thus resulting in the very desirable *supplier partnership* relationships. Having fewer suppliers means that the company can do a better job qualifying each supplier, and also, do a better job evaluating each part.

Flexibility

- **Steady Flows.** If parts and materials can be standardized to the point where there is only one version (or a predictable proportion among multiple versions), then delivery can be arranged in steady flows where "ordering" is as easy as matching the tonnage in to the tonnage out.

- **Eliminating Setup.** If the number of parts used in manufacturing is small enough, those standard parts can be permanently loaded on assembly machines or in the manual assembly pick bins. This allows all the products to be built by the same process without having to change the setup for different parts.

- **Inventory Reduction.** Inventory can be significantly reduced with fewer types of parts to stock and distribute. This will, obviously, reduce incoming or "raw" parts inventory. Parts standardization and setup elimination also encourage lean production that can drastically reduce work-in-process (WIP) inventory expense and floor space.

- **Internal Material Logistics.** The flow of parts within the plant will improve with fewer parts to order, receive, log-in, stock, issue, load, assemble, test, and reorder. Having so few parts that they can all be distributed at their points of use will avoid the space, expense, and delays of kitting parts.

- **Breadtruck Deliveries.** Low-cost standard parts can be delivered directly to all the points of use *without any overhead costs* for purchasing or internal distribution. Arrangements can be made with local suppliers of low-cost parts who will simply keep the bins full of standard parts at all the points of use, much like *breadtruck* deliveries keep the shelves stocked with bread in markets. The supplier simply bills the company for each month's usage. This is a very attractive approach for low-cost standard parts like fasteners, washers, and other hardware items. In addition to the obvious cost savings, this procurement methodology is much less likely to cause part shortages, which can stop production, even for the lack of a one cent washer.

- **Flexible Manufacturing.** Eliminating setup changes allows products to be built in any size batch. With setup reduced to zero, any quantity of any product in the family can be built. This flexibility is the key to building products to order that allows companies to build-to-order and mass-customize orders.

Responsiveness

- **Build-to-Order.** Build-to-order can eliminate finished goods inventory and let the plant build only the products that will ship immediately (for which it has orders). Another important benefit of build-to-order is that scarce parts and labor are only consumed in products that go immediately to customers, rather than being wasted on products that don't sell. Further, expensive parts are consumed closer to the time of the sale, thus lowering interest expenses.

- **Parts Availability.** In general, fewer part types used in greater quantity will mean less chance of running out of parts and delaying production. Further, standard parts are available from more sources with more total capacity (this should be a standardization criteria), so standard parts will be more available, which is especially important for rapid growth situations.

  There are special availability considerations when mass-customized products are built-to-order. Consider, for a moment, the situation when products are built to forecast: the forecast provides the information that MRP systems use to order parts in advance, so that they can be shipped to the plant before they are needed. Build-to-order environments have much less forecasting information to use for ordering parts. Stocking all parts for all possible orders incurs too much inventory expense and space.

- **Quicker Deliveries from Suppliers.** Standardization of parts and materials can accelerate deliveries from suppliers if they have fewer types of parts to order and stock. Parts designed around standard raw material sizes will be quicker to order and may even be in the supplier's own raw materials stock. Suppliers will not need to order special tools if products were designed around the standard tools that they already have.

- **Stronger Suppliers.** Standardization of parts helps part suppliers *rationalize* their product lines and allows them to simplify their supply chain management and improve their operational flexibility. This results in better delivery and reduced overhead costs and cross-subsidies, which allows them to be more cost competitive and free valuable resources to improve operations and quality, implement better product development practices, and introduce new flexible capabilities.

STANDARDIZATION IMPLEMENTATION

Standardization can be implemented by forming a standardization task force, which would include key people from Engineering, Manufacturing, Purchasing, Quality, Cellular Manufacturing, Finance, plus appropriate managers and implementers. A key member of this effort would be a "database wizard" – a designated person who has the skill and availability to quickly extract data and generate many Pareto plots from various IT systems. The standardization steps are:

- Generate interest in standardization by creating an overall Pareto chart showing the periodic consumption of *all* parts (on the vertical axis) in descending order, with only the ascending count on the horizontal axis (with no part numbers).
- Arrange training on standardization methodologies (or, at least, have all implementers read this and the next chapter) followed by workshops to standardize *each* category of parts and materials.
- Discuss any Pareto charts previously generated, including any anecdotal and documented consequences of standardization shortcomings.
- Start with the early steps: list existing parts to immediately stop the proliferation, clean up database nomenclature (e.g., for the many labels for bolts and screws), eliminate approved but unused parts, eliminate parts not used recently, and eliminate duplicate parts.
- Prioritize opportunities of what categories of parts to standardize first and start from the top of the list creating standard parts lists using the procedures presented in herein.
- Similarly, standardize raw materials, tools, features (that require special tools), and processes.
- Determine related investigative tasks to identify standardization issues, like backward compatibility, "copy-exact" policies, and implications of lean/cellular manufacture.
- Start the process of developing standardization implementation procedures. Discuss and assign tasks to change procedures or policies that may hinder standardization.
- Assign tasks to generate more Pareto plots.

- Designate people to do the above.
- Subsequent sessions would analyze new Pareto plots, specify additional investigations, start creating lists of standard parts/materials, obtain approvals, and issue the lists.
- For the standardization of expensive parts, quantify the benefits of the standardization (in the format of Figure 4-3) for specific cases, and, in general, start the process of creating a financial model to help quantify the overall cost savings and overcome resistance to some products getting what appears to be a "better" standard part.
- Issue the standardization lists and incorporate into subsequent DFM training.
- Encourage standardization through appropriate material overhead rates, prequalified standard parts, floor stock, personal display boards, spec books, and cost metrics.
- Perform *product line rationalization,* ideally as a first step if time permits, to eliminate or outsource the most unusual *products* that usually have the most unusual *parts* (Chapter 3).

The author's standardization workshop is described on page 470.

REFERENCES/ENDNOTES

1. Clay Chandler and Michael Williams, "A Slump in Car Sales Forces Nissan to Start Cutting Swollen Costs." *The Wall Street Journal,* March 3, 1993.

2. Brian H. Maskell, *Software and the Agile Manufacturer,* (1994, Productivity Press), p. 335.

3. Robin Cooper and Peter B. B. Turney, "Internally Focused Activity-Based Cost Systems," published in *Measures of Manufacturing Excellence,* Robert S. Kaplan, Editor (1990, Harvard Business School Press), p. 293.

4. Tim Stevens, "Prolific Parts Pilfer Profits," *Industry Week,* v244, n11, July 5, 1995, pp. 59-62.

5. Source: Venkat Mohan, President and COO, CADIS Inc., Boulder, Colorado.

6. George Taninecz, "Faster in, Faster out," *Industry Week,* v 224, n 10, May 15, 1995, pp. 27-30.

7. Michael R. Ostrenga, Terrence R. Ozan, Robert D. McIlhattan, and Marcus D. Harwood, *The Ernst & Young Guide to Total Cost Management,* (1992, John Wiley & Sons), p.150.

8. Marshall Fisher, Anjani Jain, and John Paul MacDuffie, "Strategies for Product Variety, Lessons From the Auto Industry," The Wharton School, University of Pennsylvania, October 9, 1992, Revised January 16, 1994.

9. Eric Teicholz and Joel N. Orr, *Computer Integrated Manufacturing Handbook,* (1987, McGraw-Hill), p. 96.

10. Tim Stevens, "Prolific Parts Pilfer Profits," *Industry Week,* v 244, n 11, June 5, 1995, pp. 59-62.

11. CADIS case study, CADIS, Inc., Boulder, Colorado.

12. David M. Anderson, *Design for Manufacturability & Concurrent Engineering,* (2008, CIM Press, 448 pages, 805-924-0100), Chapter 8, Guideline F4, "Make sure screws are standardized and have the correct geometry so that auto-feed screwdrivers can be used."

13. Robin Cooper and Peter B. B. Turney, "Internally Focused Activity-Based Cost Systems," *Measures for Manufacturing Excellence,* edited by Robert S. Kaplan (1990, Harvard Business School Press), page 293.

14. Kiyoshi Suzaki, *The New Manufacturing Challenge, Techniques for Continuous Improvement,* (1987, The Free Press).

15. Richard J. Schonberger, *Japanese Manufacturing Techniques; Nine Hidden Lessons in Simplicity* (1982, The Free Press); *World Class Manufacturing; The Lessons of Simplicity Applied* (1986, The Free Press).

5

MATERIAL VARIETY REDUCTION

Raw material variety can pose serious supply chain problems for lean production, build-to-order, and mass customization. Most companies simply accept the variety and then implement complex information systems and "supply chain management" to try to handle the situation. But too many types of raw materials can thwart spontaneity and present manufacturers with the dilemma of having to choose between stocking all types and ordering them and waiting for delivery. Fortunately, it is not hard to significantly reduce incoming material variety.

The solution is to *aggressively* standardize incoming raw materials. Fabricated materials would then be cut into various shapes by programmable CNC equipment, such as laser cutters and screw machines. Inflexible "raw parts," such as castings, moldings, stampings, extrusions, and bare printed circuit boards, can be *consolidated* into versatile parts that can be used in many products.

Of course, the first step would be to simplify supply chains with product line rationalization (Ch. 3). One example was cited in Chapter 3 of the drill bit manufacturer that reduced bar stock variety from 24 different types to only six through product line rationalization.[1]

RAW MATERIAL STANDARDIZATION

The ultimate scenario for BTO&MC is to reduce the number of raw material types within each category to one, in which case "ordering" material is as simple as matching the tonnage in to the tonnage out, so a steady flow of materials can be arranged. The resulting purchasing leverage and other material overhead savings would compensate for any cut-off waste and for some products getting "better" material than they need. As raw material variety is reduced, it becomes more feasible to use automatic resupply techniques, like *kanban,* min/max or breadtruck as will be discussed in Chapter 7. Like all standardization, material standardization minimizes the chance of mistakes from picking the wrong material.

> If materials can be standardized on one type, then steady flows can be arranged without forecasts

For materials that are to be fabricated, raw material variety can be drastically reduced by first standardizing on the absolute fewest types and grades and then cutting them on-demand instead of ordering them in many different lengths or shapes (as will be discussed below).

Standardizing raw materials uses a similar procedure for part standardization presented in the last chapter. Raw material standardization can apply to bar stock, tubing, sheet-metal, molding plastics, casting metals, protective coatings, programmable chips, and many other materials.

Bar stock and tubing. If raw materials can be standardized on one size of bar stock or one size of tubing, then computer controlled cutoff machines or CNC machining centers can be programmed to cut off required lengths from the same stock, either for on-demand use or as sources for kanban resupply (Ch. 7).

Sheetmetal. If sheetmetal can be standardized on one shape, thickness, and alloy, then computer controlled laser cutting machines can cut all the sheet metal parts needed without the setup to change sheet types.

Molding and casting. Product offerings may need a wide variety of molded or cast parts. Molding and casting operations will be more cost-effective if they can standardize on the same raw material, so that many different parts could be made in the same machine without changing over the equipment for raw materials. It even may be possible to make many different parts in the same mold, thus sharing processing time and tooling expense. Standardizing materials avoids the setup of changing materials and cleaning the equipment. If molds are designed with *fixturing* standardization, they could be changed rapidly to minimize setup time. If the raw material could be standardized, several molds could be filled with the same "pour" from a single "melt."

Protective coatings. Standardizing on protective coatings simplifies processing and makes painting and coating operations more flexible by eliminating the setup to change coating materials and clean equipment. As with parts standardization, coatings could be standardized on the "better" coating. Even if that coating *appears* to cost more, the net result would be overall cost savings, considering the process value of this standardization. The logic of this concept is discussed further in the section, "Standardization of Expensive Parts," in Chapter 4. Coating standardization could also apply to paint if the purpose of the paint is purely functional with little aesthetic considerations, for instance, for industrial equipment or inside major appliances. In fact, many industrial and agricultural products enjoy a brand recognition because of a standard paint color. For instance, farmers immediately recognize a green tractor or combine as a John Deere product.

Providing color options flexibly has been a challenge and is currently cited as one of the impediments to build-to-order for automobiles. This challenge can be solved using the principles of this book: One solution would be to have a painting gun or coating nozzle for every standard color, where nozzles not in use would be kept in a special environment to keep them from drying out. The other solution would be to dispense the desired color (even customized colors) through a versatile nozzle that is specially designed to be quickly cleaned for the next color.

Programmable chips. Many integrated circuits (ICs) can be programmed separately or in the product. For programmable chips, standardize on the fewest types of "blanks." This can greatly simplify supply chains and operations. It provides the flexibility to program these devices on-demand by on-line programming stations as they are assembled into the product. Ideally, each programming station would be dedicated to one blank device to avoid feeder setup changes. Programmable chip standardization minimizes the number of programming stations and, thus, may make it possible to program chips as they are inserted or placed onto circuit boards.

STANDARDIZATION OF LINEAR MATERIALS

Liner materials can be cut to length on-demand instead of forecasting and ordering many lengths

Standardization can also be applied to material purchased by length in reels: wire, rope, plastic tubing, cable, chain, etc. Linear material variety can be reduced in the following ways:

Cut as Needed. Linear materials can be standardized by *type* with only the length cut as needed. Available equipment can even cut and strip the ends of wires. This approach would require a dispensing machine at each point of use.

Kanban Reordering. Alternatively, wire and tubing can be cut ahead of time, but, to keep overhead low, not be given individual part numbers. The following system was developed by MKS Instruments, Inc., makers of vacuum and flow measurement/control instrumentation. Instead of issuing part numbers and treating many cut lengths as many different parts, a predetermined amount (for instance, one day's worth) of wire or tubing can be cut to the needed lengths and placed in each of two adjacent *Kanban* bins at each point of use. The kanban bins are arranged one in front of the other, as shown in Figure 7-1. When the front bin is emptied, it is sent to a central dispensing machine. The label on the bin tells the machine operator the material type, length, quantity, and the bin's return destination in the factory. After filling the bin, it is returned to the point of use and placed behind its counterpart so it can be moved forward and used when the other bin is emptied. This simple approach can eliminate hundreds or thousands of part numbers from the plant, thus lowering costs, minimizing delays, improving flexibility, and encouraging flexible operations.

Printing While Dispensing. The number of types of linear materials can be reduced further by using equipment that prints on wire and tubing as it is dispensed, so that many various colors of the material need not be ordered and stocked. Printing on wire and tubing can minimize mistakes in assembly and service, since workers do not have to memorize or refer to color codes. Using words, rather than color also solves problems associated with color blindness. Examples of printing include: "Ground"; "+12 volts"; "Supply"; "Return"; "100 psi"; etc. For international products, these codes can be printed in multiple languages, which may reduce internal variety due to labeling differences. Such printing/dispensing equipment could be utilized either at each point of use or at a central location to feed kanban bins.

ON-DEMAND CUTTING-TO-SHAPE

> Raw material could be cut-to-shape on-demand instead of forecasting and ordering hundreds of pre-cut pieces

Incoming material variety can be reduced tremendously by cutting the material on-demand to each required length or shape. This could apply to bar stock, tubing, sheet-metal, cloth, leather, wire, cable, and so forth.

This can eliminate material order size variety by arranging a steady flow of, for instance, full sheets, coil, or long bars, and cutting them by programmable CNC machine tools, such as laser cutters, cut-off machines, screw machines, CNC lathes, CNC machining centers, or less automated tools based on on-line instructions.

The cutting to length/shape could be accomplished at: (1) the first step in the process, as shown in Figure 8-1, (2) at every point of use, (3) at a local station that feeds nearby workstations, or (4) a central processing station which could be the source of *kanban* parts, as described above. Some of these modes could be combined in which material is cut at the predominant point of use and also fed to nearby work areas in addition to being a source for kanban parts.

Sheet-metal

Sheet-metal is a natural for quickly cutting-to-shape on-demand. Laser cutters, plasma cutters, and waterjet cutters can be programmed to cut, *in one machine in one setup,* the external shapes of all sheet metal parts, in addition to holes, slots, notches, cut-outs, and so forth. Cuts can be straight or curved. The laser cutter does the work of both *piercing* machines (that punch out holes and slots metal from the workpiece) and *blanking* machines (that punch out the workpiece from the sheet).

For BTO&MC, these programmable cutters do two important functions programmably: (1) they reduce incoming material variety, as discussed in this chapter, and (2) they cut parts and features on-demand as shown in Chapter 8.

By contrast, in a mass production environment, all the different sheet pieces needed for the whole plant would be cut to size on shearing machines. Often this is done by the steel supplier, but this arrangement ensures that the sheets will be cut in batches, both to spread out the setup charges and to keep them segregated for shipping. Many different sheets would all have to be forecasted, ordered, cut, put on pallets, shipped, and warehoused at the plant, which is not easy since the pallets are heavy (making stacking options limited) and they require fork lifts to move them, thus consuming a lot of floor space. Then the sheets would have to be *kitted* by directing someone to find all the sheets needed for a product, removing them from their pallets, bundling them into a kit, and distributing them to the fabrication area.

One of the author's clients, Hoffman Engineering (a division of Pentair), builds hundreds of different models of metal electrical enclosures to house factory and machine tool controls. They used to order up to 600 different sheet metal pieces, which was a logistics nightmare. When they built a flexible factory to implement BTO&MC, they followed the principles of this section and started operations with their own laser cutters which started out with only six different types of full sheets. Thus, the material variety reduction was *two orders of magnitude!*

Typically, when the author proposes these versatile machines, the "value engineers" become uncomfortable since these programmable cutters are technically not the "fastest" way to do most single tasks, nor did they *appear* to provide the most "efficient" cost per operation by mass production standards. However, the use of programmable cutters can result in the fastest *factory throughput* and lowest *total cost,* since it (1) avoids multiple operations at different machines, (2) avoids multiple transfers, (3) avoids multiple setups, and (4) avoids all the WIP inventory in between.

> BTO&MC lowers the *total cost* of operations instead of "value engineering" individual pieces and processes

Value engineers may also be concerned about material waste if pieces could not be perfectly nested into the standard sheets. But, even metal suppliers have to deal with that problem and customers will pay for it indirectly. And, metal suppliers and BTO&MC companies alike will be able to improve nesting optimization over time. Further, BTO&MC plants can make small kanban parts out of any unused material. So, in actuality, the "waste" issue may turn out to be not an issue at all.

The ultimate supply chain strategy is to feed very standard sheet metal from a coil, run it through straightener rolls, and then through a programmable shear, which could provide on-demand any size sheet needed with the least waste and most efficient nesting. Such capability could feed several programmable cutters on-demand *and* be a kanban source for other sheet metal users. Many companies may think they are "not in that league," but it may be justifiable (after aggressive sheet metal standardization) because of a combination of significant material cost savings *and* enhancing all the BTO&MC benefits.

Another concern that comes up often is standardizing on a single type of raw material, which may be the largest, widest, best grade, and possibly the thickest. It may appear to be wasteful to give all products better material than they need. However, the advantages of such a strategy can be compelling: (1) Purchasing leverage from the higher quantities result in cost savings due to economies of scale; (2) Forecasting, ordering, and waiting for shipments can be avoided by simply arranging steady flows of the standard material; (3) The logistical complications to ship, log in, track, warehouse, and retrieve multiple sheet types are eliminated; and (4) Simple automatic sheet feeders can be used to feed the laser cutters.

Multiple types of sheets would require a more costly carousel type sheet feeder. If volume and capacity balancing permitted, each material type could have its own laser cutter fed by simple feeders. But, in either case, the procurement may be more difficult. If the proportion of all the material types was constant over time, then steady flows of each could be arranged. However, it the proportions are not constant, then some kind of forecasting will be needed to predict how much of each to order.

As volume increases, it may be feasible to shift the creation of holes and slots to a programmable punch press, as long as the number of tools in the tool changer is not exceeded by the hole variety *for the entire product family or plant.*

Part of the justification in programmable equipment is the ability to perform more value-added activity, such as customizing sheet metal panels with holes and cut-outs. Since this is easy to do on the laser cutters and harder to do by customers, this represents a high-profit opportunity for the BTO&MC company.

Linear Cut-Off

Raw material variety can be greatly reduced by cutting off linear materials on-demand at the points of use or as *kanban* parts resupplied automatically to all the points of use. Linear materials include all forms of bar stock, extrusions, strips, tubing, hose, wire, rope, cable, chain, and so forth.

Linear materials that come on *reels* could be cut-to-length on-demand, for instance for metal strips, wire, rope, hose, plastic tubing, cable, chain, cork, and various rubber materials.

Programmable cut-off machines can be acquired or developed for every point of use to cut off linear material automatically on-demand. A less expensive alternative would be the programmable stop, which automatically positions a stop against which the workpiece is positioned before being cut off. The least expensive way to create this capability is to manually cut off the stock based on on-line instructions or printed instructions for each part.

For discrete linear parts, many different lengths could be cut from the longest version and either discard the excess or use it for shorter versions. A simpler version of this, that does not require any equipment, is linear material that is scored with "snap-off" features, like ball point pens and file cabinet rails.

Hoffman used these technique to reduce several lengths of many linear parts (like hinges and stiffeners) to one order length for each category. In one case, they installed a machine system that fed metal strip off a reel, roll-formed it into an "angle" shape, and then programmable cut it to-length on-demand. The total cost of this machine was less than $10,000, which is not much considering the savings it generated.

Machines are readily available that can programmable cut and strip electrical wire, which can be done at each point of use or at a central station that refills *kanban* bins with pre-cut/stripped wires.

CNC Material Standardization

Raw material variety for CNC machining centers, lathes, and screw machines can be minimized by standardizing on the input of raw material for each category and each machine. As with all raw materials, standardizing on one size and grade simplifies supply chain management to arranging steady flows of material.

Standardizing on one raw material also eliminates the cost of setup to manually change materials or the cost of multiple stock feeders. In order to reap these feeding and supply chain benefits, it may be necessary to standardize on the *largest* stock size for a family of parts and use the speed and versatile of the machinery to make any part in a family without setup or forecasting. This may involve some material waste (which would, of course, be recycled), but this could be compensated by all the cost saving benefits of BTO&MC, which will be summarized in Chapter 14.

CNC time is valuable. Don't waste it to change raw materials – *Standardize.*

Chapter 8 will show two illustrated examples in which versatile CNC machine tools cut wide varieties of parts from single stock inputs. These are illustrated in Figures 8-1 and 8-2.

Since BTO&MC is a formidable growth strategy (as will be shown in Chapter 14), manufacturing operations will grow, and, as a consequence, become more efficient. Then, instead of one screw machine making all parts from ½ inch stock, there may be several with each machine using stock more closely matched to the need of its family of parts. This could be facilitated by utilizing low-cost equipment (maybe even used or "obsolete" instead of using the latest expensive "mega-machines."

Outsourced cut-off operations may need to be brought in-house for speed and flow, as will be discussed in the next chapter.

Order Material After Receipt of Product Order

Spontaneous resupply may not be feasible for unusual or seldom-used raw materials, especially on capital equipment with inherently high diversity of materials. If material order times are less than build times, these materials could be ordered after receipt of the product order.

Materials should be *chosen* for this category, which will require a *proactive* effort, especially when companies are converting to spontaneous build-to-order.

CONSOLIDATION OF INFLEXIBLE PARTS

Inflexible "raw" parts like castings, moldings, stampings, extrusions, wiring harnesses, bare circuit boards, and so forth, can be *consolidated* into versatile common parts that can be used on many products. The problem with these parts is that their processes are inherently *inflexible* with lengthy setups, so they are normally built in large batches.

Reduce inflexible raw part variety by consolidating them into versatile shapes

---

Several different raw parts can be consolidated in a single versatile part with enough extra metal, functions, "hooks," or circuitry to be useful in many applications. If this type of consolidation is done throughout the product line, substantial reduction in raw part variety can occur. This will result in greater purchasing leverage, a fraction of the dies or molds, less dependence on forecasting (with steady flows possible), less storage space needed, and simpler internal logistics.

For example, water meter and gas regulator castings can be consolidated with extra metal included on all castings for options such as test ports. Similar logic applies to plastic moldings: all the features for many products can be molded into a few versatile plastic parts. This will have all the above advantages *plus* saving the cost of multiple molds.

Cost Added to Consolidate:

- Extra cost/part on some of the parts for extra material, circuitry, etc.
- One time cost of making the change

Cost Saved by Consolidation:

- Economy of scale savings (purchasing leverage) for ordering N times the volume at 1/Nth of the number of part types
- Tooling cost cut by a factor of N if N parts could be consolidated
- Material overhead savings on fewer parts
    - BOM/MRP expenses
    - Ordering expenses
    - Warehousing/stocking cost for materials inventory
- Setup cost saved by not having to set up the eliminated part(s); the cost savings include:
    - Setup labor reduction
    - Machinery utilization improvements
- WIP inventory savings from processing fewer part types
- Design cost of eliminated part(s) for new designs
- Prototyping cost savings from fewer new designs
- Documentation cost of the eliminated part(s) for new and existing designs
- Value of less chance of running out of multiple part types
- Value of the consolidation's contribution to flexible operations

Figure 5-1: Cost Trade-offs for Part Consolidations

Versatile "bare" printed circuit boards could have *all* the traces, vias, holes, and pads needed for several products, instead of the common practice of designing a different bare board for every product or variation. Then, steady flows of the versatile bare boards could be arranged knowing that they will be used one way or another. Printed board assembly machines are CNC machine tools and can be programmed to place or insert unique combinations of components in batch-size-of-one for BTO&MC, as will be shown in the example illustrated in Figure 8-2.

Automotive wiring harnesses are evolving toward this principle with a single wiring harness used regardless of the number of options ordered. In the old paradigm, many different wiring harnesses were manufactured so that a minimally optioned car would not have extra, unused wires. However, if total costs are considered, the results would conclude that the excess variety costs of manufacturing and inventorying multiple wiring harnesses would greatly exceed the "cost" of the unused wires.

Decisions about these part consolidations may be difficult if the cost system does not quantify total cost. The *extra* features will show up immediately and clearly as an *"extra"* cost. However, the substantial benefits may not show up anywhere in the current cost system. Until a total cost accounting can be implemented, the criteria presented in Figure 5-1 may help itemize the benefits of these consolidations and serve as a basis for quantification or, at least, for educated "leap of faith" decisions.

Consolidation Examples

Power Supplies at Hewlett-Packard. HP's staff of Operations Research Ph.D.s used total cost analysis to prove that money could be saved by utilizing one worldwide power supply for HP Laser Printers. The part cost of the universal power supply appeared to be higher, and was resisted by engineers for that reason. But when worldwide logistic costs were figured in, the universal power supply resulted in substantial *total cost* savings.

VLSI/ASIC Consolidation. Hewlett-Packard offers a wide range of specialized calculators for many niche markets such as business analysis, engineering calculations, mathematical analysis, and so forth. The book, *Product Juggernauts,*[2] discusses this example:

> "Viewed in terms of functions offered, these calculators appear substantially different. However, HP has developed common architecture, subsystems, and components to an extreme degree. This strategy often means that higher grade components find their way into low-end products. In these cases, HP spends more than necessary on parts for low-end products in the interest of minimizing component diversity. The payoff comes in the efficiencies gained in production creation across the whole product line."

Nokia also designs its cellular phones with common custom silicon to gain economies of scale by ordering huge quantities and eliminating setup changes for parts when manufacturing many models on the same line.[3]

Custom Silicon Consolidation

In general, the logic of the section "Standardization of Expensive Parts" in the last chapter can be applied to custom silicon, which can be designed to be versatile enough to be used in a wide range of products, *even if each product uses a small portion of the chip.* The increased volume for the versatile chip spreads out development and tooling costs to encourage custom silicon even on small to medium-volume products. Figure 5-2 compares the traditional analysis for ASICs (which usually discourages their use) and the total cost analysis for widespread use.

> ASICs may be justified, even for low-volume products, if they are so versatile that many products can use part of them

Traditional Analysis for Independent Application

ASICs	**Status Quo**
High NRE charge/chip	No NRE
High tooling charge/chip	No tooling charges
High Cost per ASIC in low volume	Cost/ several discrete chips
Lead time	

Analysis for Versatile ASIC with Widespread Application

Low NRE charge/chips	No NRE
Low tooling charge/chip	No tooling charges
Lower cost per ASIC in higher volume	Cost/ several discrete chips
Lead time relevant only first time	

Total Cost and Other Implications

Quality improvement when dozens of discrete chips are replaced by one ASIC. This avoids the cumulative degradation of multiple parts
Eliminate the quality costs incurred when too many components are crowded onto PC boards
Better reliability with fewer interconnections
Higher density = fewer circuit boards, maybe one, eliminating card cages
Faster performance coming from shorter signal paths for circuits in ASICs *and* other chips brought closer together, especially if circuitry can be condensed onto the same circuit board
Faster factory throughput with fewer components and maybe fewer circuit boards
Better equipment utilization and lower equipment charges
Simplified supply chain management
Better flexibility for quicker delivery and build-to-order
Scare off competition
Thwart reverse engineering
Impress customers

Figure 5-2 Decisions for ASICs
(Application Specific Integrated Circuits)

For techniques to implement standardization, see the last section of Chapter 4.

ENDNOTES/REFERENCES

1. C. Karry Kouvelas, "Getting a Grip on Product Proliferation," *APICS - The Performance Advantage,* April, 2002; Sidebar, "Successful Damage Control," p. 31.

2. Jean-Phillippe Deschamps and P. Ranganath Nayak, *Product Juggernauts, How Companies Mobilize to Generate a Stream of Market Winners,* (1995, Harvard Business School Press), pp. 35-36.

3. David Pringle, "How Nokia Thrives by Breaking the Rules," *The Wall Street Journal,* Jan. 3, 2003.

OUTSOURCING VS. INTEGRATION

CORE COMPETENCIES - the Flawed Logic Behind Outsourcing

Lately, there has been a lot written about *core competencies,* usually presenting the seemly logical argument that companies should focus at what they are "good at" and outsource the rest to someone who is "better at it." However, this cart-before-the-horse strategy results in the corporate business model being determined by a collection of "whatever everyone happens to be good at the time." This often leads to the conclusion that manufacturing is not a core competency or that contract manufacturers could "do it better."

Harvard's Clayton Christensen says that core competencies are "dangerously inward-looking. Competitiveness is more about doing what customers value than doing what you think you are good at."[1]

> The business model should determine core competencies; not the other way around

Rather, the business model should *decide* the set of core competencies that are needed and decide which core competencies need to be *developed* to support the business model. If the company wants to enjoy the benefits of the BTO&MC (Ch. 14), then it will have to make *conscious* decisions about core competencies that optimize the degree of outsourcing and internal integration.

OUTSOURCING HISTORY

For the past several years, many companies have been outsourcing more and more of their operations in order to focus on "core competencies," such as research, development, sales, and only the final assembly aspect of manufacturing.

Some companies have even outsourced manufacturing altogether. Here is an example of the logic that justifies such a move. The head of third party manufacturing for a major food company, in discussing core competencies, used the following analogy: "For example, maintaining roofs and parking lots is not a core capability of ours, so we would look to an outside company to provide this service. The same applies to manufacturing operations."[2] What this company is saying is saying is that *manufacturing is being compared importance to roof and parking lot maintenance!*

> Some companies outsource manufacturing just like roofing or paving

In the last several years, much of part manufacture and material processing was outsourced to "specialists" who were assumed to be "better at it," thinking that this would save cost and increase profits. Jeffrey E. Garten, Dean of the Yale School of Management, referred to an "obsession among global CEOs to lower production costs by outsourcing whatever they can to large-scale specialists."[3]

Some companies have outsourced production in an ill-advised attempt to minimize the money tied up in capital equipment to try to maximize return on capital, as summarized by a recent Business Week article:[4]

> "In the bid to boost return on capital and hone their core competencies, even the staid industrial giants of Germany and Japan are starting to sell off factories. They then award long-term contracts to outside suppliers – often the same companies that bought their plants."

Outsourcing Problems

Unfortunately, too much outsourcing has complicated the supply chain with more "links in the chain," more transfers, and longer lead-times, especially when there are sequential "tiers" through which materials and parts must pass on the way to the assembler. In the slow, stable days of mass production, this worked all right, since assembly could be safely scheduled in advance and there was plenty of time to order and deliver batches of parts from outsourcers. But now markets are changing so fast that the most competitive companies will need to build products on-demand and will need their parts and materials delivered quickly on-demand. Thus, the long-lead times created by far-flung supply chains will be no longer acceptable. Further, outsourcing is at odds with the inventory-less aspect of lean production and build-to-order, since outsourcing is usually a batch operation.

And in many cases, excessive outsourcing makes it hard for manufacturers to control costs and quality while optimizing profits, which often drift away from assemblers to suppliers. Furthermore, the illusion of outsourcing saving "cost" will be shattered, as will be shown below, when companies measure total cost (Ch. 12) and consider the value of responsiveness, customization, and the "opportunity costs" that are not lost with advanced business models.

Definitions: Outsourcing, Integration

Outsourcing. The dictionary definition of outsourcing is "the procurement or services or products, such as the parts used in manufacturing a motor vehicle, from an outside supplier or manufacturer in order to cut costs." [5] *The myth of saving cost from outsourcing has even crept into dictionary definitions!*

In order to help managers overcome popular hype and understand how outsourcing affects business models, a more precise definition will be needed for manufactured products:

> Outsourcing is the procurement from outside suppliers of *non-standard* parts, subassemblies, modules, or manufacturing or warehousing services.

This definition would not apply to raw materials and off-the-shelf parts (that are encouraged in Chapter 10), *which are considered to be standard,* meaning that they have no variations and have widespread use throughout the lean factory, as discussed in Chapters 4, 5, and 8. This is an important distinction since standard parts and raw materials can be procured from the outside through steady flows or automatic resupply techniques, like *kanban* and therefore support spontaneous supply chains (Ch. 7).

Integration. The word *integration* has varied and misleading definitions and is often considered outmoded or associated with the obsolete paradigm of mass production (Ch. 2). The phrase *vertical integration* usually brings to mind memories of Henry Ford's enormous River Rouge plant that had iron ore coming in one end and finished Model T's driving out the other.

However, in the rapidly changing world of build-to-order and mass customization, certain operations that affect speed, cost, variety, innovation, or quality may need to be brought in-house, in other words, *integrated,* to (1) avoid delays through long, possibly multi-tiered, supply chains, (2) mass-customize *parts* as an integral process of *product* customization, (3) maintain control over innovative parts, products, or processes, (4) assure quality first hand, and (5) minimize the total cost of all operations, especially with coordinated efforts to eliminate inventory and setup. Therefore, integration for BTO&MC would be defined as:

> *Integration refers to the production of parts or related manufacturing operations in-house or by a nearby supplier in which such operations are integrated into subsequent steps to the point where any needed parts could be quickly built on-demand based on spontaneous pull signals in any quantity without forecasts or inventory.*

A supplier that operates in an integrated mode, according to the above definition, would *not* be considered outsourcing, since the supplier would function more like an integral part of a cohesive manufacturing operation than the typical arm-length outsourcing arrangement. The section in Chapter 13, "Customer-Specific Lines," shows how suppliers can use BTO&MC principles to become an integrated part of an assembler's operations. Internal integration as part of a BTO&MC operation is discussed later.

Outsourcing and Cost Savings

Managers who favor outsourcing base their decisions on the *belief* that oursourcers are more efficient.

> Outsourcing only *appears* to save money

But, how could this lower cost? First lets "do the math." Contract manufacturers in the same country probably will have the same equipment, same processes, similar site expenses, and very likely the same labor cost. So the rational mind would ask how this could possibly save any money, especially when considering an additional layer of profits. The most likely answer is that the costs associated with that plant will *appear* to drop because corporate overhead can no longer be applied after the plant shifts to supplier status. However, that overhead will now have to go elsewhere and will appear to raise cost at other plants, thus creating more pressure "cut costs," possibly by selling them too. The result could be a fatal downward spiral, because: (1) costs are not really saved, just shuffled around; (2) operations will be disrupted by the changes; (3) more "cost reductions" will be initiated to lower the artificially rising overhead, thus resulting in more disruptions and, ultimately, the "hollowing" of the company, and (4) the company will become slower, less flexible, and less cost-efficient because of all the arguments presented in this chapter.

One common justification of selling factories to larger contract manufacturers is that their greater size allows them to lower cost with economies of scale and "pricing power."[6] However, as discussed at the end of Chapter 11, economy-of-scale is a mass production phenomenon that, in the case of multi-plant contract manufacturers, only applies to aggregated purchases for multiple plants, possibly for bulk purchases over extended periods of demand. This may be satisfactory for many low-variety mass production plants in stable markets, but building products spontaneously to-order requires that parts be pulled into each assembly plant (as discussed in the next chapter), not purchased centrally with complications arising from forecast accuracy, inventory problems, and distribution challenges. Within the same plant, contract manufacturers may not benefit from economies-of-scale if they have to buy different sets of parts for different customers.

> Outsourcing doesn't cut the biggest cost: *overhead*

Labor and materials, the focus of most outsourcing decisions, are a small part of the only "cost" that matters – *selling price,* as will be discussed in detail in Chapter 11. And the selling price contains many cost elements besides labor and materials, such as all the costs of inventory, setup, change-over, material overhead, customization, quality, product development, equipment, tooling, and distribution. Having control of the supply chain enables manufacturers to significantly reduce all of these costs in the following ways:

- Labor costs (both direct and indirect) can be greatly reduced by Lean Production[7] and Design for Manufacturability (DFM).[8]
- Quality costs can be reduced by these plus effective quality programs that can reduce quality costs from a substantial percentage of revenue[9] to the insignificant levels of six sigma (which represents the statistical goal of about three defects per million).
- Material overhead, procurement, and information system costs can be greatly reduced by standardization, product line rationalization, and automatic resupply techniques line kanban.
- Product portfolio planning, standardization, and modular design can reduce product development cost.
- Build-to-order and mass customization can reduce the costs of customization and distribution and eliminate the costs of finished goods and WIP (work in process) inventory, inventory obsolescence, setup, kitting, scheduling, tracking, and inefficient utilization of people and equipment.

In order to reap all these potential cost savings, the manufacturer must ensure that *all* costs are minimized *throughout the supply chain.* Even if it *appears* that most of a product's cost comes from purchased parts and subassemblies, total cost reduction opportunities abound at every subcontractor and part manufacturer, as discussed in Chapter 11 and illustrated in Figure 11-3.

> Outsourcing PCs to "save cost" gets further from Dell's cost advantage: BTO

In an article discussing IBM's sale of its PC assembly operations to an outsourcer, the Wall Street Journal observed that IBM often took a loss on PC sales and "hasn't been able to match the low costs of market leader Dell Computer Corp."[10] But, ironically, Dell gets its cost advantage from its ability to assemble PCs to-order and ship directly to customers from factories that are dedicated to the flexible assembly of their own products (see Introduction), not contract manufacturers that build a batch of products for one client and then shift to building another client's batch.

It may be difficult or impossible to arrange an outsourced supply chain that is proficient in lean production, DFM, six sigma quality, standardization, automatic part resupply, and on-demand built-to-order of parts. So unless the supply chain consists of extremely cooperative suppliers in close physical proximity, it will be necessary to keep in-house enough operations to gain control of cost and speed.

Outsourcing and Profitability

Some recent business books, like Slywotzky and Morrison's *Profit Patterns,*[11] recommend that companies "reintegrate the chain to capture the profit in the system." The authors state:

> "When value shifts occur up and down the value chain, the challenge is to expand the scope of the business model to include activities that were once viewed as upstream or downstream from your own position."

> Profits often migrate to outsourcers

Excessive outsourcing has had a profoundly negative effect on profits in certain industries. Personal computer manufacturers continue to struggle with very low profit margins, and yet their suppliers of microprocessors and operating systems reap the highest profits in the industry. And now, with all that profit, both Intel and Microsoft are able to expand out of their traditional roles to capture a bigger share of certain electronic systems.

The Reintegration of the Outsourcers

One of the great ironies in the outsourcing/integration debate is that as assemblers outsource more and more, the outsourcers themselves are buying up *their* suppliers to become more integrated. Lear Seating Corporation, Johnson Controls, and Magna International, major tier one suppliers of seats for automobiles, have been buying up related suppliers to be able to offer entire car interiors.[12]

> Ironically, many outsourcers are buying up *their* suppliers

Similarly, Solectron and Flextronics, leading electronic contract manufacturers, who has benefitted greatly from the outsourcing craze, have themselves been pursuing an integration strategy by acquiring some of their parts suppliers. A recent Business Week article on outsourcing summarized the trend:[13]

> "While many gurus rave about virtual corporations, Flextronics and its rivals have been quietly picking up the abandoned brick-and-mortar pieces and reassembling them into new versions of vertically integrated empires."

This may start two trends: First, reintegrated contract manufacturers may eventually become competitors as they migrate to building *and* selling their own products, as many outsourcers have begun to do including Korean's giant LG Electronics, which has been trained in BTO&MC principles.

At one time Schwinn was such a dominant manufacturer of bicycles that when asked about competition, their reply was "We don't have competition; We're Schwinn."[14] But a series of bad decisions ruined the company and led to its bankruptcy in 1992. One of those decisions was Schwinn's ill-fated attempt to cut cost by outsourcing to China. The quality was poor, so Schwinn taught the suppliers how to build bicycles. But in the process, Schwinn enabled the Chinese supplier, Giant, to become a formidable competitor.[15]

> Outsourcers may become *competitors,* especially if you teach them how to build your products

The second likely trend is that OEM assemblers, who might have shed parts operations to concentrate on "core competencies," now may want to selectively acquire, or reacquire, parts manufacturers to ensure part availability and exercise control of supply chain cost and speed. However, speed may still not be adequate if the acquired part producer is too far away, as will be discussed next.

Outsourcing and Speed

Far-flung supply chains slow down responsiveness

Outsourcing slows down supply chains if any of the following conditions exist, which are quite common for outsourcing:

(1) If suppliers are too far away, then the shipping time would slow down any attempts at rapid responsiveness and build-to-order. And even if the supplier could build batch-size-of-one parts on-demand, those parts may have to be "batched" into a daily shipment, or greater interval.

When increased border security in late 2001 caused shipping delays in the U.S., the popular press commented that this "exposed like never before the risk inherent in 'just-in-time' manufacturing." Deloitte Consulting said "some of its clients are beginning to question the fundamental tenets of just-in-time."[16] However, rather than challenging the wisdom of just-in-time, whose value been proven for decades, companies should be challenging the wisdom of far-flung supply chains and outsourcing all over the world to get the "best cost." Overseas shipping will slow down as on-going security procedures are implemented to force shippers to list contents of shipping containers;[17] if *inspections* become required, overseas shipping will encounter even more delays. Thus, as the manufacturing speed becomes more important, vertical integration and having nearby suppliers will provide another advantage – immunity from long-distance transportation delays.

Multi-tiered supply chains slow down responsiveness even more

(2) Multi-tiered supply chains slow down responsiveness even more, when each supplier must wait for parts and subsystems to arrive from *its* suppliers. Every additional tier can add days or weeks to response time.

(3) If locally outsourced parts are made in batches (as is usually the case), this would not only take too long, but it would also be contrary to the inventory-less aspect of build-to-order.

(4) If production has to wait for the outsourcers to finish other jobs, then there will be additional delays.

Outsourcing high-volatility parts can cause *product* shortages because contract manufacturers may not be flexible enough to rapidly increase production due to the above reasons. Theoretically, an outsourcer could set up another general-purpose line to meet demand spikes, but only if the order was large enough to justify the conversion and other lines were not already committed to, and set up for, other jobs. In contrast, flexible factories could shift demand to other lines without such large step-function inhibitions.

Even financial analysts are starting to realize the shortcomings of outsourcing, for instance, when they questioned the wisdom of British telecom equipment maker, Marconi, which sold off most of its manufacturing plants to a contract manufacture. Analysts questioned the long term sense of such a move pointing out that "many companies believed it important to retain control of manufacturing to ensure speed-to-market and high quality."[18]

Similarly, Hewlett-Packard is selling its PC operations to outsourcer Sanmina-SCI to focus on "core competencies," including, ironically, *supply chain management and supplier management.*[19] Any yet HP's biggest challenge is competing against Dell Computer, which assembles PCs to-order in its own factories and whose build-to-order and ship-direct model is the source of its significant cost advantage.

Outsourcing and Flexibility

Typically, contract manufacturers make parts or products in batches as specified by contracts, hence the name. They usually build or assemble several clients' products in the same facility on general-purpose production equipment. Each "run" will be set up with each customer's unique parts and procedures and possibly with unique fixtures and tooling. As in any mass production operation, the setup costs must be amortized over whatever is built, thus encouraging large batches of identical products and discouraging variety. In addition, the response time may be lengthened if the production run has to wait for lengthy setups or other jobs to finish first. Furthermore, if the outsourcer is far away, there may be more delays to wait for enough parts to fill a shipping container or truck.

Contract manufacturers like large batches of identical products

Responsiveness will be significantly delayed if the outsourcer has to wait to order and receive *its* parts and material. If the outsourcer tries to order parts and materials ahead of production based on forecasts, it will encounter the usual problems of forecasting: shortages, expediting, excess inventory, missing parts, and delayed production.

Accommodating even a moderate range of standard product variety will be difficult and mass customization will be almost impossible at the mass production contract manufacturers.

Outsourcers may argue that they can shift production to many lines to accommodate peaks and valleys of demand, but only if the lines are not busy or contractually obligated to other customers. In addition, expanding to another line would require costly and time-consuming setups; even quick change tooling and common fixturing would not avoid the setup to load each customer's unique parts.

Contract manufacturers could avoid setup changes by dedicating a line or even a plant to one customer. But the outsourcers may require linear demand (or charge for lack of it) to avoid low equipment utilization or having to change the setups of other lines, as discussed above.

To become flexible, outsourcers would need to implement BTO&MC principles, as proposed throughout this book and specifically in the section on "Customer-Specific Lines" in Chapter 13. This would include the ability to pull in materials and parts on-demand, if necessary, from its suppliers who also practice BTO&MC. Of course, for the outsourcer to be able to do this, the customer would have to cooperate by designing products around standard parts and materials (Ch. 10) that could be pulled into production (Ch. 7) and available at all points of use (Ch. 8).

Such an arrangement would require the customer to commit to enough long-term work so the outsourcer would commit to implementing the customer-specific lines or plants.

Outsourcing and Quality

In order to produce high quality products, the quality of parts and materials must be assured *throughout the entire supply chain.* Internal production is under the direct control and influence of the manufacturer. And all the aspects of quality can be assured as part of a holistic approach. Under these conditions, Six Sigma programs have the best chance of achieving six sigma quality levels.

On the other hand, it is often hard to consistently assure the quality of outsourced production. An alarmingly large percentage of the quality problems that occur are caused by outsourced parts or product manufacture. This is because most manufacturers try to control the quality of outsourced production indirectly through specifications, which are usually too simplistic to assure *all* the aspects of quality or, if specifications try to "nail down" all aspects of quality and reliability, the process becomes so cumbersome that it adds more cost, delays production, and strains relationships. Outsourced quality is even worse when suppliers are selected based on low-bidding instead of a supplier's reputation for quality production (see the section in Chapter 7 on low-Bidding).

> Many product quality problems are caused by outsourced parts

One of the common reactions to poor outsourcing quality is to dispatch teams from the assembler to "fix the problem," but this can be difficult to fix completely. The effort to try will expend money and consume valuable calendar time. This may delay new product launches or keep the assembler from to meeting demand. One of the author's clients outsourced all its printed circuit board production, but at one time the quality was so bad they had to stop the line. Being a Six Sigma company, they sent their "black belts" to analyze the data and apply six sigma principles. But, the outsourcer didn't keep enough records to analyze!

In general, outsourcing makes it harder, maybe impossible, to achieve Six Sigma quality levels because of the lack of control and disconnects caused by distances, different time zones, and varying quality cultures. Further, it is unlikely that the outsourcer will be using dedicated, one-piece flow lines since outsourcing is usually a batch operation with frequent changeovers from one assembler's products to another (see "Defects by the Batch" in Chapters 2 and "Assuring Quality with One-Piece Flow" in Chapter 8). Outsourcers that brag about "six sigma quality" operations may have only achieved that performance for their highest volume operation that has enough volume to warrant a dedicated line and enough effort to please an important client.

> Outsourcing makes it hard for product manufacturers to implement Six Sigma

The most time-consuming, expensive, and risky response to solving outsourcer quality problems is to *teach* the outsourcer how to make your products. As Schwinn found out the hard way, after they taught their suppliers how to build bicycles, the suppliers became formidable competitors.

Outsourcing and Product Development

Outsourcing manufacturing usually has the unfortunate result of separating it geographically from product development engineers. This can thwart both *design for manufacturability* and *concurrent engineering* and compromise product development success.

In design for manufacturability,[20] design engineers must design products to be easy to manufacture in existing processes, which is hard to do if the manufacturing processes are far away for two reasons: (1) Most design engineers are not very familiar with manufacturing processes, so being able to observe them and talk to plant people can greatly improve the manufacturability of their designs; and (2) Multi-functional product development teams optimize manufacturability

> Separating Engineering from Manufacturing thwarts good *Concurrent Engineering*

with early and active participation by manufacturing engineers. This is only feasible when they are working nearby in the manufacturing environment that will actually build the products being designed and, as company employees, can be assigned to actively participate on product development teams. Design for manufacturability is even more difficult when outsourcing arrangements keep changing as foolish companies chase cheap labor around the world and change vendors for a new low-bidder.

In concurrent engineering,[21] design engineers need to *concurrently* develop both products *and new processes* that do not already exist or may have to be modified from existing processes. In order to do this, it is imperative that design engineers work closely with manufacturing personnel and equipment developers from the earliest design activities through equipment/tooling development to product launch. Again, the only feasible way to do this is if manufacturing operations and the appropriate people are nearby. Separation is even worse for overseas manufacturing where the time zone difference is greater than nine hours, in which engineering and manufacturing people will not be working at the same time, thus limiting communication to one round of e-mail per day.

Sometimes companies are forced into outsourcing because of the design itself. Any time engineers design parts, modules, or subassemblies that are outside the capabilities of internal processes, they eliminate the option for internal production. If this occurs for a significant portion of the design, the company will lose control of a significant portion of cost, quality, variety, and delivery.

In the worst cases, engineers base the product architecture around a custom "mega-piece" with few, maybe only one, supplier who can build it. This means that a large percentage of the product's production will be outside the control of the OEM (Original Equipment Manufacturer). If the outsourcer is not good at DFM and doesn't work with well the OEM, costs may be high. Similarly, quality, delivery, and ability to handle variations may be insufficient. Trying to "reduce cost" or improve quality, delivery, or variety may be difficult, if not impossible, especially if *your* business is a very small percentage of *your supplier's* business.

A sad example is a robot company that developed a robot based on a combo electric motor unit that had three electric motors stacked end-to-end with three concentric shafts coming out one end. After the product was designed and outsourced, the company realized the robot cost was too high, so they went to the supplier to ask for cost reductions on the motor unit, but they were unsuccessful because (1) cost is designed into the product and very hard to remove, which is a key principle of design for manufacturability,[22] and (2) this company represented less than one percent of that supplier's business, so it could not get enough of the suppliers attention to initiate any changes.

Outsourcing and Manufacturability

Many companies have drawn an arbitrary line between "us" and "them," simplistically defining their companies as assemblers and their suppliers as part manufacturers. Usually these policies evolve from naive notions about core competency. However, such arbitrary demarcations can significantly raise cost by eliminating opportunities to optimize manufacturability. Outsourcing itself may require excessive modularization with more interface problems. This may preclude opportunities to fabricate monolithic structures on-site. For example, outsourcing strategy can dramatically compromise manufacturability when large chunks of airplanes are outsourced, thus eliminating the possibilities for a monolithic fuselage or wings.

Outsourcing and Leading Edge Processing

Not only is it important to have nearby and "on-board" support to concurrently engineer special processes, the actual production may need to be done in-house because it may be too difficult or proprietary for outsourcing (sometimes called *virtual operations*) to perform. Outsourcers may not even bid on challenging processes, since they prefer predicable jobs that use their standard processes. According to Harvard's Clayton M. Christensen, author of *The Innovator's Dilemma:*

> "When companies have to push the frontiers of performance, managerial coordination is essential. That's right - virtual operations won't work on the bleeding edge."

> Outsouring won't work "on the bleeding edge" - Clayton Christensen

> "Also, engineers don't yet understand how changes in design affect manufacturability or how subtle changes in manufacturing methods affect performance."[23]

Professor Christensen cites examples of IBM choosing to build in-house their leading-edge magneto-resistive (MR) disk drive recording heads for these reasons. He draws the same conclusion about Cisco Systems, Inc., which has so far has outsourced most of its manufacturing as well as new product development:

> "But as Cisco has moved into the most performance-demanding tiers of its markets – particularly optical networks – it is being forced to integrate, performing many product design and manufacturing activities internally. Cisco' competitors are finding that they have to become less virtual in order to compete."

WHAT NOT TO OUTSOURCE

Companies that selectively outsource must decide which products to outsource, but *most companies outsource the wrong products and parts.*

Many companies outsource their newest products and the cash cows – even the "crown jewels" – while retaining older legacy products, spare parts, and oddballs. However, the latter group consumes excessive resources and makes the least profit on a total cost basis. Ironically, many of them should have been rationalized away, using the techniques of Chapter 3.

Most companies outsource the wrong products & parts

Given the compelling reasons presented in herein, one might ask why companies outsource their cash cows while retaining the losers and the oddballs. One answer might as simple as jumping on the bandwagon "because everyone else is doing it," even though they only *think* they are saving money and don't realize the competitive consequences. Perhaps another reason is that outsourcing may be a way to avoid systematic improvements with a seemingly easy "silver bullet" solution.

Another reason might be that the new designs are "cleaner" with better documentation. Granted, older products may have poor documentation and this is a corporate *karma* that comes back to haunt companies sooner or later. But that does not mean it should dictate outsourcing strategy. Rather, the business model should determine which products should be outsourced, even if some documentation has to be cleaned up to accomplish this.

The BTO&MC business model works best when internal operations assemble products in synergistic families with standard parts flowing in or high-variety parts built on demand. This has two important implications on outsourcing strategy:

1) Outsourcers are unlikely to be able or willing to do BTO&MC as well as could be done internally.
2) Internal BTO&MC operations should not be compromised with older legacy products, spare parts, or oddballs that are not compatible with such an operation because of unusual parts, materials, processes, and procedures.

WHAT *SHOULD* BE OUTSOURCED

Savvy companies that use BTO&MC techniques products have different criteria for outsourcing with the following situations benefitting from outsourcing.

BTO&MC companies have different criteria for outsourcing

Oddball products. One of the best uses for outsourcing is to outsource "oddball" products not suitable for BTO&MC to free flexible plants from the burden of building the low-volume, unusual products that have the most unusual (and longest lead-time) materials, the least optimized processing, the most setup times, and the highest overhead demands. Some of these may be simply dropped by implementing product line rationalization (Ch. 3) because they really aren't making enough money. The remaining oddballs that still must be in the catalog should be outsourced.

One of the author's clients, Hoffman Engineering, constructed a plant to build a wide variety of electrical enclosures on-demand. The operation was so good that when competitors saw it, they realized they couldn't compete directly, so they made a deal to be the *outsourcers* for products that do not fit into Hoffman's BTO plant.

Spare Parts. Companies need to assure that spare parts are available to customers, even on products that are out of production, but *that does not mean they have to build spare parts in their own plants.* Some companies keep building them "at any cost" because of inertia or because they feel they are only ones that know how to do it. Often managers *believe* that spare parts are profitable, because their reported costs were calculated when the product was manufactured at high volume. However, many spare parts are big money-losers, especially after sales decrease for the associated products or they go out of production. Further, spare parts production can be disruptive to ongoing operations, especially if the manufacturer has implemented lean production, build-to-order, or mass customization.

Spare parts production should be outsourced except for certain parts that can be produced quickly and cost-effectively on current flexible lines. Documentation may have to be cleaned up and some transfer effort may be required. But these one-time costs will be paid back several times over by avoiding the costs and disruptions of the fire-drills caused by spare parts production. Further, focusing a BTO factory *entirely* on suitable product families will enhance the BTO&MC model and generate higher corporate profits.

Older Legacy Products. As products age, volumes decrease and that usually causes more setup changes. Supply chain costs will go up from less purchasing leverage, greater efforts to locate hard-to-find parts, enormous efforts to design and locate replacements for obsolete parts, and significant costs and risks to arrange "end of life buys" for parts that are soon to go out of production. Low volume part and materials that are different from current parts may not be suitable for automatic resupply techniques, like kanban and dock-to-line deliveries (Ch.7).

Older legacy products may no longer be compatible with current processing equipment or require excessive setup changes. Further, the fire-drills to gear up to build unusual parts would be incompatible with the operations, staffing levels, and talents of a BTO&MC plant.

One consequence of outsourcing oddballs, spare parts, and older legacy products is that their cost will *appear* to go up because outsourcing will include setup and other overhead expenses in the billed cost. However, the apparent "low" cost of in-house production of these low-volume products was *an illusion* created by conventional costs systems that failed to allocate the inherently higher overhead caused by these products (Ch. 12).

CRITERIA FOR WHAT TO OUTSOURCE

Standard parts. It may be advantageous to outsource the production of standard parts that have widespread use, no variations, and are not likely to go obsolete. If these conditions are met, these parts can benefit from mass production economies-of-scale, especially if the parts have been standardized around high-volume industry-standard parts. These can be "procured" by pull-based automatic resupply techniques, like *kanban* or steady flows (Ch. 7), since these parts will be used one way or another. Thus, these parts may be suitable for outsourced batch production and less-than-rapid shipment.

Modularity. To eliminate interface complications, outsourced production should only be allowed for parts and modules with consistent, clearly-defined interfaces that are easy to integrate.

Skill, talent, and costly equipment. If the assembler lacks special skill, talent, and costly equipment, it may take advantage of a supplier's capabilities providing all the other issues discussed herein are adequately addressed. Try to specify standard versions or suppliers' products. If these need to be customized, be sure to work together to make sure the customization can be done quickly and easily. It may be advantageous to encourage suppliers to implement mass customization techniques.

Company-specific lines. An innovative outsourcing arrangement would be for specialized outsourcing companies to set up company-specific parts production lines in or nearby assembly plants. These lines would build on-demand any part needed, be it mass-customized or standard products. The greater the range of parts, the more would be the need for the outsourcer to use BTO&MC principles. The assembly plant would get near-instantaneous delivery in a seamless integration with their operations. Of course, for this to work, the demands of product assembly would have to be satisfied by the flexible capabilities of the parts lines, which would be structured to accommodate all the products in that facility. This may require that products be designed around these part lines or that product families and flexible part lines be concurrently developed. The author has taught several part manufacturing clients (from motors to chemicals) how to provide such capability for OEMs who are ready for such an opportunity. This is discussed in more detail in Chapter 13 on implementation.

Cooperation. For outsourcing to be successful, very good cooperation is imperative in the form of supplier partnerships, where all the suppliers work closely with the assembler to optimize cost, delivery, and all aspects of quality and reliability. Outsouring proposals should be carefully scrutinized for any suppliers whose relationships have been adversarial or in any way uncooperative.

INTERNAL INTEGRATION

Tightening the Supply Chain with Selective Vertical Integration

Many articles and books have proclaimed vertical integration as "out" (as being so twentieth century!) and *virtual integration* as "in" for the twenty first century.[24] Books on about virtual integration define the *virtual organization* as "an opportunity-pulled and opportunity-defined integration of core competencies distributed among a number of real organizations."[25] But such a distribution can: (1) slow down the supply chain if suppliers are too distant or build outsourced products in batches, and (2) add unnecessary shipping, inventory, and other overhead costs, as discussed above.

The supply chain requirements for spontaneous BTO&MC are that *all* the parts and materials must be made available spontaneously. In order to accomplish this, the supply chain must be *tightened* by (a) arranging rapid on-demand delivery of parts and materials from suppliers or (b) bringing in-house any processes that have lead times or batch sizes incompatible with assembly responsiveness and inventory goals. Material cutting-to-shape, discussed in the last chapter, may be one of these processes.

The supplier base needs to be narrowed to those who can make parts to the "pull signals" of the assemblers. In fact, product line rationalization (Chapter 3) will help shrink the supplier base by eliminating or outsourcing the most unusual products with the most unusual parts and suppliers. However, suppliers may not be suitable if (1) they are too far away for adequate delivery speed, (2) they are too entrenched in batch mass production, or (3) they are unwilling to make the necessary commitments and changes.

If it is not possible to arrange suitable suppliers, certain operations may need to be selectively reintegrated in-house to eliminate order/delivery lead-times and permit batch-size-of-one flexibility. This would greatly simplify or eliminate the MRP-based purchasing normally needed to procure these parts and materials from a diverse supply base. Further, in-house production of parts on-demand can immediately feed assembly lines.

Core Competencies Revisited

Old decisions about "core competencies" will have to be revised to include supply chain responsiveness and operational flexibility as the basis of a *new* definition of core competencies. In some cases, long lead-time or large batch part production will have to be brought in (and streamlined) just to "complete the system," even if it is not considered a core competency.

> Thinking about "core competencies" will have to be revised

For example, one of the author's clients, Badger Meter, of Milwaukee, Wisconsin, found it was able to manufacture a wide variety of water meters flexibly except the printing of the face plates, which had to cope with several ways of measuring water flow plus the logo of every utility customer. So they learned how to print face plates in small quantities to complete the plant's flexible capabilities.

If a complete BTO&MC system represents the only such capability in the industry, then the assembler has a competitive advantage, which would help to justify such a move. In fact, the BTO&MC company could offer to build these parts on-demand to other companies, even competitors, thus expanding sales and widening its dominance of the industry.

Integration can improve profits because the integrated manufacturer has much better control over quality, variety, delivery, and many categories of total cost. One of the "profit patterns" that Slywotzky and Morrison identify is a selective reintegration to maximize profits:

> Reintegrate the value chain to recapture profits

> "Manufacturers are increasingly facing situations where a partial reintegration of the value chain is necessary if they are to earn an acceptable return on their investments. As value migrates from one value chain link to another, players are being forced to reintegrate in order to maintain strategic control."

When to Integrate Internally

The following specific criteria can be used to determine what is best to integrate internally:

Cost. Pure assemblers usually assume that most of their "cost" is part cost and therefore ignore opportunities to lower overhead costs. However, Figure 11-3 shows total cost pie charts through an outsourced supply chain with the revealing conclusion that each subassembly breaks out into its own pie chart (with a higher percentage of overhead) and each of those parts again breaks out into its own pie chart with most of the cost being overhead. But, with an outsourced supply chain, the assembler has little visibility or control over these overhead costs throughout the supply chain. Cost "reduction" efforts usually end up focusing on beating up suppliers; even more enlightened efforts, which try to work together to lower cost, are often stymied by lack of total cost data (Ch. 12) and resistance to change caused by suppliers not understanding *their* total cost situation.

Integration give control over *all costs*

Integration lets the assembler understand, measure, and thus control *all* the costs within its scope of operations. All the principles of this book can be applied to minimize total cost not just at assembly, but also at subassembly production and parts fabrication.

Quality. Similarly, having control over production leads to complete understanding, measurement, and control over all aspects of quality, whereas outsourced production seeks to create an acceptable specification to quality, such as defects per million. However, quality cannot really be *specified,* especially by a single number; rather, quality needs to be *assured* by meaningful quality programs. Some suppliers do have effective quality programs, but many do not and may try to meet the specs through dubious measures like screening parts, which is expensive and results in a higher proportion of parts at the "hairy edge" of not working.[26]

Outsourcing tries to *specify* quality; Integration *controls all aspects* of quality

Variety. High-variety parts may need to built on-demand in-house. The BTO&MC techniques presented herein may be the only way to quickly and cost-effectively build high-variety parts. It is more likely that these capabilities will be much more effectively implemented internally than at suppliers.

Integrated BTO is the quickest and cheapest way to build high-variety parts

Lead time. Product delivery time will suffer if suppliers cannot build parts on-demand, if the suppliers are too far away, or when there are many links in the supply chain. This may be a critical consideration when products are to be built spontaneously for rapid delivery to customers. However, lead time may be less important for capital equipment.

Availability. Internal production of parts may be able to ensure a steady supply of parts.

Size. Instead of outsourcing large modules, like airplane fuselage segments, in-house fabrication could build monolithic structures to minimize cost, weight, delivery time, flow time, shipping damage, and integration difficulties.

Interface complexity. If outsourcing forces more modularization, this may result in more interface complexity. This complexity and its problems can be eliminated by fabricating more monolithic parts in-house.

Special/flexible tooling. If special or flexible tooling is required, it would usually be better to develop and operate it in-house. As mentioned earlier and in Chapter 10, concurrent engineering of the product and tooling can produce *both* better tooling *and* better product designs – and keep improving them over time. Suppliers may be reluctant to develop such tooling without comprehensive contracts that, in turn, may not be in the best interest of the assembler. Further, such arrangements may be complicated by issues involving tooling ownership and access if the assembler wants to change suppliers or reintegrate the tooling later. In extreme cases, assemblers have been denied access to tooling *that they own* during bankruptcy or any other seizure when authorities paddle-lock the suppliers facilities and sort out tooling ownership issues later – usually much later.

> Special, flexible, proprietary tooling is best designed and operated in-house

Innovation. The higher the levels of innovation in both product development and processing, the greater the need for in-house manufacture, as Clayton Christiansen pointed out above.

EFFECT OF PLANT/SUPPLIER LOCATION ON SPEED

The effect of distance applies equally to outsourced suppliers (discussed above) and company owned plants (discussed below), many of which have been moved far from other supply chain links and customers "to save cost." However, in addition to not really saving cost on a *total cost* basis, creating great distances in supply chains slows down the flow of parts and products, in addition to eliminating the benefit of any flow in plants at either end since everything has to be batched to be shipped. This problem might have not been noticed in the mass production era of large batch forecasted production, but will become the "weakest link in the chain" for quick-response endeavors like BTO&MC.

Oceans and lean production are not compatible – Womack & Jones, *Lean Thinking*

Womack and Jones, writing in *Lean Thinking*[27] summarize it succinctly: "Oceans and lean production are not compatible." They go on to say that smaller and less-automated plants close to assembly and markets will yield lower total costs, considering the cost of shipping, the inventory carrying costs, and the cost of obsolescence when products built weeks ago no longer satisfy customers.

Shipping Disruptions

Another sobering consideration of getting parts from overseas is the vulnerability of major supply disruptions. The labor strife that caused the 10-day west coast port shutdown of 2002 was one example.
The vulnerabilities of building entire products overseas (discussed next) could be even greater. The bottom line is that companies that manufacture their products near their markets and build their parts internally or get them from nearby suppliers will be immune from overseas shipping disruptions.

Manufacturing near markets with local parts makes companies immune to overseas shipping disruptions

While its competitors were moving high-volume production to low-cost countries overseas, Timken committed $150 million to build its most sophisticated factory in Asheboro, North Carolina. Not only is the plant closer to customers, but it is flexible to be able to efficiently manufacture in small batches without lengthy setups between batches (Ch. 8). Timken's strategy is to build small runs of high-margin products in flexible domestic factories. Timken has remained profitable through that recession.[28]

OFFSHORE MANUFACTURING "TO SAVE COST"

Many companies are moving manufacturing operations overseas or outsourcing to foreign countries because they *think* it will save cost. And their primitive cost systems make such a move appear to be justified. If all they quantify is parts and labor, then moving to a "low labor rate" country labor will appear to lower labor cost. Since no other costs are quantified, it is a "case closed" decision.

> Manufacturing overseas to "save cost" rarely does based on total cost

However, when measured on a total cost basis, manufacturing products offshore for sales in the home country rarely results in a net cost savings, considering differences in labor efficiency and all the costs of shipping, quality, inventory, communications, travel, training, transferring products, support, and complete sets of equipment needed for any manufacture.

Nokia builds most of its phones in its own factories in the U.S., Germany, and at its headquarters in Finland, where average wages are 30 times more than those in "low-cost" regions. And yet, it can still be competitive manufacturing in these "expensive locations," matching the best of the Asian manufactures.[29] This is because, despite the higher labor costs, the *total cost* is less for reasons that will be discussed for the rest of this chapter.

The following discussion addresses issues affecting companies who move or expand their manufacturing operations offshore to build products for consumption at home. With the exception of startup issues, most of these issues also apply to outsourcing production offshore to contract manufacturers.

OFFSHORING THWARTS 6 OF 8 LOW-COST STRATEGIES

Offshoring has many hidden (not quantified) costs, many of which are discussed below. But since the costs are *hidden,* it is has been hard (even for the author) to persuade companies against offshoring to "save cost."

But after writing the book-length web-site, *www.HalfCostProducts.com*, it became obvious that offshoring *thwarts, compromises, or inhibits six of the eight powerful cost reduction strategies* presented, which are summarized in Chapter 11 in the section, "Strategy to Cut Total Cost in Half." Here are the six strategies thwarted by offshoring:

> Offshoring thwarts 6 of the 8 cost reduction strategies at www.HalfCostProducts.com

1) **Concurrent Engineering.** Offshoring separates manufacturing from engineering and thus thwarts Concurrent Engineering and compromises the 80% of the cost determined by the design. It is hard to do Concurrent Engineering when Engineering and Manufacturing are not even working at the same time, so what happens is that drawings are thrown over the ocean and parts come back.

   Further, transferring, supporting, and dealing with delivery and quality problems of remote manufacturing absorbs many resources in engineering, manufacturing, and purchasing whose time would be better spent *developing* low-cost products. The author has encountered several companies in which this absorbs *two-thirds of engineering resources!* And if their futures depend on new product development, then they have no future.

   After one company moved all its production from the U.S. to Asia, its *next two* new product launches were failures. One of the engineers summed up the problem as follows:

   "Concurrent Engineering:
   We had it – it worked.
   We lost it – it shows."

2) **Lean Production.** Offshoring manufacturing to distant continents increases the delivery time, which makes it impossible to pull parts for on-demand production. Even if offshore production has some flow internally, parts may still need to be batched for shipping. Further, the best lean production comes from concurrently designed processes, standardization, and an integrated continuous improvement program as part of a coordinated corporate strategy.

3) **Build-to-Order & Mass Customization,** are thwarted by long pipelines, delivery delays, batching for shipping, and difficulty doing customizations, especially if the lack of modularity does not lend itself to postponing variety. Offshore manufacturing does not support flexible use of CNC, product family lines, cellular manufacture, and other aspects of on-demand production (presented in Chapter 8). Basically, offshore contract manufacturers are predominately set up for large-batch mass production of identical parts with stable designs and predictable demand.

4) **Standardization,** which is thwarted by having to convert parts and materials to locally available versions and having to convert parts to the contract manufacturers' preferred parts lists. This would all have to change again every time production is moved to keep "chasing the cheap labor" around the world.

5) **Supply Chain Management,** which can save more money by helping product development teams and establishing vendor/partner relationships than through bidding and all the effort it takes to arrange offshoring and deal with delivery and quality issues.

6) **Quality,** which is hard to assure remotely and implement Six Sigma programs without data and control over manufacturing. Quality can not be assured by issuing a spec to contract manufacturers. *Stated* quality capabilities are probably based on high-volume products running on dedicated lines that have been refined over time, not on-demand production of low-volume/high-mix products. Six sigma can be implemented best in integrated operations as part of an overall strategy.

Total Cost Considerations for Offshore Manufacturing

Labor's percentage of the *selling price* is much smaller than people perceive because total cost is rarely quantified well. But, just because labor *is* quantified, it should not dominate manufacturing strategies. Ironically, trying to reduce a small percent of total cost can raise the following other costs even more.

Labor efficiency alone might cancel out labor rate savings, for instance if labor cost is one fifth but labor productivity is also one fifth. Unfortunately, cost systems will immediately recognize the labor cost savings but not the productivity drop, unless it can be quantified and factored into the decision making process.

Labor efficiency alone might cancel out the intended labor rate savings

Labor-intensive designs. Many decisions to move to production to low-labor-rate are based on labor-intensive designs. However, design for manufacturability (DFM)[30] can reduce labor content to the point where moving to low-labor-rate areas can no longer justified, even with primitive cost systems.

Shipping and expediting. Remote manufacturing will incur more shipping cost than manufacturing closer to customers and suppliers. In addition to the cost paid to the shipper itself, there are also costs of shipping insurance and other costs for fees, permits, duties, tariffs, compliance with import/export restrictions, and so forth. Longer shipping "pipelines" increase carrying costs for inventory in the pipeline *and* queuing up at both ends at the seaport docks and at factory shipping and receiving departments. In addition, greater distances slow responsiveness and make it harder to implement lean production, build-to-order, and mass customization. Even if a mass-producer doesn't have variety challenges, it may still have trouble responding to variations in demand and have to resort to expensive expediting to compensate for the lag between increasing customer demand and when products finally arrive from stepped up production.

Although the business case for offshore manufacturing is typically based exclusively on ground shipping, even occasional expedited air freight can seriously erode the viability of the offshore justification. Although not planned, expensive expediting may be required because of demand volatility or any of the shipping disruptions discussed earlier.

As oil cost shoots well past $100/barrel, all forms of shipping will become more and more expensive, thus making more economic sense to manufacture products close to customers and suppliers. In June of 2008, Business Week reported that American-made steel started to become *"cost-competitive because of the high energy cost of shipping steel from places like China."* The chief economist at CIBC World Markets says *"Now distance costs money."* [31]

Further, greater distances between headquarters and manufacturing result in less control. Greater distances between engineering and manufacturing compromise concurrent engineering.

Training costs may exceed projections especially if it is difficult to attract and retain workers with adequate skills. For instance, many companies that open factories in Singapore are surprised at the lack of loyalty and resulting high turnover, which, in turn, further increases recruitment and training costs in addition to the wage escalation necessary to retain valuable employees in such an environment.

Quality costs may also rise, sometimes catastrophically, because of difficulty establishing a six sigma quality culture, learning curves, and the long "pipeline" to the ultimate users which delay discovery of recurring defects (see more on the general quality discussion below).

Capacity. Needing more capacity is often the catalyst for adding a new overseas plant, but capacity can be improved in existing plants by improving equipment utilization and freeing up floor space with lean production programs that reduce inventory, batches, and setup time. By contrast, adding an overseas plant for capacity is a slow and inefficient way to add capacity, because of the above issues in general and the following ones in particular.

Every plant needs a *compete* set of equipment, whether or not justified based on capacity needs

Utilization of equipment may be way below optimal levels since every plant needs a *compete* set of equipment, whether or not justified based on capacity needs. If a company has standardized on million dollar assembly equipment or testers (which is common for printed circuit boards), then it will need complete sets of these million dollar machines in every plant, regardless of how well they are utilized.

Automation may be hard to justify when the "nominal" labor rate is low, especially when non-labor costs are not quantified. Insufficient CNC automation diminishes both quality and flexibility, which, in turn, diminishes responsiveness. On the other hand, some companies have been tempted to move automated equipment offshore, which doesn't make any economic sense, since the automated machines (a) cost the same anywhere because of world trade, (b) don't require much direct labor, and (c) would miss out on the benefits when CNC programmers work closely with engineering and manufacturing at the home office.

Cheap labor may not be good enough. Some of the first problems noticed after production is shipped offshore are (a) the tasks can *not* require skill and judgement and (b) documentation must be *complete* and *perfect*. Both of these may be taken for granted at the home factory, where skilled workers know how to build products from experience and can exercise judgement when required.

> Chasing cheap labor is an expensive way to save money

A cautionary tale about going too far afield was told at an internal conference of a large multinational company: because of the company's strict low-bidding policy, one division was forced to accept the bid of a supplier in a remote area of China where, in the words of the division's supply chain manager, "the last western person to come through there was Marco Polo!"

Cheap labor doesn't stay that way. One of the ironies of moving production overseas for the cheap labor is that the labor advantage doesn't last very long, due to rising labor cost, taxes, and local currency valuations, among other factors like changing trade agreements and local desires to force outsourcers to move "up the food chain"(see comments on tax breaks below).

A 2007 Business Week article titled "Where Are All the Workers?," said: *"The seemingly inexhaustible pools of cheap labor from China, India, and elsewhere are drying up as demand outstrips the supply of people with the needed skills."* The article also said that in Mexico, 82% of companies were having trouble filling jobs for factory workers and laborers.[32]

Moving production to chase cheap labor or opening a plant oversears is, as will be discussed in the next several points, an expensive way to save money, especially if the company is moving its own manufacturing facilities.

Plant start-up costs must be taken into account and amortized over the expected life of the plant. Just the process of starting up the plant will involve a team of ex-patriots living on expense accounts for months. New (maybe redundant) infrastructures may have to be installed for personnel, purchasing, customs, legal, and other administrative functions.

Transfer costs must be applied to every product transferred, including tooling tear-down, packaging, shipping, tooling reinstallation, documentation translation, and travel for people who may have to personally supervise the transfer at both ends.

Support costs add overhead cost to find and qualify outsourcers or arrange the startup of new manufacturing plants. Support costs continue for the ongoing administration of off-shore arrangements. These headquarters expenses are rarely subtracted from the computed savings of offshore manufacture.

The *Mechanical Engineering* article titled: "The China Road; the Outsourcing Option Isn't as Cheap or as Easy as it Looks," warns:

> "Many firms are unprepared for the amount of project management – including late-night phone calls, last-minute travel, and supply chain monitoring – needed to bring an outsourcing relationship up to speed. It may take months or even years of heavy travel to ensure that both parties understand one another. Even then, companies need strong, consistent internal processes to avoid problems." [33]

Travel costs are usually end up being higher than anticipated, especially if the plant is in an "attractive" area. Travel to plants greater than nine hours away may need to be increased to compensate for communication difficulties cause by out-of-phase work schedules.

Communication costs may be more than anticipated, especially if there is a burst of intercontinental communication for the short time that out-of-phase plants are both working.

Unanticipated/hidden costs may offset much of the expected cost savings of overseas manufacturing. Local fees, licenses, and permits may be anticipated, but in many counties, especially in Asia, it is common to have to make payments for bribes, kickbacks, favors, and protection. Having to hide such payments may create resistance to total cost measurements, which seeks to quantify and categorize all expenses, and, in extreme cases, this could encourage accounting fraud and violations of laws or company policies that forbid such payments.

The cost of moving production back home may be the *last* "hidden" cost of offshoring production, after managers realize the implications of this chapter, quantify the hidden costs with total cost measurements, or start to realize the importance of integrated manufacturing to the success of BTO&MC. The costs of moving production back home includes not only freight expenses of products, parts, and equipment, but also cost of people at both ends to arrange all aspects of the move, liquidate equipment, sell buildings, cancel leases, pay penalties, and reintegrate production back home. Re-integration is even more difficult when the outsourced products were not designed well for manufacturability but, instead, counted on abundant cheap labor for assembly and repair.

Before companies are tempted to offshore manufacturing for cost, they should consider how much money would really be saved (or lost) on a total cost basis and how all these ventures impede *real* cost reduction and the implementation of advanced business models like BTO&MC.

Offshore Traps

> Going for tax breaks focuses on *protecting* profits instead of *generating* profits

Tax breaks are often cited as an enticing incentive to move manufacturing overseas. These tax breaks can be so intoxicating that they override appropriate cautions, as exemplified by the following true story: One Fortune 50 company was going through the decision making process to select a location for a new offshore plant. One slide had list of candidate countries; someone noted that there was one country being considered he never heard of. The response was, *"For those tax breaks, we will find it!"* However, tax breaks probably will not make up for the above mentioned increase in total cost and, furthermore, they usually expire after so many years unless the company "ups the ante" and increases the commitment, for instance, by moving engineering there too.

Some home countries do not tax companies on profits made from overseas manufacturing as long as the profit does not come into the country. So, in order to keep avoiding taxes, companies reinvest the money overseas, thus increasing their commitments overseas.

The most subjective aspect about tax breaks deals with *focus,* the importance of which is emphasized throughout this book. Going for tax breaks focuses on *protecting* profits instead of *generating* profits. And often the *wrong* focus can compromise the *right* focus, as all these points illustrate.

Overhead allocation downward spiral. The actual total cost of overseas manufacturing will probably be higher than anticipated because of the above issues. Total cost measurements (Ch. 12) would predict this and prevent unwise decisions. However, without total cost measurements, these decisions may be made based on "back of the envelope" estimates of labor cost "savings."

Nevertheless, even if not quantified, all the above mentioned "extra" costs must be paid and usually end up in a big bucket called *overhead.* As mentioned in Chapter 12, most overhead *allocation* algorithms are arbitrary, and when companies have overseas plants, they will – knowingly or unknowingly – shelter these plants from all the unanticipated overhead they generate because (a) they are not quantified or (b) allocating these expenses appropriately would reveal the overseas plant was a bad decision.

So the allocation on the overseas plant is arbitrarily set to correspond to a level that supports the original business plan, *realistic or not.* But, someone must "pay the piper," usually the home country operations, whose overhead must then go up. That accelerates the transfer of more products overseas, maybe even creating demand for more plants as existing ones struggle to accept all the transfers. As domestic plants lose more products, they become less efficient, operating at partial capacity, and then their overhead rises from *their* own transfer-induced inefficiencies, thus causing even more transfers and eventually forcing domestic plant closures. Continuing this pattern results in a downward spiral (some companies call this the *death spiral*) until there is no domestic production left to absorb the excess overhead. By the time all production moves overseas, all the excess costs of overseas production will be known, but by then it might be too late to make up for the competitive disadvantages caused by excessive costs and other considerations discussed next.

Competitive Considerations for Offshore Manufacturing

Product delivery. As with outsourced overseas production, internal production far from customers slows delivery to customers and makes Lean Production, Build-to-Order and Mass Customization more difficult, maybe impossible, to implement. In addition to shipping time, more delays may be caused as shipments are batched to fill shipping containers. Delays may also be caused by slower production due to lower productivity, quality problems, or other unanticipated problems caused by overseas manufacturing.

> Overseas production poses major challenges for delivery, quality, innovation, control, and strategic focus

Primitive transportation infrastructures, such as poor roads, air, and rail service can further delay both incoming parts and outgoing products. Recently, cross-border shipments into the U.S. have been delayed by security checks.

Parts delivery. Too much distance from suppliers can slow down responsiveness and make it difficult to pull parts and materials from suppliers. Vendors need to be close to Engineering to help design products. Vendors need to be close to manufacturing for quick deliveries and optimal interactions.

Supply Vulnerabilities. Jeffrey E. Garten, Dean of the Yale School of Management, raised the issue of the vulnerabilities of moving too much production to one country, in this case, China. His concern was that the world economy is becoming "dangerously vulnerable to a major supply disruption cause by war, terrorism, social unrest, or a natural disaster."[34]

Quality. Quality may suffer if the new plant has not established an effective quality culture. Japanese "transplant" factories in the United States often spend up to a year training workers before the plant opens. Quality problems are more of a concern when quality is not a inherent element of the local culture. High turnover can also cause quality problems and limit effects of training efforts. Even with the optimum quality culture in place, starting up any new plant has a certain "learning curve" phase, which also applies every time another product is transferred. Quality may also suffer if recurring defects are produced overseas and not detected until hundreds of defective products are discovered at the end of the long transcontinental "pipeline."

Design Competitiveness. Separating engineering from manufacturing discourages interaction between engineering and manufacturing and makes it difficult for manufacturing people to participate on product development teams, which is one of the keys to design for manufacturability and concurrent engineering.[35]

Control. It may be harder to exercise effective control of overseas operations because of time zone separations, languages, cultural differences, lack of face-to-face communications, and the "out of sight; out of mind" syndrome.

Communications. Communications may be less than optimal due to different languages, communication equipment, and time zones, which, if over nine hours, means the plant and home office may never be working at the same time.

Local cultural differences may create unexpected problems related to quality, deliveries, holidays, communications, and treatment of local workers, which has tripped up many companies lately.

Technology introduction delays. New manufacturing technologies are usually introduced in the home plants first and overseas plants last. Such implementation delays may defer realizing technology benefits and temporarily make new technology plants incompatible with old technology plants.

Customs and regulations. There may be unanticipated difficulties with customs and regulations from both the home and overseas countries. One high tech company located their "high volume" circuit board assembly factory in Singapore, but later found out that U.S. national security restrictions prohibited the latest microprocessors from being shipped in or out of the plant. Trade wars can disrupt or penalize overseas production and shipping.

Liability. American companies can face liability risks based on actions of their overseas joint-venture partners.

Strategic focus. Setting up and operating overseas manufacturing plants consumes a lot of effort and may distract companies from other improvement programs, such as BTO&MC.

When Offshore Manufacturing Does Make Sense

The preceding discussion addresses the problems of locating production overseas *to save cost* on parts or products shipped back to the home country.

On the other hand, overseas manufacturing *does* make sense when those factories are close to customers and sources of supply, as is the case with "transplant" automobile factories set up in the U.S. by Japanese, European, and Korean automobile companies. In fact, the American automobile industry has been transformed by these transplant factories who now even design products in America for American tastes.[36]

This corresponds with the philosophy of BTO&MC which is *local production for local consumption.* This author predicts that this will reshape world trade, as will be discussed next.

WORLD TRADE IN THE ERA OF BTO&MC

The role of distance has a crippling effect on customer responsiveness at the distribution end of the supply chain. Even if the materials end of the supply chain is tightly coupled with close proximity and one-piece flow based on pull signals, the operation will probably not be able to provide the optimal customer responsiveness if finished products have to be shipped across oceans. Shipping delays will cause a significant competitive disadvantage for any company trying to take advantage of the strengths of BTO&MC. This is especially true for markets that require fast responses: rapid replenishment of stores and dealers; individually mass-customized goods; internet sales; and on-demand part manufacturer for BTO assemblers.

> BTO&MC favors local production for local consumption

Build-to-order and mass customization will favor local procurement, local production, and local distribution to ensure speed throughout the entire supply and distribution chains. And, as pointed out through this book, supply chain speed is what allows the BTO&MC company to build products on-demand and ship them to customers or stores as fast as the batch manufacturer can ship forecasted inventory, but without the costs of inventory or the outage vulnerabilities of forecasting.

Thus, as build-to-order and mass customization grow, exporting should diminish in both directions. In developed countries, the mass marketers who mass-produced products overseas will gradually evolve or be replaced by BTO&MC companies who build mass-customized and standard products on-demand for quick shipment to domestic customers.

> As BTO&MC grows, exporting should diminish in both directions

Developing countries will be able to raise themselves above the low status of being the "cheap labor" provider with all of its problems of workplace abuses, "sweat shop" allegations, and population dislocations and disruptions. In fact, the current form of "globalization" has not helped many poor countries, according Joseph Stiglitz, author of *Globalization and Its Discontents.*[37]

In its place, developing countries will to evolve to the dignified status of designing products suitable for their own cultures, building them in local plants by a proud local workforce, and distributing mass-customized or standard products quickly to local consumers. And in both developed and developing countries, companies, and ultimately consumers, will benefit from avoiding the costs of transoceanic shipping, inventory, obsolescence, tariffs, and duties plus minimizing macroeconomic problems of international currency fluctuations and balance of trade deficits.

> "Cheap labor" outsourcing countries will evolve to the dignified role of making their own goods on-demand

ENDNOTES/REFERENCES

1. "How Great Companies Stay Great;" Interview with Clayton Christensen; *Business 2.0,* October 2003, p. 36.

2. Maria A Ferrante, "Outsourcing: Gaining the Manufacturing Edge," *Food Engineering,* 69 (4): 87-90, April 1997.

3. Jeffery E. Garten, "When Everything is Made in China," *Business Week,* June 17, 2002.

4. Rete Engardio, "The Barons of Outsourcing," *Business Week,* August 28, 2000, p. 177.

5. Outsourcing definition from *The American Heritage Dictionary of the English Language,* Third Edition (1992, Houghton Mifflin Company).

6. Olga Kharif, "PC Companies Look Beyond the Box," *Business Week online,* January 15, 2002.

7. The seminal book on Lean Production, *Lean Thinking,* by James P. Womack and Daniel T. Jones (1996, Simon & Schuster) concludes, based on years of benchmarking, that converting to lean production will: double labor productivity all the way through the system (for direct, managerial, and technical workers, from raw materials to delivered product); cut production throughput times by 90 percent; reduce inventories in the system by 90 percent; cut in half errors reaching the customer and scrap within the production process; cut in half job-related injuries; cut in half time-to-market for new products; offer a wider variety of products, within product families, at very modest additional cost; and reduce capital investments required to very modest, even negative, levels if facilities and equipment can be freed up or sold." Further, the authors content that "Firms having completed the radical realignment can typically double productivity *again* through incremental improvements within two to three years and halve again inventories, errors, and lead times during this period."

8. David M. Anderson, *Design for Manufacturability & Concurrent Engineering; How to Design for Low Cost, Design in High Quality, Design for Lean Manufacture, and Design Quickly for Fast Production,* (2008, CIM Press, 448 pages; 805-924-0200)

9. Phillip Crosby, *Quality is Free* (1979, Mentor Books).

10.William M Bulkeley, "As PC Industry Slumps, IBM Hands Off Manufacturing of Desktops," *Wall Street Journal,* January 9, 2002, p. B1).

11. Adrian J. Slywotzky and David J. Morrison, *Profit Patterns, 30 Ways to Anticipate and Profit From Strategic Forces Reshaping Your Business,* (1999, Times Business/Random House), pages 116-123, "Reintegration."

12. Carliss Y. Baldwin and Kim Clark, "Managing in the Age of Modularity," *Harvard Business Review,* September-October 1997.

13. Rete Engardio, "The Barons of Outsourcing," *Business Week,* August 28, 2000, p. 178.

14. Sandra D. Atchison, "Pump, Pump, Pump at Schwinn," *Business Week,* August 23, 1993, p. 79.

15. Sydney Finkelstein, *Why Smart Executives Fail, And What You Can Learn from Their Mistakes,* (2003, Portfolio/Penguin Group). pages 43 and 74.

16. Jeffrey Ball, "Chrysler Averts a Parts Crisis," *Wall Street Journal,* September 24, 2001, p. B1.

17. "Shipping industry in Hong Kong May Fail U.S. Test," *The Wall Street Journal,* January 3, 2003, p. A6.

18. Madeleine Acey, "Marconi Sells Plants; Shares Rise; Outsourcing Deal Will Net Jabil Circuits $4 billion; *FT Market Watch,* London, January 12, 2001

19. Mike Tarsala, "H-P plans to Outsource more PCs." *CBS Market Watch,* January 17, 2002.

20. Anderson, *Design for Manufacturability & Concurrent Engineering.*

21. Ibid., Chapter 3, "Concurrent Engineering."

22. Ibid., Chapter 1, "Design for Manufacturability."

23. Clayton M. Christensen, "Limits of the New Corporation," *Business Week,* special issue on The 21st CenturyCorporation, August 28, 2000, p. 180.

24. John A. Byrne, *Business Week*, August 28, 2000; Table: "What a Difference a Century Can Make," page 87.

25. Steven L. Goldman, Roger N. Nagel, Kenneth Preiss, *Agile Competitors and Virtual Organizations,* (1995, Van Nostrand Reinhold);page 205, "Virtual Organization Characteristics."

26. Anderson, *Design for Manufacturability & Concurrent Engineering,* Chapter 10, Design for Quality, p. 199.

27. James P. Womack and Daniel T. Jones, *Lean Thinking, Banish Waste and Create Wealth in Your Corporation* (1996, Simon & Schuster), p. 224

28. Adam Aston and Michael Arndt, "The Flexible Factory; Leaning heavily on technology, some U.S. plants stay competitive with offshore rivals," *Business Week,* May 5, 2003, pages 90-91.

29. David Pringle, "How Nokia Thrives by Breaking the Rules," *The Wall Street Journal,* Jan. 3, 2003.

30. Anderson, *Design for Manufacturability & Concurrent Engineering.*

31. "Breaking Point," *Business Week,* Cover story on "Oil & the Economy," June 9, 2008, pp. 22 - 24.

32. Peter Coy and Jack Ewing, "Where Are All the Workers?," *Business Week,* April 9, 2007, pp. 28 - 30.

33. Alan S. Brown, "The China Road: The Outsourcing Option Isn't as Cheap or as Easy as it Looks," Cover story, *Mechanical Engineering,* March 2005, pp. 36 - 40.

34. Jeffery E. Garten, "When Everything is Made in China," *Business Week,* June 17, 2002.

35. Anderson, *Design for Manufacturability & Concurrent Engineering.*, Chapter 3, "Concurrent Engineering."

36. Joann Muller, Kathleen Kerwin, and David Welch, "The New American Auto Industry," *Business Week,* July 15, 2002, pp. 98- 106.

37. Joseph E. Stiglitz, *Globalization and Its Discontents,* (2002, W.W. Norton), 282 pages; Book review by Michael J. Mandel, "Where Global Markets are Going Wrong," *Business Week,* June 17, 2002, p. 17.

7

SPONTANEOUS SUPPLY CHAIN

The two basic principles of for spontaneous, on-demand production of mass-customized and standard products are:

Automatic resupply of parts and materials without forecasts or inventory (the subject of this chapter). Without this ability to be *pulled* spontaneously, production would have to wait for ordered parts and materials to arrive, which would not be spontaneous or on-demand.

On-demand production of the parts from standard materials that are always available at all points of use (the subject of the next chapter). For spontaneous, on-demand production, parts would have to be built on-demand from standard materials, if they can not be procured spontaneously.

Spontaneous supply chains play a key in BTO&MC to avoid the need to generate forecasts, count inventory on-hand, generate purchase order inputs through MRP (Material Requirement Planning) systems, place purchase orders, wait for parts to arrive, expedite those that are late, receive (and maybe inspect) materials, warehouse, group into kits for scheduled production, and distribute within the plant. BTO&MC can avoid all these costly and time-consuming steps with a *spontaneous supply chain,* which is able to *pull* in materials and parts on-demand. It is highly unlikely that this can be accomplished merely by asking existing supply chains to deliver all your current parts on-demand.

The prerequisites to the spontaneous supply chain were presented in the last chapters: Product line rationalization to eliminate or outsource the most unusual products that have the most unusual parts (Chapter 3); Standardization to reduce the variety of parts and materials to the point where they can be pulled into factories (Chapter 4); On-demand cutting-to-shape to reduce incoming raw material to a few standard sheets, bars, tubes, and so forth (Chapter 5); and selective vertical integration to eliminate any remaining long-lead time parts and materials (Chapter 6).

The above list of prerequisites, and the steps presented in this chapter, will require some time and effort. Fortunately, these steps will generate cost and throughput savings in their own right. Combined with the overall benefits of BTO&MC (Chapter 14), these steps will provide even more returns.

Spontaneous Resupply

The concepts presented herein are labeled as the *resupply* of parts and materials as opposed to *procurement* or *purchasing*. This is to emphasize that most of supply chain management is a *resupply* function of parts and materials that have been procured before. Therefore, the role of procurement/purchasing departments should focus on arranging spontaneous *resupply* of *standard* parts and materials instead of *purchasing* a proliferated list of parts and materials.

> Instead of purchasing parts for every order, Purchasing's new role will be to implement automatic resupply

In a build-to-forecast batch environment (mass production), Purchasing's role was to order parts based on forecasts and MRP data and then expedite part shortages due to forecasts being off, bill-of-material errors, inventory count errors, and late deliveries. In BTO&MC environments, Purchasing's new role will be to:

- Encourage and maybe drive standardization of parts and raw materials for current products and help new product development teams design around *aggressively* standardized parts and materials.
- Identify standard raw materials and parts that will be available *throughout the life of the product*. Arrange breadtruck replenishments.
- Arrange steady flows of standard parts and raw materials; nurture supplier/partner relationships with the focus on delivery.
- Establish kanban and pull signal arrangement with suppliers.

Most of the spontaneous part and material resupply will be automatic or manually triggered by production personnel, not by MRP and purchasing, thus saving that overhead cost.

For the 25 finalists in Industry Week's annual Best Plants rating, 55.6% have implemented pull systems compared to the average of 18.6% for all 2,511 manufacturers surveyed.[1] *The best plants used pull systems three times more than the average plant.*

Best plants' use of pull systems was 3 times the average

Prerequisites to Spontaneous Resupply

1) **Aggressive standardization** parts and materials, which in many cases means that some products will get a better part or a stronger material, but any perceived "extra" cost would be much smaller than the total cost value of the spontaneous BTO model. Such standardization will:
   - get the volumes high enough for frequent deliveries
   - assure wide-spread use for assurance that (a) the standard parts/materials will be used one way or another and (b) wide-spread use allows spot shortages to be covered by other bins in the factory, which is even more important for long-lead-time parts.
   - get the part counts small enough to be able to ensure spontaneous availability at all points of use with respect to (a) arranging "dock-to-line" deliveries to all points of use and (b) space at the workstation

2) **Product Line Rationalization,** which eliminates or outsources the most unusual *products*, which generally have the most unusual *parts and materials*. For product that survive rationalization, keep them out of flexible plants.

There are two aspects of the spontaneous supply chain: raw material resupply and part resupply.

RAW MATERIAL RESUPPLY

Raw material variety reduction. Too many types of raw materials can thwart spontaneity and present manufacturers with the dilemma of stocking all types of materials or ordering them and waiting for delivery. The effective procedures presented in Chapter 4 should be used to aggressively standardize raw materials. Standardization reduces the incoming variety which helps enable the spontaneous supply chain needed for BTO&MC. Purchasing leverage and other material overhead savings compensate for some products getting "better" material than they need. Standardization can also protect a manufacturer from shortages, which are more likely to occur if production depends on many unusual materials.

Raw materials could be automatically resupplied using the following techniques (in order if increasing variety):

Steady Flow of Standard Raw Materials

The ultimate scenario for spontaneous BTO is to reduce the number of raw material types within each category to one, in which case steady flows can be arranged for each standard raw material. Ideally, there should only be only one type of each material. Then forecasting multiple types would be unnecessary and "ordering" would be as simple as matching the tonnage in to the tonnage out; in other words, the incoming flow of the standard raw materials would be equal to the monthly consumption of the plant. These will be used one way or another. Multiple types of materials in each category would allow the same spontaneity if the ratio is constant or predictable and usage is segregated (if not, material changeover would have to be quick and hopefully automatic).

> The ultimate resupply scenario is a steady flow of one standard material for each category

Min/Max Stacks

In the "min/max" technique, often used for raw material like sheet metal, material is consumed until the stack reaches the "min" (minimum) level, usually marked on the rack or wall. This triggers a reorder of the material to bring it up the "max" (maximum) level without the usual purchasing costs. Price and delivery arrangements, based on average usage data, could be negotiated on a long term basis for greater purchasing leverage.

> When materials get to the "min" level, reorder up to the "max" level

Material Cut-to-Length/Shape

Instead of ordering every possible shape, raw materials can be *cut to-shape* on-demand from the longest version or standard sizes by programmable CNC equipment, such as laser cutters, programmable shears, or from less automated tools directed by on-line instructions displayed on monitors. This subject was addressed in more detail in Chapter 5 and shown on the illustration in Figure 8-1.

> Cut to shape on-demand, instead of ordering every possible shape

Linear Cut-Off

Raw material variety can be greatly reduced by cutting off linear materials on-demand at the points of use or the cut-off station could be a *kanban* source for parts resupplied automatically to all the other points of use. The cut-off machine could be fully programmable or a worker could position the material up against a programmable or manual stop. Linear materials include all forms of bar stock, extrusions, strips, tubing, hose, wire, rope, cable, chain, and so forth.

Kanban

For raw materials that come in bins rather than stacks, the *kanban* resupply technique could be utilized. It may be that the *source* is one of the cut-off operations mentioned above (for more on kanban, see the illustrated discussion below in "Part Resupply").

Strategic Stockpiles

Until the above techniques can be implemented, it may be necessary to have strategic stockpiles of certain materials. The manufacturer could use *selective* stockpiles to temporarily compensate for any parts or materials that cannot be "pulled" or for temporary availability problems on standard materials. Stockpile ordering would have to be based on some kind of forecasts, but if the material was standardized, then the forecast would be easier to make for the *aggregated* demand for all consumption.

Order Material After Receipt of Product Order

As mentioned in Chapter 5, spontaneous resupply may not be feasible for unusual or seldom-used raw materials, especially on any products with inherently high diversity of materials. If material order times are less than build times, these materials could be ordered after receipt of the product order.

Materials should be *chosen* for this category, which will require a *proactive* effort, especially when companies are converting to spontaneous build-to-order.

PARTS RESUPPLY

The typical response when suppliers are asked to deliver parts just-in-time to their customers' pull signals is to keep building the parts in large batches, try to stock enough in *their* finished goods inventory, and meter them out "on-demand." A special-case variation of this approach is the Dell model where suppliers warehouse their parts next to the assembler's factories (see Introduction). Not many

> When asking to pull parts on-demand, most suppliers will try to dish them out from inventory despite all the problems

assemblers are big enough and powerful enough to force their suppliers into such an arrangement.

However, this is not really a pull-based supply chain and does not qualify for the definition of *spontaneous build-to-order.* Parts availability would depend on assemblers' forecasts, which are becoming increasingly less accurate, and the supplier's inventory, which is costly to carry and prone to obsolescence.

PART RESUPPLY STRATEGY

Part resupply strategy depends on the variety of the parts. At one end of the spectrum, very standard parts that are used in almost all products could arrive in a steady flow like standard raw materials.

At the other end of the spectrum, parts with high variety would be *built* on-demand using the techniques presented herein including flexible CNC fabrication and manual equivalents.

In between, there are several other strategies such as *kanban* for somewhat standardized, medium-variety parts and breadtruck resupply for small, low-cost commodity parts such as fasteners. Inflexible parts, such as casting and plastic parts, can be consolidated into versatile parts that can be used in many products. Parts could be automatically resupplied by the following techniques (listed in order, with the easiest first):

Steady Flow of Parts

As with standard materials, steady flows could be arranged for *very* standard parts, which would be used one way or another. The criteria for steady flow of parts would be standardization and widespread use.

Breadtruck Resupply

> Breadtruck suppliers simply keep the bins full and bill for whatever is consumed. Thus, no overhead.

The easiest and "lowest hanging fruit" in material logistics is the *breadtruck* (sometimes called "free stock") delivery system for small, inexpensive parts, like fasteners. Instead of counting on forecasts to trigger an MRP system to generate purchase orders, all the "jellybean" parts can be made available in bins at *all* the points of use. A local supplier is contracted to simply keep the bins full and bill the company monthly for what has been used, much like the way bread is resupplied by the breadtruck to a market.

All the MRP/purchasing expense is eliminated and this type of delivery can assure a constant supply of parts, thus avoiding work stoppages. Being off the forecast/MRP system, the supply of these parts can be assured for "forecast-less" operations such as BTO&MC. Typical parts suitable for breadtruck deliveries are fasteners, hardware, and almost any small, inexpensive part.

As companies become more agile, they may include slightly more expensive and slightly larger parts into the breadtruck system. The more expensive parts may incur some inventory carrying cost, but that should be outweighed by savings in purchasing, material overhead, expediting, and avoiding work stoppages.

Criteria for Breadtruck Deliveries:

- A reliable supplier can be contracted. Many suppliers welcome such business and want to perform well, since they usually get all the business for their categories of parts and raw materials.
- Parts can be distributed at *all* points of use, without cluttering assembly areas with too many parts. Of course, part standardization will help achieve this goal.
- Parts are small enough and cheap enough so that sufficient parts will always be on hand. Bin count can be set high enough to preclude any chance of running out.
- Parts are not likely to go obsolete or deteriorate while waiting to be used.
- The breadtruck parts are not so "attractive" as to create a significant pilferage problem, since, generally, companies do not correlate part consumption with product sales. However, making breadtruck parts freely available for R&D prototypes and factory improvements may encourage innovation and the use of standard parts.
- Manual reorders are not anticipated to occur. The supplier should be in a continuous improvement mode and be constantly adjusting bin count to correspond to prevailing demand. The factory should alert the supplier about any anticipated "spikes" in demand.

Kanban

In *kanban* resupply, parts with limited variety are made, maybe in batches, and resupplied automatically to replenish parts bins as they are consumed. This is one of the many *pull systems* used to "pull" parts into assembly operations.[2] The resupply is automatic once the pull signal gets to the supplier. There are many simple ways to do this without complex information systems such as MRP or ERP. Thus, kanban resupply avoids the uncertainly of forecasting, the cost of purchasing, and the cost and risks of inventory.

> *Kanban* can automatically resupply standard parts, even parts made in batches

Kanban works best for semi-standard parts without too much variety, which would increase work-in-process (WIP) inventory and clutter assembly stations with too many part bins. Kanban parts can be made in mass produced batches, but part manufacturers may have to implement setup and batch size reduction to be able to economically make batches small enough for kanban deliveries. *Do not let suppliers refill kanbans from inventory.*

In-house CNC machine tools and programmable cut-off machines can be kanban sources when they are not making parts on-demand, thus increasing their utilization and helping to justify such equipment

The principles of kanban can be best explained using the *two-bin* system as illustrated in Figure 7-1, which shows two rows of part bins which are set up for resupply.[3] Initial assembly starts with all bins full of parts.

When the part bin nearest the worker is depleted, the full bin behind moves forward, as shown by the empty space in the illustration. The empty part bin then is returned to its "source," which could be the machine that made the part, a subassembly workstation that assembled the part, or a supplier. The source fills the bin and returns it to this assembly workstation behind its counterpart which is still dispensing parts.

The beauty of Kanban resupply is that the system ensures an uninterrupted supply of parts *without forecasts or high-overhead-cost ordering procedures*. The number of parts in a bin is based on the highest expected usage rate and the longest resupply time. The size of each bin is determined by the bin quantity and size of the parts. For large parts, some companies use two-truck kanbans, in which parts are drawn from one truck trailer while the other trailer goes back to the supplier for more parts. Alternate systems include kanban squares for larger parts and a two-card system where the cards travel (or are faxed or e-mailed) back to the source instead of the bins. Electronic equivalents can also be utilized.

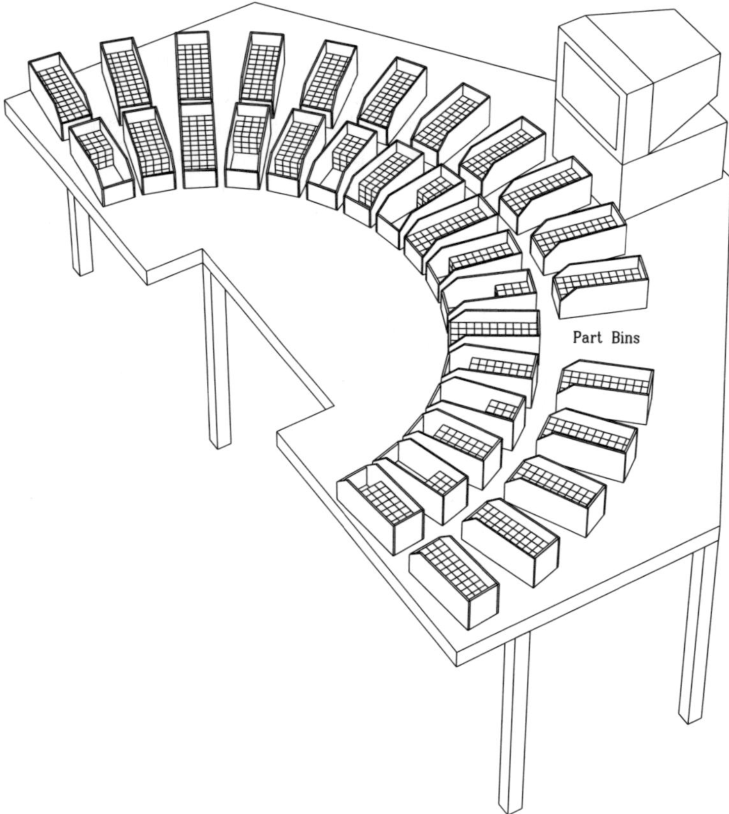

Figure 7-1: Kanban Part Resupply

In order for Kanban systems to work, there must be enough room to dispense all parts at *all* the points of use. This, again, emphasis the importance of part standardization.

Yasuhiro Monden, in an updated version of his classic *Toyota Production System,*[4] states that "the kanban system's most remarkable feature [is] its adaptability to sudden demand changes or exigencies [urgencies] of production." This is exactly what is needed for build-to-order environments, which are based on demand, not on planned production schedules.

Spontaneous Build-to-Order of Parts

For parts that are too varied for kanban, the assembler or the suppliers themselves would need to implement spontaneous build-to-order so that *they* could actually build on-demand to their customers' (the assemblers') pull signals. This is the only way to supply mass-customized parts on-demand for mass-customized products. Parts can be made on-demand in-house or by nearby agile suppliers.

> Parts may need to be built on-demand using BTO&MC techniques

It may *appear* that spontaneous build-to-order of parts may cost more than mass production, but in reality, a complete BTO&MC operation is very cost-effective when measured on a total cost basis, as discussed in Chapters 11 and 12.

Parts Made On-Demand by Suppliers

Hopefully, it may be possible to find suppliers who can implement these techniques to make your parts on-demand in response to your pull signals. Pull signals need to be initiated early enough and response time needs to be quick enough so that parts arrive without causing assembly delays.

Spontaneous BTO of parts may require the development of supplier/partner relationships in which suppliers establish the ability to build parts in any quantity on-demand. The distance to the supplier must not be so great so that part delivery delays product delivery.

Be wary of suppliers that are "pulling" parts from inventory because of the risk of not having enough (which they may blame on *your* inadequate forecasts) and higher than necessary cost for inventory carrying costs.

Parts Made On-Demand In-House

In order for spontaneous build-to-order to work, *all* parts must be available on-demand. If there are any key parts that are not suitable for kanban and no supplier can build them *and ship them quickly enough* to your pull signal, then you might have to bring those operations in-house, as discussed in the last chapter on selective vertical integration (Ch. 6).

Flexible Processing

Regardless of the sources of parts, spontaneous part manufacturing operations must be able to make parts on-demand efficiently in a batch-size-of-one mode without setup or inventory. CNC programmable machine tools and flexible assembly can produce a high variety of parts without setup costs and delays from standard raw materials, as will be shown in the next chapter. Similarly, manual assembly can be made flexible as will also be discussed in the next chapter. This may require concurrent engineering of product families, parts, and processing to eliminate all setup changes, as will be discussed in Chapter 10.

Strategic Stockpiles

Until the above techniques can be implemented, it may be necessary to have strategic stockpiles of certain parts. The assembler could use *selective* stockpiles to temporarily compensate for any aspect of the supply chain that cannot be pulled or for temporary availability problems on standard parts.

Order Parts After Receipt of Product Order

Spontaneous resupply may not be feasible for unusual or seldom-used parts, especially on capital equipment with an inherently high diversity of parts. If parts order times are less than build times, these parts could be ordered after receipt of the product order.

Parts should be *chosen* for this category, which will require a *proactive* effort, especially when companies are converting to spontaneous build-to-order from pseudo "build-to-order" operations.

SUPPLIER LEAD-TIME REDUCTION

As BTO&MC implementations continue to decrease the product build time, there will come a time when long-lead part delivery times will become the critical path and start to delay product shipments. *Before* this happens, the process of shortening supplier lead times needs to begin. For parts and materials with long lead-times:

> Shorten supplier lead times *before* they become the critical path

- Identify parts with long lead times. Plot in Pareto format for all parts and then by categories as appropriate. Within each category, identify various suppliers.
- Explore ways to minimize the lead time (listed in order of the easiest first):
  - Standardize parts and raw materials and focus lead-time reduction efforts on standard part and material lists. For new designs, eliminate parts and materials that currently have excessive lead times. Standardization efforts result in fewer part types with higher order quantities, which makes lead-time negotiations more focused and, ultimately, more successful.
  - Investigate and understand reasons for long supplier lead times, for instance:
    - Inherently long processing time with many steps
    - Too much supplier outsourcing with many links in *suppliers'* supply chains
    - Large batch mass production with lengthy setups
    - Parts shipped from inventory, which may be fast if in stock but much too slow if not in stock
    - Order entry problems or too many manual steps to process the order

    - Parts ordered are not suppliers' most popular products; consider shifting to those that are.
    - Customizations are performed in the reactive fire-drill mode.
    - Your order may wait in a queue before starting.
    - Your business is too small a share of your supplier's business.

  - Negotiate to improve lead-times: apply pressure; offer incentives; make delivery a key condition of doing business; show them your rapid operations to emphasize the importance of faster delivery.
  - Shift demand to product variations with the fastest suppliers.
  - Look for faster suppliers; first, let current ones know their business is at risk over delivery.
  - Work with suppliers to improve their lead times. Most delivery delays are caused by paperwork delays, setup, and processing inefficiencies. Share with them your successes and techniques. Encourage suppliers to embark on lean production, build-to-order, and mass customization programs. Send a team to train suppliers or pay for outside experts to train them and help them implement BTO&MC. The resulting improvements to supply chain will be well worth the effort.
  - Establish vendor/partner relationships with fast lead times as a key criteria.
  - Bring production in-house or at least under the control of the assembler through supplier/partnerships, alliances, acquisitions, or mergers. Certain processes may not be currently considered a "core competency," but may complete the "missing link" in a flexible operation and support the business model.

- As a last resort, carry a stock of *selected* long lead time parts, the size of which would be based on sales forecasts *and how standard they are.* As with any forecasting paradigm, this approach is dependent on forecast accuracy, which may not be very accurate, so the risk of running out has to be balanced against inventory carrying costs and obsolescence risks.

DOCK-TO-LINE PART DELIVERIES

Dock-to-Receiving-to-IQC-to-Warehouse-To-Kitting

In most plants incoming parts go through a slow and expensive procedure starts in the receiving department (where they are logged in), then to the incoming quality control (IQC) department (where they are inspected), then to the raw material warehouse (where they are inventoried), and then the kitting department (where they are counted and grouped into batches).

Dock-to-Line Deliveries.

To be truly agile, incoming parts and materials must flow directly to all points of use without the steps listed above; this is called *dock-to-line delivery.* This is sometimes referred to using the more common, but less accurate phrase, *dock-to-stock,* which technically means parts go to some kind of internal warehouse, which may include incoming inspection, before being distributed to the line.

Dock-to-Line means deliveries go directly to all points of use

Dock-to-line may be more easy to implement after implementing part and material standardization, product line rationalization, breadtruck deliveries, and kanban resupply of appropriate parts. Freeing up floor space by inventory reduction efforts will make room for internal distribution at all points of use. Lean environments require much less raw materials inventory than batch oriented operations, so there will still be a net reduction in floor space requirements after implementing lean production. In addition, part warehouse space may now be more available.

Dock-to-line deliveries can be either triggered by purchase orders that come from MRP systems or automatic pull signals like kanban deliveries.

In order for dock-to-line to work, *quality must be assured at the source* by suppliers whose processes are so in control that their customers (the assemblers) do *not* need to inspect incoming parts, which costs money, takes time, and may reduce flexibility to the point where parts have to be kitted for production batches.

Dock-to-line deliveries can be an essential part of a lean production program or may be instituted primarily to save cost and improve throughput. Industry Week's "Best Plant" survey of the 25 top performing candidates indicated that 68% of suppliers deliver parts to the point of use in the plant.[5]

The Problems with Incoming Inspections

The big paradigm shift required for dock-to-line deliveries is the elimination of incoming inspections of parts and raw materials. Incoming inspections are impractical for two reasons: time and effort. Just-in-Time deliveries, as the name suggests, should be *just in time.* Having to go through incoming inspections, usually at central receiving stations, would cause too many delays for a fast-moving lean environment. JIT deliveries may be smaller and occur more often than the traditional large order that is delivered infrequently. Consequently, inspecting many small orders would be very inefficient because of the inspection setup which include finding getting up to speed on quality standards and procedures, setting up and calibrating test and inspection equipment, and dealing with problems via MRB (Material Review Boards).

Ensuring Quality at the Source

But incoming inspections cannot simply be eliminated without some way of assuring that the incoming parts and raw materials will have adequate quality. If a manufacturer simply dictates new standards for incoming part quality from suppliers, the suppliers may respond by shifting inspection from the manufacturer's receiving to the end of the suppliers operations. This may screen out bad parts but at too great a cost in money and agility. In addition, when companies try to achieve quality by rejecting out-of-spec parts from a "wide" bell curve, the result is that the parts that *do* pass will have a high proportion close to the "hairy edge" of not working, which may cause more worst case failures.[6]

Suppliers, internal and external, need to become good at assuring quality *at the source.* If part manufacturing and raw material processing is sufficiently *in control,* quality will be assured by the *process,* not by subsequent inspections. Statistical Process Control (SPC) is a proven tool for assuring quality by process controls.[7] Even though SPC is firmly founded on statistical principles, its implementation does not require the proverbial Ph.D. in statistics. Control charts, available from the American Society of Quality (ASQ), have the statistics built into the charts so that factory workers can use them by literally filling in the blanks and performing some simple arithmetic computations.

Certification

Suppliers that can prove that their processes are in control and, thus, can deliver good parts directly to the line are *certified* by the manufacturer. Similar certifications may also be applicable for raw material suppliers to allow them to ship metals, plastics, and chemicals directly to the points of use without incoming inspections.

LOW-BIDDING

Going for the low-bidder is something we management consultants have been trying to discourage for years, but now it has seen a resurgence just because it is easy to do on the internet. This comes at a time when many purchasing functions are under heavy pressure to "use the net" and "get into e-commerce."[8]

> Low-bidding finds the cheapest parts –
> *and that is the problem*

Bidding for the cheapest parts is not only an ineffective way to achieve *real* cost reduction, but it can substantially raise less-obvious costs and compromise other important goals like quality and delivery, which are especially important for BTO&MC. Quality usually takes a back seat when buying decisions are focused on purchase cost, especially in bidding situations. Some say that quality "standards" can be set for all bidders, but it is a dangerously naive assumption that quality can be assured simply by setting a single metric. Further, focusing cost reduction efforts on part bidding distracts attention from *real* cost reduction opportunities, which are addressed throughout this book.

> There is more to procurement than price – Dell Computer

Dick Hunter, Vice President of Fulfillment and Supply Chain Management for Dell Computer says that on-line auctions are no "silver bullet:"

> "Auctions and exchanges have fueled the thinking that price is everything. But there is more to procurement of materials than just price. Quality, service, responsiveness, and the willingness to improve common processes also are very critical to driving down the total cost of materials."[9]

The Cost Reduction Illusion

In old-paradigm companies, "cost reduction" efforts are focused primarily on parts and materials (hereafter called parts) because that is all that most cost systems are able to quantify, besides labor. Many manufacturers, especially the U.S. automobile companies, beat up their part suppliers for repeated cost reductions. And now internet web-sites are able to conduct competitive bidding auctions to offer suppliers the "lowest cost" parts.

> Low-bidding parts may seem "the cheapest" at the expense of quality, deliver, and flexibility

But before manufacturers fall for a magic elixir, they should consider how part costs really would be lowered under such pressure. One assumption is that either purchasing agents have naively offered to pay too much or that cavalier part makers have been gouging their customers. While this may have been true in sleepy industries of the past, it is rarely true in today's dynamic marketplaces.

> How do bidders lower their prices to get the work? You may not like the answer!

Another assumption is that supplier inefficiencies can be somehow be corrected after a supplier "wins" a contract at a lower-than-usual price. However, soon after a supplier wins a bid, it is expected to deliver the goods, and there will not be time to implement any meaningful cost reduction program, like the eight strategies presented at *www.HalfCostProducts.com.* Thus, without a real means to lower costs, the supplier will either have to cut its margins (which will be resisted from the corner office all the way to Wall Street), cut corners (as discussed below), or do the same to thing to *its* suppliers, who will have the same difficulty achieving real cost reductions. Further, if suppliers are either making disappointing profits or struggling to reduce costs, they will not be very receptive or cooperative with BTO assemblers' needs for on-demand part delivery, which compromises their responsiveness and real cost reduction efforts.

> Without a *real* way to lower cost, bidders will either cut margins or cut corners

In some cases, suppliers will temporarily lose money to "buy into the business" with the expectation of raising costs later, once they are "in." And there are even suppliers out there whose strategy is to bid jobs at zero profit and plan to make all their money on expected change orders.

In other cases, low-bidders "win" because they don't understand the problem and then are ultimately unable to deliver at all. In other cases, winning bidders are "vapor" companies, whose goal is keep bidding down until they win, and then patch together a "virtual" network of alliances to somehow fulfill the order. This phenomenon came out in an in-house seminar, when he asked if anyone had noticed any problems with on-line competitive bidding. Within seconds, the purchasing manager was jumping up and down waving both hands in the air. They discovered that one bidding competitor was working out of an apartment and its strategy was to win the auction and then figure out later how to deliver the goods! He also said that a corporate dictate to do part bidding was alienating their valued suppliers with whom they had good relationships.

> Bidders may be *vapor* companies that will figure out fulfillment only if they win

The Cost of Bidding

Bidding keeps Purchasing so busy *managing* the bidding process that they will not be able to help their teams develop products, assure availability, and set up vendor/partnerships. The *Toyota Product Development System* book sums up the cost and wasted resources of bidding as follows:

> *"Searching the globe for the lowest cost means managing very large numbers of suppliers as well as introducing a steady stream of new suppliers into your system. These suppliers are unfamiliar with your requirements and demand a great deal of attention to get up and running. While administering complex contracts, managing global bidding wars, and overseeing the constant introduction of new suppliers into the process, U.S. automakers must maintain mammoth purchasing organization, deal with incredibly cumbersome and slow sourcing processes, and live with constant variation of supplier performance in the development process."*[10] – all to "save cost!"

Even after all that effort, problems often arise because low-bidders may not understand the problem or may be cutting corners, which raises other costs such as quality, expediting, delayed launches, warranty costs, or the costs of recalls.

The biggest cost of bidding may be value of the resources it takes away from product development to support bidding (at both the customer *and* vendor) to do the following: first, update/change documentation, CAD files, materials, tooling, and processing; second, complete transfers; and, third, deal with new or ongoing problems related to ramps, delivery, quality, or getting up the learning curve, sometimes through many iterations. All the above problems are much worse when offshoring to another continent.

Pressuring Suppliers for Lower Cost

Pressuring suppliers for drastic part cost reduction is what J. Ignacio Lopez de Arriortua tried at GM, which not only failed to generate real, lasting cost savings, but also alienated its supplier base and drove the best suppliers to its competitors. The suppliers that remained put their best people on Ford and Chrysler projects and withheld their newest developments from GM, since

> Beating up suppliers on cost alienates the best ones

Lopez was using proprietary supplier information to press all suppliers for lower prices. Further, the so-called "savings" in purchasing cost caused severe cost to be incurred elsewhere. An ill-fitting ash tray from a new low-bidder caused a six-week shutdown in one Buick plant! In another incident, managers had to beg for help from a supplier Lopez rejected because of a 5% "savings" from the low-bidder whose parts failed quality tests 50% of the time.[11]

> *The "cost-cutting campaign of Lopez, whose heavy-handedness drove away many of the company's best suppliers – and, perversely, may have helped raise GM's total costs."*[12]

When Rubbermaid first encountered cost pressures from powerful retailers, its first response as a "leading" company was to make sure its customers understood the necessity of price *increases!* When they realized that they really had to reduce prices, they tried what didn't work for Lopez and got the same alienation of the supply base, according to the largest research project ever devoted to corporate failures, *Why Smart Executives Fail, and What You Can Learn from Their Mistakes:*

> "With little talent in cutting costs in-house, Rubbermaid looked to shift responsibility elsewhere. Suppliers were prodded to knock down their own prices, alienating some of the best, low-cost vendors in the process."[13]

> Bidding creates a standoffish relationship that inhibits the *real* source of cost reduction: *cooperation*

Bidding creates a standoffish relationship between buyers and sellers that inhibits cooperative cost reduction, which, as will be proven next, are the key to *real* cost reduction. An extensive study that analyzed deficiencies in the American automobile industry concluded this about the effects of bidding on suppler relations:

"A key feature of market-based bidding is that suppliers share only a single piece of information with the assembler: the bid price per part. Otherwise, suppliers jealously guard information about their operations, even when they are divisions of the assembly company. By holding back information on how they plan to make the part and on their internal efficiency, they believe they are maximizing their ability to hide profits from the assembler." [14]

By contrast, *"Philips Consumer Electronics Company begins its product design process by sharing all relevant information with key suppliers."* Said one supplier, "*This kind of interaction allow us to give PCEC the maximum possible value for their money. Yet it is rare with other customers."*[15]

The Value of Relationships for Cost Reduction

Another common assumption is that if suppliers know they will have to bid, they will implement effective long-term cost reduction efforts. However, the most successful real progress in cost reduction in supply chain management has come from long-term relationships where manufacturers work together with suppliers.[16, 17]

The book that launched the lean production movement in the U.S., *The Machine That Changed the World,* notes that in lean production companies, suppliers *"are not selected on the basis of bids, but rather on the basis of past relationships and a proven record of performance."*[18] Honda's criterion for selecting suppliers is the *attitudes* of their management.[19] As a philosophy-driven company, Honda feels it is easier to *teach* product and process knowledge than to find a technically-capable supplier with the right attitudes, motivation, responsiveness, and overall competence.[20]

> Suppliers should be selected on the basis of past relationships and a proven record of performance, not bids

In BTO&MC much of the real cost reduction opportunities are not just at the assembler or at the supplier, but rather *in their relationship*. In a thorough study of Japanese lean manufacturers, *When Lean Enterprises Collide,* Robin Cooper states that "it is no longer sufficient to be the most efficient firm; it is necessary to be part of the most efficient supplier chain." The key to accomplishing this is *inter-company cooperation*, summarized by Cooper as follows:

> "The blurring of organizational boundaries becomes critical as competition intensifies because it not only reduces the time it takes the entire supplier chain to bring out new products with increased functionality but also allows quality to be improved while reducing cost."

Cooper recommends partner companies "create relationships that share organizational resources, including information that helps improve the efficiency of the interfirm activities." [21]

Create relationships that work together to improve efficiency

Such inter-company cooperation offers significant cost reduction opportunities, especially if suppliers can build parts on-demand for BTO assemblers; then *both* avoid all the cost and risk of parts inventory in addition to minimizing many categories of overhead for procurement, material overhead, expediting, warehousing, internal distribution, and so forth.

However, switching suppliers every time a competitor drops its price is incompatible with this strategy and can jeopardize ongoing relationships. A Fortune magazine analysis of dot-com failures summarized the failure of an on-line bidding site:

> "For the bulk of spending, corporations have long been moving in precisely in the opposite direction, establishing deep relationships with a few favored suppliers in a 'total cost' approach. Under this approach, price is but one of a host of criteria, which include quality, cycle time, service, and geography." [22]

The same article also revealed some realities about the purchasing process that question how welcome bidding would be for typical buyers:

> "In retrospect, say analysts, most B2B efforts betrayed pronounced cluelessness about how industrial buying actually works. Start with the supposition that purchasing managers would be thrilled to take bids online from dozens if not hundreds of suppliers each vying to be the lowest bidder."

Another article that proclaimed B2B auction sites as "yesterday's darlings," said that:

> "Many companies just weren't willing to dump the networks of suppliers they had built up over the years and do all their buying through a new, unfamiliar medium." [23]

Cheap Parts – Save Now; Pay Later

Actually, this phrase should be, more precisely: *save a little now, pay a lot later.* Many times trying to save money on purchase cost has the unintended effect of driving up other costs many times the assumed savings, like the old English adage: *penny wise, pound foolish,* or the more colloquial *"you get what you pay for."*

> Cheap parts can cause an explosion of other costs and problems

Cheap parts are usually just that – *cheap parts* that usually earn the stereotypical image of poor quality, which will add significant cost in the plant and cost even more if bad products get out, not to mention hazards to life and limb and loss of corporate reputations.

Even though quality disasters may look like infrequent anomalies, these costs must be included in the company's cost of quality, not just written off (literally and figuratively) as a one time "charge." Programs that aim to improve quality should be justified on their ability to prevent *all* quality costs, from an accumulation of many to "the big one."

Ford's enormous problem with tires was not surprising coming from an industry historically obsessed with bidding on part cost. An Industry Week article described the procurement process for tires at Ford:

> ". . . five tire manufacturers participated in an auction earlier this year in which an initial bid was set by Ford and the tire suppliers then reverse bid downward to capture the business. 'Twelve hours later they were still bidding,' says Brian Buersmeyer, Ford's e-business planning manager. 'The suppliers kept lowering the cost. The market tension that created was dramatically different than the traditional buying processes.' " [24]

In the 1990 J.D. Power rating of automobile reliability, Mercedes-Benz received the top rating. But by 2003, Mercedes' rank slipped to 26 out of 37 cars ranked.[25] One of the reasons for the drop in quality was cited by European analysts:

> "Executives of what then was Daimler-Benz grew worried about escalating production costs in the early 90's. Executives then made a policy decision to start trimming costs by notching down specifications for many components."[26]

Reduce Total Cost Instead of Focusing on Cheap Parts

In addition to the cost ineffectiveness of part bidding and its detrimental effects on relationships, there is the compelling argument that other cost strategies provide much greater opportunities for real cost reduction, like those presented at *www.HalfCostProducts.com.* Chapter 11 presents many ways to minimize *total* cost; Chapter 12 presents some easy ways to quantify it.

An Industry Week cover story pointed out the disappointments of auction-based exchanges and how they distracted focus away from programs that promise real promise:

> The cheap part focus distracts companies from focusing on *total cost*

> "Meanwhile, executives looking for big-time cost reductions could be in for major disappointments. Perhaps worse, the infatuation with auction-based first-generation exchanges threatens to sidetrack supply-chain management initiatives that offer the greatest promise for long-term results." [27]

There are enormous opportunities to reduce total cost throughout the supply chain, without any negative consequences, by eliminating the costs of setup, inventory, and obsolescence and substantially reduce the costs of quality, distribution, and material overhead.

The Value of High Quality Parts

Receiving high quality parts is especially important to BTO&MC operations because:

(1) Dock-to-line deliveries count on "quality assured at the source" so that incoming inspections are not necessary and parts can go straight to all the points of use. This not only saves on the cost of incoming inspections, but also enables spontaneous resupply techniques.

(2) One-piece-flow operations are more sensitive to failed parts "looping" back and disrupting the flow.

(3) Testing large batches of identical parts is not compatible with flexible production.

(4) Having part quality assured at the source plus the continuous quality feedback of one-piece flow will enable BTO&MC plants to assure quality by process controls.

Of course, there are suppliers that practice *kaizen* continuous improvement and *can* provide both high quality products at a low price. But because they value cooperation and have a "big picture" orientation, these companies would naturally align with customers who value long-term relationships instead of participating in the bidding process.

Another related trend is becoming apparent: The best suppliers are shunning B2B auctions. Philip L. Carter, professor of purchasing at Arizona State University, Temple, and executive director of the Center for Advanced Purchasing Studies (CAPS) concludes that:

> "Manufacturers that are tempted to source key parts and materials through a trading exchange may find it difficult to connect with the most innovative, quality conscious vendors, since many likely will boycott the auction bidding process, viewing it 'as a margin-squeezing play,' "[28]

The Value of On-Demand Part Delivery

For BTO&MC manufacturers to build products *spontaneously,* suppliers must be able to deliver parts on-demand. Since the best BTO&MC manufacturers will be building products without forecasts or inventory, their suppliers will also have to build parts without forecasts or inventory – in other words, suppliers will also have to practice BTO&MC techniques. Relationships will not evolve if suppliers are selected based on purchase cost and suppliers are switched often. A Business Week article summarized the views of Michael R. Katzorke, senior VP for supply-chain management at Cessna Aircraft Company:

> On-demand delivery comes from supplier relationships, but don't count on it if you keep switching suppliers

> "For companies pursuing lean-manufacturing initiatives – including the use of pull signals to trigger supplier shipments – it is critical to integrate suppliers into the manufacturing process. 'But how am I going to do that if one guy has the part today and somebody else will be making the part tomorrow?' "[29]

Ironically, suppliers who practice BTO&MC principles will actually provide *lower* cost not only for purchases but also for several categories of material overhead. And since these BTO&MC suppliers will probably have a healthy network of BTO&MC customers, they will start losing interest in "old paradigm" customers that want them to bid on purchase cost.

Another problem with procuring parts though competitive bidding is that usually the bid is for a *batch* of parts, which will probably be built in a batch and then drawn from inventory with all the problems of forecast inaccuracies, inventory costs, and obsolescence risks.

Low bidders will probably prove to be disappointing with respect to supply chain coordination, adjusting for demand variations, quality issue resolution, customization, bar-coding, labeling, radio frequency ID tags, shipping bin reuse, and help with customer service at the user level. All these services may be inadequate if either the low bid is forcing suppliers to eat into their margins or the suppliers are too stretched trying to satisfy the bid commitment.

Product Development

A key theme of *Concurrent Engineering*[30] is multifunctional teams with active and early participation from suppliers, who are in the best position to help design the parts they will be making. However, suppliers will not participate in early design team efforts if they do not have some assurances that they will get some business out of it. Michael R. Katzorke, senior VP for supply-chain management at Cessna Aircraft Company said: "With auction bidding, alignment is damaged. And integration into design and manufacturing is out the window."[31]

> Suppliers will not help develop products unless they foresee future work

No Such Thing as Commodities in BTO&MC

Even companies that value relationships still justify competitive bidding for "commodity" parts and materials. However, for BTO&MC, *there are no commodities* – even the most basic parts must be made available spontaneously, so the *service* of supplying these on-demand becomes a very important aspect of supply chain management. And accomplishing this will require supplier/partner relationships, not switching suppliers on every procurement for a "lower" purchase cost.

> There are no commodities in BTO&MC

Companies like GE brag about how much money they think they have "saved" by bidding on commodity parts like nuts and bolts, but the fact still remains that they are still *purchasing* them, with overhead costs that must be far exceeding the cost of the fasteners for material overhead and distributing a year's worth of fasteners to GE divisions.

Even for high-cost purchases, B2B transactions still don't reduce material overhead. Dave Oppenheim, e-business director for Cessna, summarizes the situation for big-dollar items:

> "So it still takes 45 buyers to conduct their business. The people aren't going away. The work isn't going away. And cycle time is not being reduced, even after spending millions of dollars on these B2B tools." [32]

Instead of competitive bidding, there are many ways to lower the total cost of "commodity" parts and materials. For low-cost parts, the strategy would be to *eliminate* the much greater overhead costs, such as all the costs of forecasting, bills-of-material, MRP, purchasing, shipping, receiving, and internal distribution. One of the authors favorite lecture props is a bag of three screws that one company *issued* to the assembly area. Not only did the issuance cost hundred of times the cost of the parts, but the assemblers complained that if they lost one of those screws, they had to fill out forms to get one replacement screw!

Kanban, breadtruck, and min/max resupply techniques, discussed in this chapter, can automatically resupply parts and materials and make them always available in all points of use and in the process *avoid considerable overhead costs.*

ENDNOTES/REFERENCES

1. David Drickhamer, "Aim High; How Industry Week's Best Plants Measure Up," *Industry Week,* October 2002, pp. 67-70. This article is available online at http://www.industryweek.com/CurrentArticles/asp/articles.asp?ArticleId=1321.

2. Yasuhiro Monden, *Toyota Production System, An Integrated Approach to Just-in-Time,* Second Edition (1993, Industrial Engineering and Management Press, IIE), Ch. 2, "Adaptable Kanban System Maintains JIT Production," and Ch. 3, "Supplier Kanban and the Sequence Schedule Used by Suppliers."

3. This figure was created for the author's seminars and first published in copyrighted class handouts. It was then published in *Agile Product Development for Mass Customization,* by David M. Anderson (1997, McGraw-Hill).

4. Monden, *Toyota Production System,* p. 27.

5. Industry Week, *The Complete Guide to America's Best Plants,* (1995, Penton Publishing).

6. David M. Anderson, *Design for Manufacturability & Concurrent Engineering; How to Design for Low Cost, Design in High Quality, Design for Lean Manufacture, and Design Quickly for Fast Production,* (2008, CIM Press, 448 pages), Chapter 10, "Design for Quality."

7. Robert Amsden, Howard Butler, and Davida Amsden, *SPC Simplified,* (1989, Quality Resources, New York, NY).

8. Philip L. Carter, professor of purchasing at Arizona State University, Tempe, and executive director of the Center for Advanced Purchasing Studies, was cited in *Industry Week* (February 12, 2001, p. 43) as observing that "Many purchasing organization are under heavy pressure form the corporate brass to implement some form of e-commerce."

9. John H. Sheridan, 'Proceed with Caution," *Industry Week,* February 12, 2001, pages 38 - 44. One of the cover stories on Manufacturing Exchanges.

10. Morgan & Liker, The Toyota Product Development System, Chapter 10, "Fully Integrate Suppliers into the Product Development System, , p. 200

11. For more on the counterproductive effects of this type of "cost reduction" see page 37-38 of *"The Connected Corporation,"* by Jordan D. Lewis (1995, Free Press).

12. "Smart Partner," *Business Week*; review of *The Connected Corporation* by Jordan D. Lewis.

13. Sydney Finkelstein, *Why Smart Executives Fail and What You Can Learn from Their Mistakes,* (2003, Portfolio/Penguin), p. 62.

14. James P. Womack, Daniel T. Jones, & Daniel Roos, *The Machine that Changed the World, The Story of Lean Production,* 1990, Harper Perennial, p. 142.

15. Jordan D. Lewis, The Connected Corporation; How Leading Companies Win Through Customer-Supplier Alliances (1995, Free Press); Chapter 4, "Practices for Joint Creativity," p. 74.

16. Yasuhiro Monden, *The Toyota Production System,* Second Edition (1993, Institute of Industrial Engineers).

17. James P. Womack, Daniel T. Jones, & Daniel Roos, *The Machine that Changed the World, The Story of Lean Production,* (1990, Harper Perennial), Chapter 6, "Coordinating the Supply Chain."

18. Ibid., p. 146.

19. Jeffrey Pfeffer and Robert I. Sutton, *The Knowing-Doing Gap; How Smart Companies Turn Knowledge into Action,*(2000, Harvard Business School Press), p. 23.

20. John Paul MacDuffie and Susan Helper, "Creating Lean Suppliers: Diffusing Lean Production through the Supply Chain," *California Management Review,* Summer 1997, pp. 118-150.

21. Robin Cooper, *When Lean Enterprises Collide,* (1995, Harvard Business Press), Chapter 9, "Interorganizational Cost Management Systems."

22. Jerry Useem, "Dot-Coms: What Have We Learned?" cover story, *Fortune,* October 30, 2000, p. 92.

23. "Lessons from the Dot-Com Crash," cover story, *Fortune,* October 30, 2000, p. R8

24. *Industry Week*, August 21, 2000, p. 39.

25. Lee Hawkins, Jr. "Finding a Car That's Build to Last," *Wall Street Journal,* July 9, 2003, page D1.

26. John O'Dell, "Even Mercedes Hits a Few Speed Bumps," *Los Angeles Times,* July 13, 2003, pages C1 and C4.

27. John Sheridan, "Proceed with Caution," *Industry Week,* February 12, 2001, p. 38.

28. Ibid., p. 43.

29. ibid. p. 40.

30. Anderson, *Design for Manufacturability & Concurrent Engineering*, Chapter 2, "Concurrent Engineering."

31. John Sheridan, "Proceed with Caution," *Industry Week,* February 12, 2001, p. 40.

32. Ibid., p. 44.

ON-DEMAND LEAN PRODUCTION

The ability to build mass-customized and standard products on-demand is the payoff for lean production programs. Lean production[1] has become a very popular program, but lean production efforts are usually confined to within Manufacturing Departments and existing MRP-based build-to-forecast scenarios.

This chapter shows how to implement *on-demand lean production,* which depends on a simplified, *spontaneous* supply chain (as discussed in the previous five chapters) that can deliver parts and materials spontaneously without forecasts, purchase orders, part purchasing delays, or warehoused inventory of parts and materials. When this supply chain is coupled to flexible batch-size-of-one lean production, the result is *on-demand lean production.*

In the phrase *batch-size-of-one*, the quantity "one" refers to the smallest order quantity anticipated, which could be one product or one case or one pallet. Batch-size-of-one capability means that setup delays have been eliminated or reduced to the point that production facilities can *efficiently* build any size order on-demand without having to batch orders together to spread out the setup delays.

In general, there are two types of lean production: *replacement* and *spontaneous build-to-order,* which is the subject of this book. In replacement lean production, common parts are built ahead and available to be pulled into assembly from *kanban.* When part supplies are consumed, more parts are built to *replace* them. Replacement lean production becomes less feasible with too many types of parts or too much variability in demand. If replacement lean production is not feasible for all parts, then the common parts would be made available through kanban and the overly numerous parts with variable demand would be built on-demand from standard raw materials by CNC (Computer Numerically Controlled) machine tools or manually from on-line instructions.

The most important capability of on-demand lean production is *the ability to build products quickly and efficiently in a batch-size-of-one.* In order to do that, all setup must be eliminated including any delays to kit parts, find and load parts, position workpieces, adjust machine settings, change equipment programs, and find and understand instructions.

PROBLEMS WITH SETUP

Mass Production deals with setup by accepting it as a "necessary evil" and then spreads it over as many products as possible in *batches* or *lots* to minimize the set-up charge per part. For decades, industrial engineers have used formulas to try to calculate the "Economic Order Quantity" (EOQ). However, manufacturing in batches drastically raises costs and lead times because of the following considerations:

> Mass production accepts setup as a "necessary evil" and wants to spread it over large batches

- **Setup costs.** The cost of setup includes all the labor to change fixturing/molds/dies, change parts, change tools, change programs, change workpiece holding means, and ensure the first part off the new setup is good. Once there is a dominant setup, then the cost of any setup change would be *both* the cost of changing that setup *and* the cost of returning it to the original setup.

- **Lower machine tool utilization.** While setup changes are being made, expensive machine tools may be idle, thus lowering machine tool, and maybe plant, capacity.

- **Throughput times.** Batching parts slows throughput because at each step the batches wait in line for setup changes and the processing time of all the parts in the batch.

- **WIP inventory.** Batches *create* WIP inventory, whereas BTO *eliminates* Work-in-Process inventory, thus speeding throughput, enabling quick exposure of quality problems, and saving inventory carrying costs, which average 25% of the value of the parts per year (Ch.2). And in some companies, WIP inventory *carrying costs exceed profits!* Eliminating this cost alone may justify lean production, BTO and mass customization programs.

- **Defects.** Parts made in batches might be made with recurring defects, which may not be discovered until after hundreds of defective parts have been made. See the section, "Defects by the Batch," in Chapter 2.

- **Space.** Batched parts occupy much more space than a single piece flow, especially if batches are so heavy that fork lift aisles are needed.

- **Disruptions.** "Rush jobs" can cause major disruptions to scheduled production by adding two additional setups for every affected part. "Treasure hunts" may be needed if there are any part shortages due to errors in forecasting, bills-of-materials, or inventory counts. Womack and Jones conclude that treasure hunts "are the distribution equivalent of the expediting always necessary in batch-and-queue production operation."[2]

- **Flexibility.** Because of all of the above, batch and queue manufacture is not flexible and therefore does not support build-to-order and mass customization.

For a treatise about the shortcomings of setup and mass production, see Chapter 2.

SETUP & BATCH ELIMINATION

If successive products are to be unique and different, there cannot be *any* significant setup delays to get parts, change dies and fixtures, download programs, find instructions, or any kind of manual measurements, adjusting settings, or positioning of parts or fixtures. For a plant to mass-customize or spontaneously build products to-order, *all production setup must be eliminated* (not just the low-hanging fruit or "as much as you can"), except for standard materials and kanban parts. The words "setup" and "elimination" are defined as follows:

> Change tooling quickly or eliminate the need to have to change it

"Setup" definition. *Setup* is the sum total of *all* the tasks to change over equipment and materials when shifting from one product or part to another. It also includes all tweaking after changes, warm-up times, and post-setup run-in until the first of a new batch is deemed acceptable. Setup costs should include any quality costs occurring because of setup changes or any shortfall from the *kaizen* continuous quality improvement possible with dedicated lines.

If equipment has a *dominate* setup, then running anything different will include *two setups:* both the setup to change over to the odd product *and the setup to return the equipment to the dominant configuration.*

"Elimination" definition. Many lean production efforts aim to *reduce* setup as much as possible in as many operations as possible, given the scope of the setup reduction efforts. However, the "tipping point" for flexible operations is the capability for an entire line to be able to produce any product/part in the family without change-overs. The definition of setup *elimination* includes elements of depth and breath. At each work station, setup is reduced to the point where it is feasible to build anything in the family, anytime, in any quantity. Such setup elimination applies throughout the line so that the whole line can build any product in any quantity.

Eliminating High-Setup Situations

Eliminate Setup with Dedicated Lines. A high-volume line can avoid setup if it is dedicated to one product and management avoids the temptation to "fill the line" by adding another product, which would add *two* setups, one to change to the new product and another to change back to the high-volume product. Medium-volume products can present a dilemma: the volume may not seem high enough to justify their own dedicated mass production line, so they are run together with other medium-volume products or stuffed into the high-volume line, with setup changes between every product change. The solution would be to eliminate setup by creating small dedicated lines in which "setups" are permanent.

> Dedicated lines have no setup changes

Since these smaller dedicated lines do not need the capacity of high-volume lines, they can be equipped with smaller *and less expensive* machines, either "low-end" new equipment or used machinery, which can be procured at a fraction of its original cost. Such equipment may be technically "obsolete," but it still may be adequate for these dedicated lines. In fact, even the *perception* of obsolescence can lower the cost further on the used equipment market. It may be possible to rectify quality shortfalls of used equipment because, as dedicated equipment, they can be refined to do one job well through *kaizen* continuous improvement efforts.

These medium-volume dedicated lines may not have "ideal" utilization and line balancing, but this is all right as long as they can produce the output because: (1) The smaller, slower machines used to complete cells cost less than the mega-machines that must be keep running at high utilization; (2) These lines eliminate expensive setup and inventory costs while reducing material overhead; and (3) There is a significant value to "completing the picture" for BTO&MC to enable the optimal customer satisfaction, competitive advantage, growth, and profits. Remember, when it comes to cost or any related metric, the only cost that counts is *total cost.* Many companies have compromised their overall company performance in the pursuit of specific metrics, such as utilization or productivity.[3]

Medium-volume dedicated lines each require a certain amount of floor space, but this should be made available by other efforts to eliminate work-in-process inventories, which can cut plant floor space needs in half.[4, 5, 6]

Implement product line rationalization to eliminate or outsource low-volume products that are the cause of many small batches and excessively high setups (Ch. 3). The remaining oddball products should be *kept out* of flexible plants, or at least segregated to the point where they do not disrupt BTO&MC operations or overburden the overhead charges to much more efficient flexible lines.

Avoid proliferation. Resist temptations to mix dissimilar products into flexible facilities through mergers, acquisitions, and filling under-utilized lines.

Discourage small orders of inflexible products. For the products that survive product line rationalization and cannot be made flexibly, discourage small orders by:

- Making sure the sales force understands setup cost and flexibility issues for multiple-product orders. The sales force needs to understand the real customer push-back tolerance – will they really shift a big order to a competitor if one low-volume product is not available? Finally, the sales force needs to understand the big picture – that company-wide flexibility can offer more customer satisfaction than making deals "complete at any cost"

- Shifting sales commissions from volume to actual profit, not profit margins (see "The Margin Trap" in Chapter 3).
- Charge all products their *actual* setup charges, not averages and not just canned "standards."
- Applying all overhead charges to every product. Low-volume products have higher-than-average overhead charges, and so will be under-reported when overhead is allocated (averaged).
- Outlawing orders that are incompatible with strategies for dedicated or flexible lines. Start with *profiles* that identify "red flags" for potentially incompatible orders for scrutiny. For any product red-flagged, require approval signatures from Operations, Supply Chain Management, Product Development, Distribution, and so forth. As the profile develops, it can automatically block inappropriate orders.

Eliminate tool/fixture setup. Design the product/processes to eliminate the need for tooling changes for:

- **Dies and molds.** Design castings, forgings, stampings, and moldings with *versatile shapes* that can be used in many products, maybe all products in the family. Any "extra" cost for material will usually be paid back many times by cost saved in fewer dies/molds to build and the cost/time to keep changing them. Very versatile dies/molds coupled with inexpensive machines might eliminate die/mold change entirely.

- **Flexible tooling.** Families of parts should be concurrently designed with versatile tooling and fixtures, that are versatile enough so that they never have to be changed.

- **Unique tools.** Eliminate tool change setup by designing parts around common tools (cutting tools, bending mandrels, punches, etc.), ideally one tool that never has to be changed.

- **Cutting tools.** If multiple tools are required, designers must keep tool variety well within tool changing capacity *for the whole family*. Ideally, eliminate cutting tool changes with versatile multi-purpose tools can do all the operations.

Versatile fixtures can eliminate positioning setup

Tool plate setup changes can be eliminated for printed circuit board by either developing versatile tool plates that can handle all products processed or laying out all circuit boards on the same size *panel.* Tektronix builds all its boards on 12" by 18" panels, which may contain one large board or several smaller boards. Any cut-off waste would be more than offset by the value of eliminating setup changes. Nokia circuit boards all have the same dimensions to avoid board size set up changes.[7]

CNC to Eliminate Machining Setup. CNC machine tools are very versatile tools to eliminate setup since many operations can be done by multi-axis machine tools without having to reposition the workpiece or move it to another machine. CNC machining centers can perform a wide range of operations such as machining, drilling, tapping, and so forth. The more operations that can be done in one operation, the fewer times the workpiece needs to be moved and set up. In fact, a key DFM and quality principle to ensure tight tolerances is to make sure all critical dimensions are cut in the same machine tool in the same clamping.[8]

Therefore, it is desirable to maximize the amount of dimensional variation done with CNC. However, physical setup must be eliminated, for example, for workpiece positioning and tool changes. Products may need to be designed for CNC to completely eliminate setup.

Doing all operations in one versatile CNC eliminates a lot of moving and setups

Setup elimination can be accomplished by: standardizing raw workpieces to avoid material setup; make fixturing flexible to eliminate positioning setup or changing fixtures; designing parts with standard fixturing geometries; quickly and automatically changing CNC programs; specifying standard cutting tools within tool changing capability; and keeping hole punch variety within punch press capabilities. Automatic material feed and ejection are nice to have, but these tasks can be performed manually by an alert workforce. Sheet metal nesting optimization can improve over time.

For unusual and low-volume parts, using a CNC machine to "hog out" parts may lower *total* cost if it avoids (a) stocking a high variety of low-volume material shapes or raw parts and (b) complications, delays, and high cost in the supply chain for too many low-volume castings, forgings, or moldings. To make the right decisions on flexible use of CNC, total cost must be used to include machine time, material cost, *and* all related overhead costs.

Program/Instruction Setup. Eliminate setup delays to program or load programs for CNC machine tools or any other computerized instructions. Programming setup can be eliminated by downloading machine tool programs on-demand or generating them on-the-fly. Digital labels like bar codes can be wanded to identify the product after which the program database should be able to download the programs almost instantaneously.

Process Variable Setup. Develop processes, concurrently with product development as appropriate, to standardize on the variables that are the most time-consuming to change. For heat treating, oven brazing, or baking, standardize on furnace or oven temperature, since th7at would cause a long setup change to change; instead, vary the speed through a furnace or, better, design all products in the family to use the same processing. Standardizing process variables also eliminates the chance of using the wrong setting. This is an example of the Japanese quality assurance concept of *poka-yoke,* or mistake proofing. Poka-yoke can be designed into concurrently engineered products and processes.[9]

> Eliminate process setup by standardizing on the hardest to change variables

SETUP REDUCTION

- **Design to minimize setup.** When designing or redesigning parts to minimize setup, designers must understand all the current or projected setups, moves, repositioning, zeroing, and machining steps in order to design parts and processing for minimum setup. Then they need to actually design the product/process to minimize setups (Ch. 10).

- **Consolidate inflexible parts.** There are two techniques to deal with inflexible parts that have unavoidable setup and therefore can only be made in batches. Inherently inflexible parts (like castings, moldings, extrusions, stampings, and bare printed circuit boards) may need to be *consolidated* into very versatile standard parts that can be ordered in a steady flow with confidence that, because of their versatility, they will be used one way or another. Part consolidation is discussed in Chapter 5. Parts that qualify for *kanban* resupply *can* be made in batches if the combination of setup time, run time, and delivery time is short enough to keep the bins full (Ch. 7).

> If parts are standardized, they can be distributed to all points of use, thus avoiding kitting

- **Distribute parts at all points of use to eliminate kitting.** If part variety is too excessive to allow distribution at all points of use, then enough parts for a *batch* would be assembled into a *kit* which is put together in the raw materials warehouse, delivered to the assembly area, and then distributed to part bins or automation machines. This *kitting* is a set-up which will inhibit flexibility. Techniques to eliminate kitting are *aggressive* standardization (Chapters 4 and 5) and the concurrent engineering of products and processes that minimizes the number of different parts at any assembly station (Chapter 10).

- **Minimize the number of times parts are moved and set up.**

  - Minimize processing steps through design simplicity and combining parts, which eliminates the need to machine both interface surfaces for every combination. Combined parts often result in a net reduction in the total cost and flow time.

  - Perform processing in the fewest number of machine tools, moves, and setups. Even turning a part over to machine the other side could be take considerable time to turn over, reposition, and zero in. Making all cuts in the same setup assures that *all* those dimensions are made to machine tool tolerances (usually +/- .001" or better).

- **Manual Processing Setup.** All of the above setup elimination strategies can apply to manual processing. But a setup that applies uniquely to manual assembly is finding and understanding instructions. This setup can be eliminated by displaying instruction on on-line monitors that instantly and clearly show what is to be done at that work station to any product being worked on (see Figure 8-3 at the end of this chapter).

- **Shipping Setup.** Shipping may cause finished goods to be batched if they have to wait to fill a truck or shipping container. As batch size goes down in the factory, batch shipping could negate those gains. The solution would be an outbound flow with more frequent shipping in smaller trucks or the new transportation paradigm of "less than full container load" shipping using trucks that open on the sides to allow partial shipments to be loaded and unloaded independently.

QUICK CHANGEOVER TECHNIQUES

A key ingredient of setup and batch reduction is *quick changeover techniques,* which leading plants are implementing. For the 25 finalists in Industry Week's annual Best Plants rating, 89.6% have implemented quick-changeover techniques compared to17.4% for the average of all 2,511 manufacturers surveyed.[10] *The best plants used quick-changeover techniques five times more than the average plant!*

Best plants' use of quick-changeovers was 5 times the average

Many manufacturers have successfully implemented quick tooling changeover programs based on "quick die change" techniques, which applies to all types of tooling, not just dies. Shigeo Shingo developed the methodology called "Single Minute Exchange of Dies" (SMED) for Toyota.[11] Clever die and mold mounting geometries have been developed to facilitate quick changeovers.[12] Conveyors and carousels, that were first applied to moving parts and products, are now being applied to moving dies quickly in and out of presses and molding machines.[13] SMED principles are as follows:

Identify internal and external setups:

- Internal setups are performed when the machine is stopped.
- External setups can be performed while the machine is running.

Convert internal setup to external setup

While the machine is running:

- Line up the right people, parts, tools, and tooling, which should be kept in a good state of repair.
- Prepare tooling off-line; e.g., inspect tooling; lubricate dies; preheat if necessary.
- Bring everything close to the machine and position hardware for a quick change.
- Install near the machine appropriate racks, rails, bolsters, etc.
- Fasten tooling to off-line turrets or intermediate jigs, cassettes, or pallets.

Streamline setup operations

- Shorten transportation time; position the most frequently used tooling closest to machines.
- Organize tooling storage, identification, and labeling.
- Standardize on tooling geometry so all tooling is the same height and has the same mounting holes; if this is not inherent, add standard spacers and adaptors as necessary.
- Install self-jigging features and geometries for workpieces and tooling. For rectangular workpieces, arrange to position all workpieces against a back wall and the right of left wall. If this is a uneven part like a raw casting, the workpiece would need 6 location bosses, typically arranged as 3 on the bottom, two on the back, and one on the side.

Workpieces and tooling can also be quickly positioned with V-shaped projections or centering clamping devices, such as lathe chucks or centering vices.

Another technique is tooling pins and holes in standard locations. To avoid overconstraints and tolerance problems, utilize reamed holes and pairs of round and "diamond" pins, which can be used to locate workpieces to tooling in addition to the tradition application of locating tooling to machine tools.[14]

Intermediate jigs, cassettes, or pallets to accomplish the above and mount to the machine through standard mounting geometries.

- Locate parts visually with visual locating marks, cameras, and CNC positioning algorithms.

- Utilize quick fastening techniques to secure tooling, such as bolts in T-slots, quick-turn bolts, split thread bolts, pear-shape washer slots, C-shape washers, swing away clamps, cam locks, toggle clamps, and so forth.[15]

- Consider inserts. Too many different molds or dies might cost too much to build and take too much time to change over. If only a portion of the mold or die is unique to each product, then custom *inserts* could be substituted into standard cavities in the molds or dies. The length of constant cross-sections could also be varied with inserts. Clever styling could camouflage the seams caused by the inserts. These techniques could enable changes to happen quicker, since the bulk of the tooling could remain fixed.

- Consider one-use flexible molds and dies. It may be possible to make single-use molds and dies, possibly modeled after the old Linotype machines which cast lead blocks of letters for each line of newspaper type, and then melted them down for the next usage. Patterns for such molds, or the molds themselves, could be made by rapid prototyping or machined from a free-machining material by CNC equipment and then recycled for the next use.

FLOW MANUFACTURING/ONE-PIECE FLOW

Spontaneous build-to-order and mass customization depend on flow manufacturing for efficient lot-size-of-one production. If setup can be eliminated or reduced enough to eliminate the need to manufacture in batches, then parts, sub-assemblies, and products can flow one piece at a time. One-piece flow may be essential when building to-order a wide variety of standard or mass-customized products. It eliminates much of the waste of batch-and-queue manufacturing: waiting, interruptions, overproduction, extra handling, recurring defects, and other non-value-added activities. The one-piece flow aspect of flow manufacturing is the key to rapid throughput, little or no WIP inventory, and rapid quality feedback.

> Best plants' use of flow production was 3.5 times the average

For the 25 finalists in Industry Week's annual Best Plants rating, 77.8% have implemented continuous-flow production compared to the average of 22.7% for all 2,511 manufacturers surveyed.[16] *The best plants used continuous-flow production three and a half times more than the average plant.*

One-piece can lower costs as a freestanding program. Joseph Day, Chairman and CEO of tier-one automotive supplier Freudenberg-NOK, said "In my 10 years of experience and 8,000 Kaizens, I've never found a process where one-piece flow wasn't lower-cost than a batch environment."[17]

Flow manufacturing is achieved by: eliminating all setup; eliminating batches, scheduling, kitting, and any accumulation of products or parts in the line; eliminating WIP inventory with one-piece flow; and arranging spontaneous delivery of parts and materials (Ch. 7) which are delivered to all points of use "dock-to-line" without the delays and cost of incoming inspection. To accomplish this, suppliers will need to be *certified* to assure quality *at the source.*

> Without setups, waiting, and batches, production can *flow*

Another aspect of flow manufacturing is the dedicated *cell* or line, which may be arranged to build any variation within a product family without any setup. Each cell has a compete set of "right-sized" inexpensive machines. Utilizing older machines that have been made "obsolete" by centralized "mega-machines" is a very cost-effective way of building complete cells. The fact that cells consist of dedicated equipment occupying dedicated space often makes this incompatible with outsourcing any of the cell's parts, unless the outsourcer is nearby strategic partner willing to make this type of commitment.

Dock-to-line deliveries directly to *all* points of use is a key element of flow manufacturing, lean production, build-to-order, and mass customization, as discussed in Chapter 7.

Assuring Quality with One-Piece Flow

One-piece flow has a distinct advantage for assuring quality. First, flow manufacturing eliminates the possibility that recurring defects may be built into several batches before being caught at a downstream inspection step. Second, people working in flow manufacturing look for any visible deviation as each part is handed to "its customer" (the next station). Further, if the part doesn't fit or work in the next operation, the feedback will be immediate leading to quick correction of the problem.

> Flow provides instant feedback and prevents "defects by the batch"

In flow manufacturing, parts may be manually handed to the next station, which may be very close, thus eliminating the need for mechanized conveyance or fork lifts, whose aisles may occupy as much space as the production line.

Six Sigma quality improvement programs will be more effective and have a greater chance of achieving six sigma quality levels when integrated manufacturers: implement one-piece flow; eliminate machining setup[18] and manual positioning; standardize parts (Chapter 5) to eliminate the chance of picking the wrong part and eliminate setups to change parts; and continuously improve dedicated lines (discussed earlier and in Chapter 13).

On the other hand, Six Sigma implementations will be more difficult, and quality levels lower, for outsourced production because most outsourcing is scheduled in batches with fewer opportunities for one-piece flow, more setup changes, and few opportunities for continuously improved dedicated lines.

U-Shaped Lines

U-Shaped lines provide visual control, problem awareness, & let workers help each other

Rather then laying out "lines" in a literal straight line, it may be advantageous to create a *U-shaped line* which bends the line into the shape of a "U" for the following reasons:

- **Visual control.** Everyone in the line can see the whole operation, enhancing *visual control,* thus resulting in greater group ownership, continuous improvement (kaizen), and problem solving. Visual control can be further enhanced with clearly visible *andon* (warning or status) lights and display boards.
- **Problem awareness.** When everyone in the line works close together, problems at all stations will be heard by the entire line, thus leading to faster problem identification and resolution.
- **Helping out.** If one worker gets behind, nearby workers can help out, which requires that workers have the necessary skill and training.
- **Skipping steps.** Having work stations closer together makes it easier to process orders that skip steps, since the part can quickly be handed to the appropriate station.

Grouping Products, Subassemblies, and Parts into Families

With enough volume, certain products, subassemblies, or parts would have enough volume for their own dedicated lines and, thus, avoid setup changes, allow a one-piece flow, and continuously improve quality. But there may not be enough consistent demand for a dedicated line.

Synergistic grouping into *families* or *platforms* lowers the volume threshold for "dedicated" lines and also avoids the problems of matching demand to the capacity of multiple lines.

A *family* would be a group of products, subassemblies, or parts that *can spontaneously build all variations on-demand* without any significant setup changes with all parts and materials always available. The two main criteria are:

1) All *processing* setup changes are eliminated or quick enough to allow efficient on-demand processing, including machine setup, tool changes, downloading CNC programs, locating/zeroing workpieces, clamping workpieces, changing materials, and finding and understanding instructions.

2) All parts and materials needed for the product family must be *automatically distributed* to all points of use. To satisfy the on-demand capability, materials for the family must be available spontaneously without forecasts, purchasing, or waiting for parts. Instead, parts and materials should be resupplied spontaneously with steady flow, *kanbans,* breadtruck, min/max, and on-demand fabrication of parts.
   This will require standardization *across the whole family.* Depending on the proliferation of the group, incorporation of standard parts and materials may have to substituted or wait for the next generation of the family.

The *scope* of the family will be determined by anticipated product demand and the capacity of the grouped line.

If the line is populated *predominantly* by high-capacity machines that exceed demand, then the group may have to be large enough to fill the line. If it is not full, avoid the temptation to "fill" the line with anything incompatible. Once there is a predominant setup, adding anything else requires *two setup changes* – one to change it and another to change it back.

If the line is populated by *few* high-capacity machines, then the effect of local undercapacities would be overwhelmed by the value of the business model (see "right sized" machinery discussion below) and the group could be narrower.

Combining the build-to-order of standard products with the mass customization of custom products will increase the volume of the operation and thus allow the family to qualify for larger line or be more focused.

If they meet the above criteria, *other* products, subassemblies, and parts may qualify for inclusion in the family, even if they are not part of the currently stated "product line."

For existing products, it may be necessary to *convert* a group of products into a family, including the following actions, with implications noted:

- converting to "better" grades of fasteners and materials. This can be a global "better than" substitution without formal change orders

- converting to standard bolt *lengths*, in which case most bolts would be longer. Each longer bolt situation in the family needs to be checked out for clearances, interferences, and safety. This would involve waivers or minor change orders

- converting to standard bolt *diameters*. Larger bolt diameters would have to be checked for clearance and material thicknesses in adjacent materials. Smaller bolt diameters would have to be checked for strength (bolt *and* bearing surface)

- converting to thicker standard material stock sizes, such as plate, sheet metal, bar stock, etc. Product weight gain needs to be calculated and approved, but this might be offset by system weight savings from more efficient structural elements (see next to the last section in Chapter 9), corrugating flat sheets, more support members for large plates, and so forth. Drawings of mounts may need to be changed. Guidelines for new products should recommend designing all mounts on the *same side* of the sheet or bar stock, to allow the stock thickness to accommodate future standardizations.

- converting inflexible parts (stampings, castings, forgings, moldings, extrusions, bare circuit boards, etc.) to versatile shapes. It may be necessary to check versatile shapes for all usage scenarios. Change orders may be required for all versions.

Steps that are too difficult now may have to wait for the next generation design of the product family.

Products that *cannot* be grouped into any family should be considered for elimination or outsourcing through product line rationalization. If these "non-groupable" products are necessary for product line completeness, then they may have to be redesigned (or changed) to accommodate standard parts, standard materials, optimal integration, spontaneous supply, and/or setup/batch/inventory elimination. If neither of these options are possible, those products need to be moved to their own plant or area with *support demands and overhead allocations are separated from BTO operations.*

Product family grouping will be also discussed in Chapter 13.

Cellular Manufacture

Flexible operations work best with dedicated cells or lines for every family. Cells can be permanently configured so that within a product family, all setup has been eliminated. This strategy work best with many simpler dedicated machines instead of a single "mega-machine," unless the mega-machine is flexible enough to handle a family and the family is large enough to justify its expense. In some cases older or "obsolete" machines may be used to provide a complete set of machines for the cell; this was one of the solutions covered in Eli Goldratt's novel, *The Goal.*[19] Remember that speed or capacity may not be as important as flexibility to support BTO&MC.

Cells are structured to make a family without setup changes

Total cost analysis must be used to take into account all related overhead costs in addition to the usual material and processing cost. In some cases, cells may be installed even if the cell alone cannot be justified by traditional analyses, but if the cell completes a valuable plant capability like build-to-order and mass customization. The guiding strategy for cell design is flexibility and setup elimination.

For the 25 finalists in Industry Week's annual Best Plants rating, 74.4% have implemented cellular manufacturing compared to the average of 19.3% for all 2,511 manufacturers surveyed.[20] *The best plants used cellular manufacturing almost four times more than the average plant.*

Best plants' use of cells was almost 4 times the average

Line/Cell Layout. Structure a flexible line to assemble all variations of the product on-demand. Feed the product line with *cells* that make all versions of subassemblies and custom parts. Feed these cells with (a) materials cut to length/shape on-demand and (b) custom parts that are fabricated on-demand by CNC. Arrange for spontaneous resupply of standard parts to the cells *and* the line through steady flows, kanban, min/max, and breadtruck. If there is too much part variety for these automatic resupply techniques in flexible cells or lines, it may be necessary to initiate change orders to convert proliferated parts to standard part values, as mentioned above.

Right Sizing. Cells usually have smaller capacity than the mass production lines that many production machines are designed for, but the cell will need a *compete set* of machines. The solution is smaller capacity "right-sized" equipment.

Right-sized equipment was one of the solutions covered in Eli Goldratt's novel, *The Goal.*[21] Dedicated lines of low-cost equipment are especially relevant for eliminating setup for inherently inflexible processes that use dies and molds. Remember that speed or capacity may not be as important as flexibility to support BTO.

A good source of right-sized equipment is the *used machine* market. The best bargains can be found for "obsolete" machines that, ironically, have been replaced by "mega-machines" or hard automation by mass producers.

The quality of these used machines can usually be raised to acceptable standards since (a) continuous *kaizen* improvements can raise the quality when the settings are constant, (b) the machines may be used at lower speeds and feed rates than they were designed for, and (c) used machines that are only needed for certain functions can have those functions certified when purchased, even when unneeded functions are not working (which lowers the cost even further).

The machines in cells may not have "balanced" capacities or *takt* times and may not even be running all the time, but they may be a valuable part of a BTO system, especially if they complete any "missing link" in on-demand capabilities.

Cells consist of dedicated equipment occupying dedicated space feeding one part at a time. Therefore, cells are incompatible with batch-oriented outsourcing, unless the outsourcer is nearby strategic partner who can supply parts on-demand and frequently enough to support the line.

Simplifying the product line first through product line rationalization (Ch. 3) can make cellular implementation easier, as was the case of the drill bit manufacturer cited in Chapter 3:

> By eliminating the very low volume product line, the company was able to set up a simple kanban system between finished goods and the manufacturing cells, which eliminated the need to operate a complicated, computer-based work order system.[22]

Machine Maintenance

In sequential one-piece flow, when one production machine breaks down, the whole line will go down, unless there are redundancies. Therefore, proactive equipment maintenance is important to prevent unexpected production interruptions. A good TPM program should assure this.[23] Inventory buffers may give an allusion of protection, but are costly and may still require special measures, like overtime, to recover.

Equipment maintenance can be more responsive and less costly with standardization of all equipment replaceable parts: belts, motors, fuses, controllers, and so forth.

Line Balancing

Ideally, to achieve optimal machine tool and work station utilization, one-piece flow lines should be balanced so that the time to do the required tasks at each station, called the *takt* time, is fairly constant throughout the line.

If takt time at each station is equal to line capacity, arrange the process into sequential line. If takt time does not equal station capacities, but does not vary with products: (1) upgrade appropriate capacities or find faster machines to achieve balance, (2) group machines/stations into series/parallel paths to achieve better balance, perhaps three of one feeding two of another, or (3) don't worry about overcapacities if those machines are not expensive, keeping in mind that the entire system can provide high value and support the business model.

If takt time varies with different products, (1) make stations/machines flexible enough to share workload, (2) sequence jobs to compensate for imbalances (alternating heavy and light workloads at an overtaxed workstation), or (3) size the line based on the most expensive machine and provide excess capacity for the less expensive machines.

Another way to balance lines is to make selected stations become kanban sources, so that they make kanban parts during times when they have excess capacity.

LEVELING PRODUCTION

Flow manufacturing, build-to-order, and mass customization work best with a fairly constant or *level* production output. Critics of one-piece flow cite demand fluctuations as justification for building lots of inventory. However, most irregularities in factory workload are artificially induced.

Artificially Induced Irregularities.

Raw material comes out of the ground in a steady flow. Most products are ultimately consumed in a steady flow. But in between, there are many flow irregularities. Most irregularities in factory workload are artificially induced.[24] The following is a list of sources of irregular factory workload *that can be eliminated:*

> Most irregularities in factory workload are artificially induced
> – Womack & Jones, *Lean Thinking*

- **The "business cycle."** According to Womack and Jones, author of *Lean Thinking,* half of the effects of downturns are caused by working off excess inventories and half the upturns are caused by building up inventories for anticipated upswings in demand.[25]

- **Production goals and quotas** for the end of the month, quarter, and year. Graphs of production output at many companies look like the familiar "hockey stick" with fairly level production for almost all of the measurement period but with a dramatic upswing in production toward the end of the measurement period. In fact, Dell Computer begins its fiscal year on February 3 in order to be out of phase with other companies so that it will be able to *buy* when most companies are desperate to *sell* to make their monthly, quarterly, and yearly goals.

- **Sales Promotions,** instigated by store managers to meet their revenue goals and quotas. Of course, nonlinear sales will then translate into nonlinear factory production to replace the items sold in the promotion. Look as the different approaches practiced by large retailers. Wal-Mart emphasizes "everyday low prices" which levels production and supply chains to support their mastery in this area. On the other hand, sales events and temporary discounting is disruptive to production and supply chains.

- **Periodic Promotions** can induce non-linear orders of regular items, even if the promotion was only intended to liquidate over-stock inventory. This situation is compounded when customers hold off purchases until the expected sales events, which, in turn, exaggerates nonlinear demands on production.

- **Capricious Sales Promotions,** unrelated to quotas or overstock situations, where store managers simply want to create more floor traffic. One tool supplier to Home Depot lamented that the store chain ran a promotion and then had to order six times the usual amount of tools without warning!

- **Quantity discounting and "deal making"** encourage large batch shipments to achieve "economies of scale," which will be debunked in Chapter 11. This is *doubly* counterproductive since is causes the producer to work overtime to deliver large batches *and* causes the customer to incur inventory carrying costs once the batch is received. During an in-house seminar at a telecom company, production managers were anguishing over orders for 100,000 telephones, so the author challenged such practices pointing out that no phone company could possibly sell that many telephones anytime soon, so the best approach would be to manufacture the telephones at the same rate as they were sold.

> Big deals force factories to work overtime and customers to carry excessive inventory

- **Lack of *dealer* confidence of product availability,** leading to unnecessary inventory in factories, stores, dealers, on-line retailers, and other resellers.

- **Lack of *customer* confidence in product availability,** leading consumers to "stock up" when they feel insecure.

These artificial induced irregularities can and must be eliminated by implementing performance measurement systems based on "the big picture," educating customers (end users, stores, suppliers, and assemblers) on why leveled production is in everyone's best interest, and building consumer confidence that products will actually be delivered in a steady flow as promised. To discourage large batch shipments, manufacturers can offer discounts for level orders or penalties for disruptively large orders, both of which make sense from a total cost standpoint. If this supports corporate leveling strategies, it should be built into *firm* pricing policies to avoid this turning into a concession just to "make the sale."

Seasonal Products

Although lean/flow production works best for level demand, spontaneous build-to-order techniques can accommodate seasonal demands better than stockpiling finished goods inventory, with all the associated inventory cost and risk of building too many or too few of the "right" products.

The prerequisite would be to first try to level demand with off-season incentives based on actual differences in total cost. For the remaining seasonal variations, seasonal products can be manufactured on-demand with a minimum of part inventory and risk, using BTO&MC principles. Again, if seasonal discounts support corporate leveling strategies, it should be built into *firm* pricing policies.

General strategies include the following:

- Developing industry-specific supply chain and operational strategies optimized for seasonal products.

- Reducing labor content and maximizing plant output with design for manufacturability.[26]

- Reducing part/material variety with standardization (Ch. 4 & 5) so that supplies will be available for more product variations from fewer parts bins.

- Selectively build parts or subassemblies ahead of time that are (a) so standard they will be used one way without the risk of not being sold or going obsolete, (b) low value to keep inventory carrying costs low, (c) small and stackable enough not to take up too much space, and (d) high labor or machine content, so off-season product would free up capacity during the peak

- Building versatile "vanilla" platforms off-season and then add the "flavors" on-demand as sales come in (see Postponement section in Chapter 9), either stocking various modules (flavors) or, if necessary, building them on-demand from standard parts. This will result in the minimum inventory costs, compared to inventorying an vast array of complete products that may not be the variations that sell this season.

- Achieving 100% *equipment* utilization during the peak season by eliminating change-over setup delays using the principles presented herein. Until that is possible, make predictable low-volume products ahead of time so that equipment can mass produce the high-volume products during the peak season with fewer setup changes. Maximize utilization during the peak season with optimal equipment maintenance including a round of preventive maintenance before the peak season.

- Achieve 100% *people* utilization by eliminate changeover tasks, shortening remaining changeovers, automatically delivering parts to all points of use, eliminate firefighting and wasted efforts, and provide instructions that are quickly displayed and easily understood.

- Expanding production during the peak season by operating the existing equipment for two or three shift a day and seven day a week. The extra labor could come from overtime (e.g. two extended shifts), temporary/seasonal workers if skill and experience demands are not too high. If the permanent workforce has to grow, find work for them during the off-season.

- Outsource routine, stable, well-documented parts to contract manufacturers not affected by your seasonal cycles.

- Building out-of-phase seasonal products in the same plant, such as combing operations for snow removal with lawn mowing products, snow mobiles with water sports and off-road products, and air-conditioners with furnaces, as is done by United Technologies Carrier.

- Leveling *part* production by standardizing on parts and modules that can be used on out-of-phase seasonal products, such as the same motors for furnaces and air-conditioners or the same engines, power train elements, and electrical systems for both snow removal with lawn mowing products, *even if manufactured in different plants or for different companies*

- Optimizing *fabrication* flexibility and output with CNC machine tools and versatile fixturing

- Optimizing *assembly* flexibility and output with mechanization, flexible automation, and on-line manual assembly instructions (as will be discussed and illustrated below)

- Optimizing synergies between operations in northern and southern hemispheres (that are inherently out-or-phase for seasonal products) for parts light enough and small enough to be shipped between hemispheres.

- Freeing up people and machine time by *rationalizing away* product variations that waste valuable resources, generate little profit, and, during the peak season, may force the company to turn down much more profitable orders. At the minimum, defer all demanding, high-setup products to the off season.

CAPACITY

Short-Term Capacity Challenges. Line capacity is usually determined by *peak* demand. Equipment is then sized and installed based on peak demand estimates or experience, adding extra capacity to compensate for non-value-added time for setup changes, maintenance, rework, and troubleshooting for gearing up to build unusual products, new product launches, quality problems, and so forth.

In build-to-order and mass customization, less equipment would be needed for the same *plant output* capacity because of the elimination of setup and the downtime caused when attempting to manufacture seldom-built products which can be rationalized away (Ch. 3).

> Eliminating setup and delays raises utilization which, in turn, raises capacity

When companies wrestle with equipment purchases, they start to appreciate the need to minimize the occurrences of *artificially induced peaks* discussed above. In addition, an optimal strategy for seasonal products can reduce the need for extra capacity specifically for seasonal products, as discussed in the previous section. Even if equipment purchases are not on the foreseeable horizon, freeing up capacity can postpone or eliminate further equipment purchases that may be needed for growth.

Mixed Model Production. After reducing artificial peaks, specific line capacity needs can be reduced by making *other* lines flexible enough to handle peak demand by shifting production from line to line. Mazda resolved such a dilemma by moving production of the popular Miata from its Hiroshima plant to the Hofu plant which was underutilized with large sedan production. Because of manufacturing flexibility, Mazda could combine production of a niche sports car with a family sedan in the same plant to balance output at these two plants.

In general, if production can be shifted to other lines, then capacity planning can be based on *total plant output* (or total company output) instead of having to add a cushion to the capacity to each line.

In an age when most car manufacturers are offering huge discounts, rebates, and subsidized financing to "move the metal," *some* cars are actually selling *above* list price because the manufacturers' production lines are flexible enough to economically build niche market vehicles. A Wall Street Journal article reported that, for flexible manufacturers, "instead of needing to sell 200,000 or more of one vehicle to make money, a car maker can make a profit selling 40,000 to 50,000 each of four different models."[27]

Nokia builds 9 different products on the same production lines

Nokia has designed its own nine factories to be flexible enough to quickly switch production from one model to another. Nokia's Salo factory builds nine different products on the same production lines. Even its top-of-the-line handset, with far more parts than a standard model, flows down the same lines as the most basic cell phones.[28]

This capability will also help manufacturers satisfy unexpectedly high demand of new models. Without it, companies are sometimes put in the pathetic paradox of turning away customers for one product while closing other plants.

> Some companies can't build enough of one product, but are closing other plants at the same time

If demand sometimes exceeds daily capacity for a line, build-to-order companies can prioritize production into delivery categories, such as next-day, two-day, or time available within the week, and charge accordingly, either a premium for next day or a discount for slower delivery.

For short-range peak demand beyond capacity of a line or plant, overtime can be used to catch up. At Toyota plants, the nominal work shift is eight hours with up to two hours of spontaneous overtime to be worked to meet unexpected demand.

Long-Term Capacity Challenges. For projected long-range capacity shortfalls, permanently outsource the *least* efficient or *least* compatible products and operations. Assign efficiency ratings and work from the bottom of the list.

Capacity can be expanded cost-effectively by using low-end or used equipment, by creating dedicated lines for medium volume products that used to be built in batches, as discussed in Chapter 13. Removing these products and their setups from the expensive lines will increase *their* utilization and make it more likely that they will also be able to run without setup changes.

For all production lines, unexpected loss of capacity can be avoided with preventive maintenance (TPM) and quality assurance programs, like TQM (Total Quality Management), Six Sigma, and process controls, to avoid interruptions and products looping back for rework.

RESULTS: SETUP ELIMINATION/BATCH-SIZE-OF-ONE FLOW

Setup elimination leads to batch-size-of-one flexibility which leads to one-piece-flow which leads to WIP elimination.

Eliminating Setup Itself can decrease throughput time to approach the "labor standard" (the actual time spent to fabricate and assemble a product) and eliminate setup delays on expensive equipment. This will improve machine tool utilization, save setup labor costs, and eliminate inspection time and scrap costs verifying the first parts made after such setup changes.

In addition, Batch-Size-Of-One Flexibility can allow *dock-to-line* part delivery, thus eliminating inventory carrying costs of raw parts inventory (floor space, administration, interest, etc.) and the cost and delays of incoming inspection and logging in parts. It can also eliminate the penalties of kitting: the labor cost to kit; the floor space for kitting area; the production delays and expediting costs when kits are short; and the waste or restocking costs of excess parts left over.

In addition, one-piece flow can improve quality with rapid feedback to catch and correct quality problems fast, eliminate fork lifts (including the labor, equipment, and floor space for the aisles), and foster psychological flow,[29] improve job satisfaction, relieve boredom, and encourage continuous improvement.

In addition, eliminating WIP inventory can eliminate WIP inventory carrying cost, which is equal to 25% of value of inventory per year (Ch.2), and cut floor space needs in half.[30, 31, 32] This is especially important in times of growth, but floor space savings should *always* be assigned a value to encourage more efficient utilization of space.

BUILD-TO-ORDER STRATEGY

The overall Build-to-Order strategy is to *standardize all that you can and procure or build the rest on-demand.*

A) Standardization works best with *aggressive* standardization in which:

- *Raw material variety* is reduced to the absolute minimum. When this standardization approaches *one* type for each category of material, then the material no longer needs to be ordered based on forecasts, but, instead, steady flows can be arranged based on aggregated sales, as described in Chapter 7 – in other words, just match the tonnage in the tonnage out.
- *Parts* are standardized enough to allow then to be spontaneously pulled into production with automatic resupply techniques such as kanban, min/max, or breadtruck, discussed in Chapter 7.

These parts may be pulled from in-house operations or suppliers who can build them in small enough batches for kanban resupply.

Aggressively standardizing usually means that most products will get "better" materials and parts than they need, but the total cost savings and the value of the business model will exceed any perceived "extra" cost for the parts and materials.

B) On-Demand Manufacture or Procurement. For everything else that can't be standardized:

- ***Build* the remaining parts on-demand** (from standard raw materials) using CNC machine tools with versatile fixturing and machine tool programs that are instantly loaded or generated.
- ***Assemble* the products on-demand** using flexible assembly automation machines (Figure 8-2) or manual assembly stations fed by kanban bins and directed by on-line instructions shown on computer monitors (Figure 8-3).

Processing Shift

BTO companies will be shifting to CNC programmable processes and away from inflexible processing such as castings, moldings, stampings, forgings, extrusions, and bare circuit boards, that make many discrete parts with many discrete tools *unless* these can be very versatile shapes that are (a) versatile enough to be used in many applications as is or (b) can be fabricated or assembled into many final versions by flexible CNC processes

HOW BTO & MASS CUSTOMIZATION WORK

The following examples were created to show mass customization principles for fabricated parts or products (Figure 8-1) and electronic products (Figure 8-2). The author creates perspective illustrations, like these, for BTO&MC client because they show, on one page, the flow of materials and information through easily recognizable equipment. Further, three-dimensional drawings showing the actual equipment are more meaningful than two-dimensional block representations to a broad audience. These can be drawn in 3D CAD solid model software. In the following discussions, bold words refer to labels on the illustrations.

The process starts with a dialog with the customer in which customer queries are quickly answered. This rapid dialog is the only dialog (the only two-way arrows in both figures) in BTO&MC, as opposed to the traditional practice of many lengthy inquires back and forth with Engineering, Procurement, and Manufacturing departments. These inquiries usually start with "Can you do this?," followed by "I don't know, but I'll find out," and then followed by "We can't do that, but we can do this," and so forth and so on. This process can go on for days, weeks, or even months.

In BTO&MC, various "what if" scenarios can be explored instantly, complete with price and availability quotes, using **configuration software**, called "configurators," which were first developed to ensure order accuracy for standard products and have now evolved to the point where they are the perfect order entry tool for mass customization (as will be discussed in the next chapter). Configurators can be on a salesperson's laptop computer or on the company web-site, in which case customers could "design" their products and place their own orders. As is done by Cisco, customer input for mass customization may include body scans for clothing or foot scans for shoes.

When the customer has optimized the configuration and approves the order, the order information is sent by **modem input to the factory** where it enters the **order entry database,** which accepts the information and converts it into various data packets that go (1) to **on-line assembly instruction** monitors, which tell workers how to assemble each product, and (2) to the parametric **CAD/CAM work station.** This is an automatic or semiautomatic computer that accepts customer order data and turns it into *parametric CAD* drawings, which are drawn with "floating" dimensions that accept the customer's data and then stretch all the part drawings, which also stretches the assembly drawings. Finally, this station automatically translates these drawings into **CNC Programs** for the CNC equipment.

BTO&MC for Fabricated Parts or Products

The model illustrated in Figure 8-1 shows how to build-to-order standard of mass-customized fabricated parts or products. Hoffman Engineering has adopted this model to build a wide variety of standard and customized sheet metal electrical enclosures.[33] In this illustration, solid lines show the flow of parts; dashed lines show the flow of information.

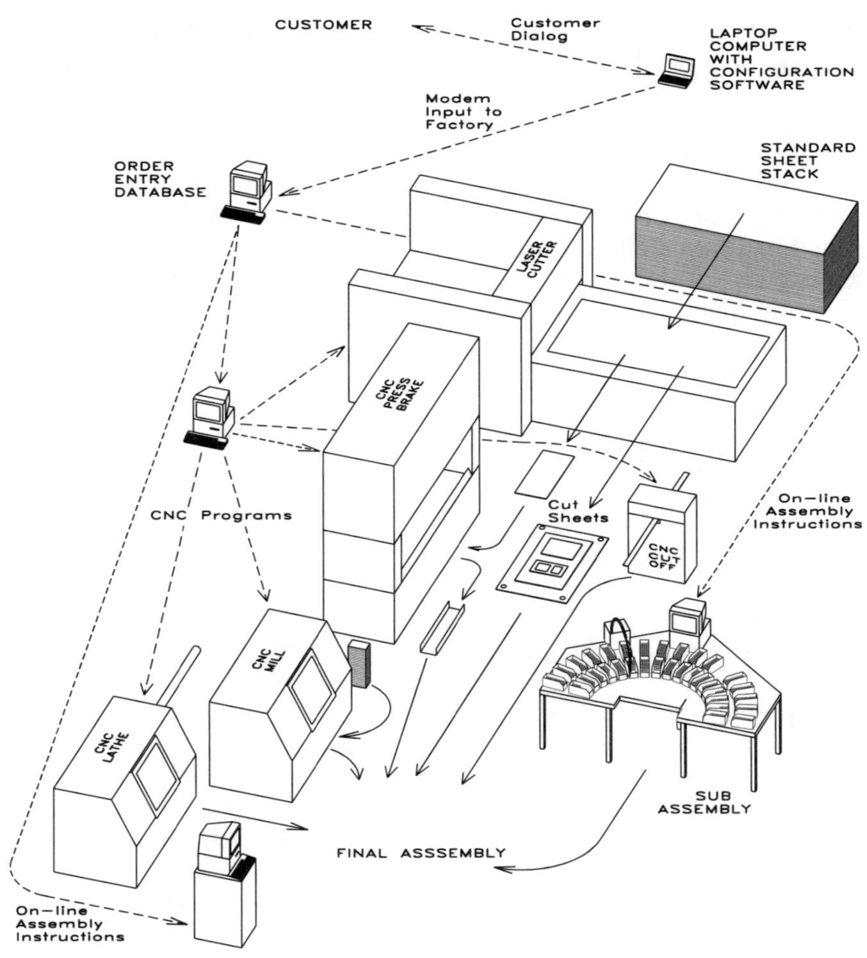

Figure 8-1, BTO&MC for Fabricated Products

The actual production starts when sheet metal from the **standard sheet stack** is fed into the **laser cutter.** *Aggressive s*tandardization of raw material is the key to minimizing material overhead, providing raw material on-demand, and building any product in any batch size in any order. As mentioned earlier, a single standard raw material type would be ideal because "ordering" would be as simple as matching the tonnage in to the tonnage out of the plant and arranging a steady flow of materials. Multiple types of raw material might be all right from a procurement standpoint if the proportion of each type was fairly constant. More than one type would require either an automated multiple sheet feeder or a worker who loads whichever sheet is specified by on-line instructions. The cost savings in material overhead and less material handing (including avoiding multiple sheet feeders) should easily compensate for any cut-off waste or products getting "better" material than they need. And with experience, waste will decrease as CNC programs get more efficient at nesting various shapes. Extra material can also be utilized as a source of supply for *kanban* parts, if they can be designed to use the same material. If usage is high enough for a *single type* of sheet metal, it could procured at much low cost in bulk and fed from a coil, which would also incur less waste between parts.

The **laser cutter** performs all the sheetmetal cutting including the outside shape and all holes, notches, and cutouts. This may not appear to be the "fastest" time per hole or the most "efficient" cost per operation by mass production standards, but operating without setup changes or WIP inventory will result in the fastest *factory throughput* and lowest *total cost.* If hole quantity warrants, a programmable punch press could be used, as long as the number of tools in the tool changer is not exceeded by the hole variety *for the entire family or the entire factory.*

The output of the laser cutter is a set of **cut sheets** for each product. Some will go through the **CNC press brake** for bending, and the rest may go to welding (in the case of Hoffman cabinets) or straight through to final assembly.

Milled parts are made from a standard blank (ideally only one as shown) in the **CNC mill** (machining center) which machines any variations for this product family. Similarly, the **CNC lathe** (or CNC screw-machine) makes all the turned parts for the family, ideally from a single size of bar stock.

Many factories can use several **CNC cut-off** machines, which programmably cut off bars, tubes, coiled sheet, or any long part that also has shorter versions. Usually, these are used for linear parts that are too simple to justify an expensive machine tool. Each time these are employed, they may reduce part variety from several types to one. Used for several part families, they can significantly reduce plant material overhead costs and provide these parts efficiently on-demand without the need to forecast and order them. A semiautomatic version would use a programmable "stop" against which a worker would place the material prior to manual cut-off. In the manual version of this concept, a worker would cut the part on-demand according to the length displayed on a monitor. Hoffman Engineering has several of these for hinges and various brackets, one of which is fed from a reel, roll formed, and cut to length on-demand; the cost of the latter was less than $10,000.

The **subassembly** workstation allows manual BTO&MC by manual assembly of kanban supplied parts according to instructions displayed on the monitor. One standard fastener that is dispensed through the autofeed screwdriver accomplishes all fastening. **Final assembly** is also directed by a computer monitor which gives appropriate instructions for each product. Figure 8-3 shows an enlarged view of manual BTO&MC.

BTO&MC for Electronic Systems

The model illustrated in Figure 8-2 shows how to mass customize an electronic system, which could be any type of computer, instrument, communication device, audio/visual equipment, electronic game, small appliance, and so forth.[34]

For ease of illustration, this model shows a small instrument that can be customized with unique software and many different combinations of meters, dials, cases, and various internal parts. However, the BTO&MC principles described here apply to all electronics products. This model is structured to be very "do-able" in that it is does not depend on high volume or special automation. The only programmable machine tools used are the most common in industry: circuit board assembly equipment and CNC machining centers. The final assembly is manual, which was selected for this example because most companies may not have the volume or the budget for the equipment necessary for all automated flexible assembly. Of course, all the principles demonstrated here can be applied to automated operations. All products are built-to-order with automatic *kanban* part resupply, discussed in Chapter 7.

Figure 8-2 shows the overall flow of information and parts for the mass customization operations. Notice the variety of assembled products on the output cart, shown in more detail in Figure 8-3. A wide variety of standard BTO products or mass customized variations can be built through the following:

- Various combinations of different parts, automatically resupplied by kanban in Figure 8-3.

- One of three cases, depending on the space requirements of the internal parts, also resupplied by a dual kanban system.

- A printed circuit board (PCB) assembly based on a single versatile bare board that can be assembled with different components for various functionality. The assembled PCB is delivered on a simple conveyor belt from the PCB assembly area.

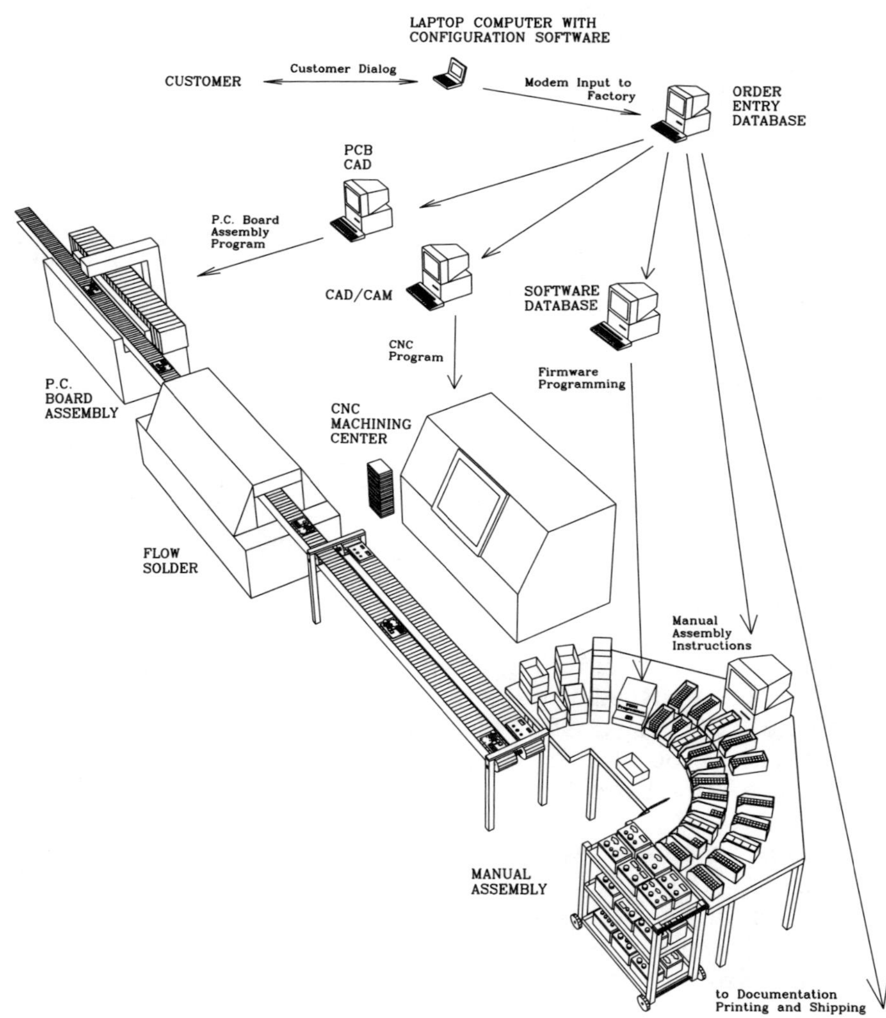

Figure 8-2, BTO&MC for Electronics

- Customized software loaded into a standard firmware blank and then inserted into circuit board.
- A custom machined front panel milled from one standard blank. Appropriate holes and slots are milled for various combinations of dials, meters, ports, switches and controls. The custom front panel is delivered on a simple conveyor belt from the CNC milling machine.
- Rapidly downloaded assembly instructions displayed on computer monitors based on information from the product database file server.

BTO&MC for Manual Assembly

Contrary to popular beliefs put forth by journalists continuing to write about the "computer revolution," not all flexible operations must be performed by automated computer controlled machine tools. In fact, the same principles apply to manual assembly or part fabrication.

Figure 8-3 shows an enlarged view of the flexible manual assembly station illustrated in the last figure.[35] Parts are automatically available through kanban bins, as shown. All fastening can be accomplished by one screw type, which is dispensed through an auto-feed screwdriver. Instructions are created/updated by the order entry process. Displayed instructions are changed whenever operators wand bar codes on the incoming circuit board.

Manual Assembly Instructions. Finding assembly instructions or a drawing is a setup, which would inhibit the flexible manufacturing of mass customized products. To eliminate that setup, manual assembly instructions can be displayed on a computer monitor shown in Figure 8-3.

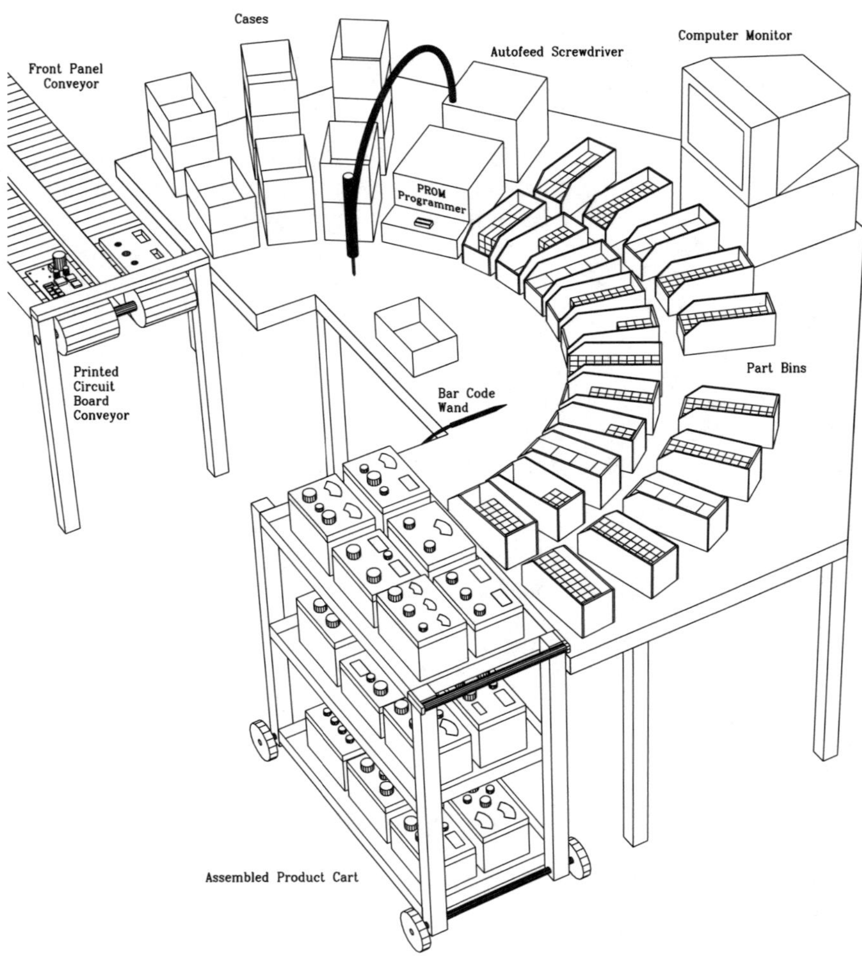

Figure 8-3, Manual Assembly for BTO&MC

Displays for manual assembly instructions can be generated from two-dimensional CAD drawings, 3D CAD solid models, drawn illustrations, or digital photographs that can be touched up to emphasize certain points. Important processing sequences could even be animated, based on either virtual imagery or live action video. At the very minimum, the monitor can display text instructions that could be exported from appropriate documents in word processing or database applications.

Instructions can be made available in multiple languages for multilingual workforces or to ease the transition if pilot production needs to be transferred to a plant in another country.

After performing each operation, the worker would advance to the next instruction by pressing a "next" or "page down" button on an input device. The input device could be a foot pedal to keep hands free for assembly work. Keeping track of the timing of these steps can provide good statistical data for subsequent task analysis.

Instructions related to printed circuit boards can use CAD views of the board in the appropriate level of assembly. Components being assembled could be highlighted by a different color on a color monitor.

ENDNOTES/REFERENCES

1. James P. Womack and Daniel T. Jones, *Lean Thinking; Banish Waste and Create Wealth in Your Corporation,* (1996, Simon & Schuster).

2. Ibid., p. 83.

3. Many a company have compromised their overall company performance in the pursuit of specific metrics, such as utilization. Large mega-machines are hard to justify and, once purchased, must be keep running at high utilization to pay them off. This is one of the lessons of Eli Goldratt's novel, "The Goal," where a plant was maximizing the utilization of it's biggest machine, keeping it busy making parts that were not even ordered, but the plant in danger of being closed because it was losing sales and money. See *The Goal; A Process of Ongoing Improvement,* by Eliyahu M. Goldratt, 1984; 1992 second edition, North River Press.

4. Jones, Daniel J., "JIT & the EOQ Model: Odd Couples No More!," Management Accounting v72, n8, Feb 1991, pp. 54 - 57.

5. Richard J. Schonberger, *World Class Manufacturing, The Lessons of Simplicity Applied,* (1986, Free Press), p. 83.

6. Ibid., pp. 229-236.

7. David Pringle, "How Nokia Thrives by Breaking the Rules," *The Wall Street Journal,"* Jan. 3, 2003.

8. David M. Anderson, *Design for Manufacturability & Concurrent Engineering; How to Design for Low Cost, Design in High Quality, Design for Lean Manufacture, and Design Quickly for Fast Production*, (2008, CIM Press); Chapter 9; Guideline P14: "Design machined parts to be made in one setup (chucking)."

9. Ibid., Chapter 10, "Design for Quality," Section 10.9, "Poka-Yoke (Mistake Proofing)."

10. David Drickhamer, "Aim High; How Industry Week's Best Plants Measure Up," *Industry Week,* October 2002, pp. 67-70. This article is available online at http://www.industryweek.com/CurrentArticles/asp/articles.asp?ArticleId=1321.

11. Shigeo Shingo, *A Revolution in Manufacturing, The SMED System,* (Portland, OR, Productivity Press, 1985). A summary of SMED principles has been written for shop floor personnel: *Quick Changeover for Operators; The SMED System,* (1996, Productivity Press).

12. Kiyoshi Suzaki, *The New Manufacturing Challenge, Techniques for Continuous Improvement,* (New York, Free Press, 1987), Ch. 3.

13. Kiyoshi Suzaki, *The New Manufacturing Challenge; Techniques for Continuous Improvement,* Video program (Dearborn, MI, Society of Manufacturing Engineers).

14. Anderson, *Design for Manufacturability & Concurrent Engineering*, Chapter 8, Guideline A3, "For critical alignment of parts, use round/diamond pins."

15. Productivity Press Development Team, *Quick Changeover for Operators; The SMED System,* (1996, Productivity Press); Chapter 6, Streamlining All Aspects of the Setup Operation."

16. David Drickhamer, "Aim High; How Industry Week's Best Plants Measure Up," *Industry Week,* October 2002, pp. 67-70. This article is available online at http://www.industryweek.com/CurrentArticles/asp/articles.asp?ArticleId=1321.

17. Russ Olexa, "Freudenberg-NOK's Lean Journey," *Manufacturing Engineering,* January 2002, pages 34 - 45.

18. Anderson, *Design for Manufacturability & Concurrent Engineering,* Chapter 9, Guideline P14, "Design machined parts to be made in one setup."

19. Eliyahu M. Goldratt, *The Goal,* Second Revised Edition, (1992, North River Press).

20. David Drickhamer, "Aim High; How Industry Week's Best Plants Measure Up," *Industry Week,* October 2002, pp. 67-70. This article is available online at http://www.industryweek.com/CurrentArticles/asp/articles.asp?ArticleId=1321.

21. Eliyahu M. Goldratt, *The Goal,* (North River Press, Second Revised Edition, 1992).

22. C. Karry Kouvelas, "Product Proliferation," *APICS - The Performance Advantage,* April 2002, pp. 26 - 30.

23. Several books on Total Productive Maintenance are published by Productivity Press (www.productivityinc.com) and the Society of Manufacturing Engineers (www.sme.org).

24. Womack and Jones, p. 87, "Is Chaos Real?"

25. Ibid. p. 88, "Do we really need a business cycle."

26. Anderson, *Design for Manufacturability & Concurrent Engineering.*

27. Jonathan Welsh, "A New Status Symbol: Overpaying for Your Minivan; Despite Discounts, More Cards Sell Above the Sticker Price," *Wall Street Journal,* July 23, 2003, p. B1.

28. David Pringle, "How Nokia Thrives by Breaking the Rules," *The Wall Street Journal,"* Jan. 3, 2003.

29. Mihaly Csikzentmihalyi, *Flow: The Psychology of Optimal Experience* (1990, Harper Perennial).

30. Jones, Daniel J., "JIT & the EOQ Model: Odd Couples No More!," Management Accounting v72, n8, Feb 1991, pp. 54 - 57.

31. Richard J. Schonberger, *World Class Manufacturing, The Lessons of Simplicity Applied,* (1986, Free Press), p. 83.

32. Ibid., pp. 229-236.

33. Chris L. Conway and Jerome M. Tiry, "Mass Customization Comes to Hoffman," *Agility & Global Competition,* Vol.2, No. 2, Spring 1998; Guest Editor, David M. Anderson; pages 16 - 23. A summary of this article is presented on-line at www.build-to-order-consulting.com/hoffman.htm.

34. This figure was created for the author's seminars and first published in copyrighted class handouts. It was then published in *Agile Product Development for Mass Customization,* by David M. Anderson (1997, McGraw-Hill).

35. Ibid.

9

MASS CUSTOMIZATION

Mass Customization[1] is the ability to quickly and efficiently build-to-order customized products. It uses all the techniques presented so far for the build-to-order of *standard* products and extends that to *custom* products. These products can be customized for individual customers or niche markets, such as versions optimized for certain market segments, industries, regions, or countries.

Mass customization quickly and efficiently builds-to-order customized products

Prerequisites for mass customization are acquiring the ability to: (1) quickly and efficiently build products on-demand, as discussed in the last chapter, and (2) procure materials and fabricate parts spontaneously, as discussed in Chapters 3 through 7. Without these techniques, part and material delivery can present a dilemma for companies since, except for big-ticket capital equipment, mass customization customers may not wait for parts to be ordered and delivered. And mass customization customers are too unpredictable for production to count on having just the right parts in forecasted parts inventory.

Mass customization has not yet reached a critical mass in industry because: (1) there are very few benchmark companies doing it well; those that do may not get much publicity and, when the press does discover them, it will be after years of "building up the flywheel" as Jim Collins points out in *Good to Great,*[2] (2) most of these companies have only *experimented* with mass customization and not extended these principles to their entire line of standard products, and (3) writers of books and articles don't understand enough about manufacturing and supply chains to tell readers how to implement it.

Definitions

Build-to-Order is defined as the on-demand production of "standard" product variations identified with pre-defined configurations and designations.

Mass Customization is defined as products built using the on-demand production capabilities of BTO for product variations that are, in part, specified by:

- modular count variations, which could be pre-defined and validated ahead of time;
- modular combinations not specified as "standard" products, for which certain allowable combinations could be pre-validated or unique combinations would have to be validated for the order and then added to pre-validated list;
- unique dimensions, which may involve Engineering-to-Order (ETO) efforts, except for pre-defined "templates"that automate the customization from parametric CAD through automatic CNC program generation.

Awkward, and Expensive, Customization Attempts

Many companies are trying to better satisfy customers by building what they want, but are not doing it very well. Most companies today are trying to offer variety to customers in the mass production heritage by layering new "exception" activities on top of existing inflexible processes, but this is not mass customization.

In mass production companies, Manufacturing really wants to build a single product in high volume (for economies-of-scale), but Marketing would like to please customers with many "variations" of standard models or even new and different products. This can be expensive and time consuming for Engineering and Manufacturing when the product line is not designed to easily accept those changes. Many companies are customizing by Engineering Change Orders, a very inefficient process.

Some companies offer customers variety by employing "custom engineering" departments, but this high overhead activity can still be inefficient if the product and production processes are not designed to readily accept the custom engineering. If cost accounting systems do not quantify overhead adequately, inherently high overhead activities, like custom engineering, will probably be subsidized by the standard products, thus distorting decision making on customization. Some companies are able to do custom products somewhat quickly, but sacrifice cost and control. These approaches are shown on the "reactive" side in Figure 9-1.

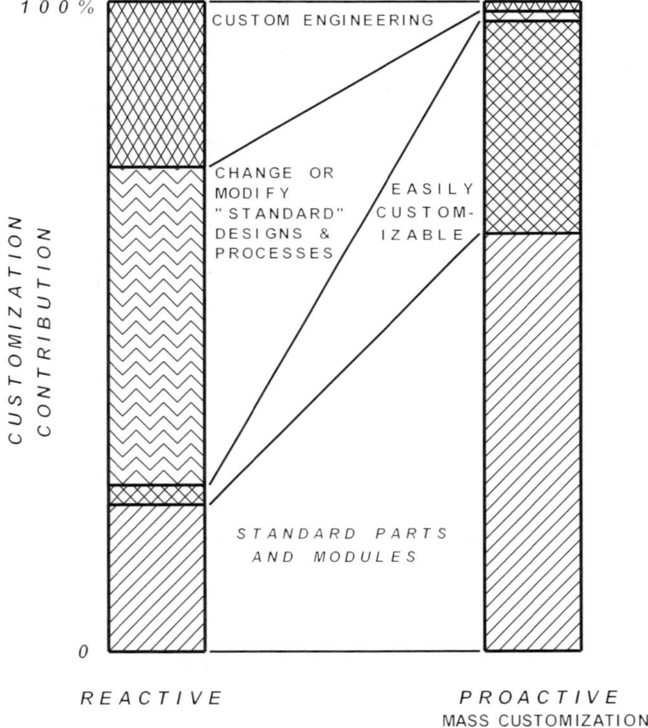

Figure 9-1: Customization Contribution for Reactive and Proactive Modes

Thus, the goal is to achieve *timely and efficient* customization of products which is *mass customization* (as shown at the "proactive" side in Figure 9-1) with a much higher proportion of standard parts and modules and "easily customizable features," which will be discussed throughout this chapter.

The Buzz about Mass Customization

There has been a lot of hype lately about mass customization, but much of it is just *customization.*

> The is a big difference between customization and *mass* customization

Many journalists and authors jump at the chance to say something about the latest management trends, but fail to comprehend the potential of mass customization because they don't understand the principles of on-demand production and spontaneous supply chains as uniquely presented in this book. Most writers still limit their view of mass customization as merely assembling *modules* to customer orders, although they are unable to explain how these modules can be always available for on-demand assembly and how materials can be always available for on-demand module production. Although modularity is an important technique, mass customization is much more than modular assembly.

Similarly, many consulting firms like to toss around the latest buzzwords and may even make a vague reference to mass customization or add the phrase on a list of "specializations" on their web-site. But in reality, these may just be a new label to fairly conventional practice.

> Mass customization is much more than bolting modules together

Another cause of misunderstandings about mass customization is the benchmarking approach: find "successful" companies who are doing anything *called* mass customization (by writers who don't understand it), superficially analyze them with plant tours or case studies, extrapolate "the possibilities," and then try to apply this to different industries. Further, because of their background and experiences, they view what they see "through the lens of mass production." [3]

> Most MC examples cited are trivial, experiments, or special cases

The main shortcoming of the benchmarking approach is that there are simply not that many companies doing mass customization very well, and most of those are not glamorous products from household-name companies, so very little is written about them. Most of the examples touted are either (a) somewhat trivial, (b) just experiments that have not worked their way into a whole company strategy, or (c) special cases, such as *assemble-to-order* personal computers (as exemplified by Dell, as discussed in the Introduction) which are inherently modular products made by very large operations that have enough clout to force part suppliers to warehouse forecasted parts inventory *at their expense and risk* next to their assembly plants. These companies are successful because of their size, their direct sales model, and lack of finished goods inventory which make them better than the competition. But, that does not mean that they are practicing the full vision presented in this book.

Misconceptions range from buzz-word PR to over-hyping that MC can customize anything

The worst aspect about the buzz about mass customization are the misconceptions. At one end of the spectrum, company PR departments or ambitious consulting firms use the label to juice up a company image or broaden the market for conventional supply chain management or lean production offerings. At the other end of the spectrum, mass customization is over-hyped to imply that a mass customization plant can build any oddball variation without any limits or constraints. In reality, mass customization can quickly and efficiently build any variation *within the capabilities that the mass customization system was established to build.*

What Mass Customization Really Is

Real mass customization is actually *manufacturing* custom products (not just assembling modules) quickly and efficiently in quantities as low as one. The best way to visualize this is to imagine a customer placing an order for one customized product. An imaginary plant would be constructed and equipped to fabricate all the parts and assemble this product, as if the product ordered was so good that a mass production plant would be built to make thousands of them (the only way to keep cost low in mass-production thinking). The factory fabricates and assembles this one product and ships it out that day, and then, in a few seconds, magically

Imagine a plant that magically retools itself for each product reconfigures itself and its supply chain to produce the single product ordered by the next customer. This idealized conceptually illustrates the potential of mass customization.

Now to bring this fantasy into reality, apply the techniques of this book to achieve the "magical" transformation of a manufacturing facility in seconds to build different products at "mass production efficiencies" – actually better, without all the inventory carrying costs and other inefficiencies.

First of all, the factory is not equipped with typical mass production machines: expensive, specially-designed, "hard" automation that cannot be used on other products. Rather, the equipment consists of less-expensive, readily-available CNC machine tools that don't go obsolete and can be reused again and again. Not only can these machines do much of the customization automatically, but they can do it efficiently in batch-size-of-one without setup. Even manual assembly can be reconfigured in seconds with spontaneous part delivery (Ch. 7) and on-line assembly instructions (Ch. 8).

Instead of parts and materials appearing magically in the above example or coming from boundless inventory in traditional mass production, supply chains are simplified by product line rationalization (Ch. 3), part standardization (Ch. 4), material variety reduction (Ch. 5), and selective vertical integration (Ch. 6) to form a spontaneous supply chain (Ch. 7).

HOW TO MASS-CUSTOMIZE

Despite all the attention given to modularity, there are really *three* ways to mass-customize products: *modular, adjustable,* and *dimensional* customization. Mass customization works best when the optimal combination are used, not just the one that is obvious or easy to comprehend. Each of these will be briefly defined and then discussed at length after this summary:

> There are 3 ways to mass-customize: modular, adjustable, and dimensional

Modular Customization. The most obvious form of modularity is the literal "building block modularity," which can customize a product by assembling various combinations of modules. In electronics, modules could include processor boards, power supplies, plug-in integrated circuits, daughter-boards, and disk drives. In software, code could be written in modules (objects) that can be integrated into various combinations. For vehicles, modules could include engines, transmissions, seats, audio equipment, tires, wheels, and so forth.

Adjustable Customization. Adjustments are a *reversible* way to customize a product, as in mechanical or electrical adjustments. Adjustments could be infinitely variable. Discrete adjustments, or **configurations,** would represent few choices, such as those provided by electronic switches, jumpers, cables, or discrete software controlled configurations. Unlike the following category, adjustments and configurations are reversible. These adjustments and configurations make the product *customizable* by the factory, distributor, store, or user.

Dimensional Customization. Dimensional customization involves a *permanent* machining, cutting-to-fit, mixing, or tailoring. Dimensional customization could be infinite or have a selection of discrete choices. Examples of discrete dimensional customization would be hole punching or soldering selected electronic components onto a printed circuit board. Examples of infinite dimensional customization would include machining customized shapes on CNC machine tools, tailoring clothing, drilling holes in bowling balls, grinding eyeglasses, mixing of paints or chemicals, cutting wire or tubing, and so forth.

The National bicycle unit of Matsushita mass-customizes bicycles[4] using all three methods: The *modular* contributions to the customization were the wheels, pedals, drive chain, derailleur, seat, handle bars, and controls. The *adjustable* contributions were adjustments for the height and angle of the seat and handle bars. The *dimensional* contributions were the lengths of the frame tubing, which were cut-to-fit, after measuring the customer on an adjustable dealer fixture.

Firmware could be any of these three categories depending on the application. Programmable integrated circuits could be: (1) programmed and treated as modular parts, (2) configured or reconfigured in nonvolatile (flash) memory, or (3) permanently programmed in the product, as in "Programmable Read-Only Memory" (PROM).

Modular Customization

As mentioned before, most of the attention on mass customization has gone to modularity because it is easy to comprehend and because dimensional customization is hard to comprehend without a thorough understanding of flexible manufacturing processes. In addition, modularity may be part of an outsourcing strategy, as is popular now in the automobile industry where the assemblers outsource well-defined modules such as seats, instrument control panels, suspension systems, fuel systems, and so forth. Personal computers have some obvious modularity: monitor, keyboard, motherboard, memory, disk drives, modems, video controller, and option boards. For audio systems, customers can choose their favorite amplifier, tuner, CD player, and speakers and simply connect them together. Dell and Gateway derived significant competitive advantage by assembling to-order personal computer modules.

Modularity. Modularity is a good strategy to concentrate functions most likely to change into one module. If the modules can be replaced in the field, this can be a key element in repair and upgrade strategies.

In product development, it can allow parallel development after the modular architecture, interfaces, and standards are established. It also allows the use of off-the-shelf modules, which can simplify product development, assure reliability with proven modules, and take advantage of the suppliers' economy of scale. The off-the-shelf approach would be ideal for standard modules with little or no variety. On the other hand, too much variety of off-the-shelf modules can cause supply chain complications, especially if module demand is hard to forecast.

However, modular systems can be more difficult to design than the equivalent integrated systems.[5] In addition, adding interfaces for modularity in electronics can add cost, slow down signal speed, and may create reliability problems if it adds more low-voltage mechanical interfaces.[6] For more on the *design* of modules, see Chapter 10.

Modularity as a Customization Tool. There are many aspects to decisions about modularity as a means to mass-customize products. It can be useful *as part of* a mass customization strategy. For widely used standard modules, outside suppliers can deliver steady flows to avoid forecasting and inventory complications. If modules have significant variety, then the outsourced model may bog down in forecasting difficulties, out-of-stocks, overproduction, excess inventory, obsolescence, and delays unless the supplier is nearby and very cooperative (Chapter 6). Therefore, modules with considerable variety may need to be made on-demand in-house using all the mass customization techniques of this book.

One alternative to customizing through modularity is flexible *fabrication* of integrated designs instead of assembly of modules. For instance, in the electronic system example in the last chapter, various electronic functionality can be *manufactured* by programmably placing all the components needed for any set of functions onto a versatile bare board, which, because of its standardization, can flow into the factory without forecasts or inventory.

Another alternative is to consolidate many electronic functions into very versatile VLSI integrated circuits or custom silicon that provide the required functions for all product variations. This is the strategy Hewlett-Packard uses to provide a various functionality for a wide range of calculators. All calculators have the same versatile VLSI chips and access the functions they need. Although this might seem wasteful, the standardization of the VLSI chips provides economies-of-scale and simplifies the supply chain.[7]

Virtual Modularity. Another variation of modularity would be *"virtual modularity"* which does not limit modules to the obvious physical building blocks. Drawings of virtual modules (known in CAD as components, symbols, or blocks) could be combined and "assembled" in CAD systems; in fact, some CAD systems, like Pro/ENGINEER, use the *assemble* command to assemble components.

For printed circuit boards, certain groupings of circuitry and their components (integrated circuits, resistors, capacitors, etc.) could be a *virtual* module in a CAD system, perhaps segregated by CAD layers. Combining various virtual modules could customize PC boards with a minimum of special design work. Practically all printed circuit board manufacturing equipment can programmably place and solder these virtual groups of components onto standardized bare boards that are versatile enough to have all the holes, pads, vias, and traces for all customization scenarios.

Hidden Modularity. In some markets, customers would prefer the *appearance* of an integrated product. Internal or *hidden modularity* could be transparent to the customer, but still allow the assembler to benefit from modularity. Products could be assembled from various choices of modules, but then be assembled into what appears to be integrated products. Electronic systems could be sold as an integrated "system," but with a high degree of modularity inside the cabinetry. The manufacturers could assemble standard modules inside of integrated enclosures.

If product image or uniqueness is important, then modularity could be hidden. The visible portions could be unique with functional modularity inside.

Adjustable Customization

Adjustments can be automatic as with shavers such as the Gillette Sensor razor, the Schick Tracer brand, and the Braun Flex Control electric razor that always keeps the razor head at a ninety-degree angle to the user's face.[8] Matsushita has developed a washing machine that can automatically choose one of 600 washing cycles.[9]

Adjustments can be manually set by the plant, dealer, or user. Examples of user-controlled adjustments include: seat positions in cars or office chairs; bicycle seat and handlebar height and angle; the "pump" tennis shoes; and all the adjustments on machine tools.

There may be an optimal balance of automatic adjustments and manual adjustments: basic users may want automatic adjustments, but advanced users may want to be able to configure the product to meet their advanced needs. Many PC software applications are installed with the most popular *default* settings, which can be reset by more advanced users.

The ideal way to set adjustments would be through programmable software settings, which could be automatically programmed in at the factory or dealer, upgraded by modem, or programmed by users. Lutron Electronics converted 40 different lighting control products into a single programmable "smart" microprocessor-controlled product that can be programmed by the user.[10]

Discrete adjustments, or *configurations,* can be made by switches, jumpers, special cables, or automatically through software settings.

One caution about adjustability: Make sure that adjustments are not *coupled* in such a way that one adjustment will affect another. Another problem with multiple adjustments might be worst-case incompatibility. Designers must be sure that no adjustment will be incompatible with any other adjustment to the point of degrading product performance. Too many interconnected adjustments may frustrate dealers or users and, in extreme cases, might result in an adjustability "Catch 22," where several adjustment attempts have rendered the product unusable to all but an expert.

A significant advantage to adjustable customization is that the adjustable features can be mass-produced without having to forecast the choices, build many versions, and distribute a great variety of products. That was the strategy for adjustable eye glasses, invented by Joshua Silver for developing countries who don't have the optometry infrastructure taken for granted in developed countries. The focal length of the glasses can be varied by controlling the amount of silicon oil between flexible and rigid membranes. The user does this through a removable pump that can be reused to adjust the focal length again over time.[11]

Dimensional Customization

Dimensional customization is accomplished by a *permanent* dimensional change such as machining, cutting-to-fit, mixing, or tailoring. This has great potential, although hard to comprehend without understanding how manufacturing equipment can perform such customizations.

On-demand dimensional customization has the distinct advantage that it does not have be forecasted, built, shipped, stocked, and distributed with risk of obsolescence from overproduction or delays from underproduction, as might be the case of modules made in the mass production paradigm.

Dimensional and adjustable means may be best for wide ranges of customization whereas modular customization would be best for standard modules.

Perhaps the earliest application of dimensional customization is house paint. Decades ago, paint companies realized that stocking every color and texture of house paint in every city would be impossible. Consequently they standardized the raw material (white paint) and provided each store with the means to quickly mix any color using pre-formulated mixtures of a few primary color pigments. Today, some paint stores can even customize the color to match a photograph or chip of old paint.

Other examples of dimensional customization are the widely available house gutters which are fabricated on-site from standard sheet metal, which is fed off a coil through a choice of rollers that form it into the desired shape and then cut the length to fit the house. The Peerless Saw company has developed a computer-controlled laser system to cut custom tooth profiles on cutting saws.[12]

Dimensional customizability can be performed manually, but care must be taken to ensure consistency and plan operations to avoid set-up delays. For instance, tubing could be cut to the required length by a worker who reads the dimension on a computer monitor and then manually positions the tubing in a cut-off saw.

Many dimensional customizations can be performed automatically by computer numerical controlled (CNC) equipment, such as machining, hole punching, laser cutting, wire cutting, and circuit board assembly. This *programmable customization* is a valuable set of tools for quickly customizing parts, modules, or products. The programs that control these machines can be changed instantly, without a set-up change, thus permitting flexible manufacturing in batch-size-of-one.

For *programmable assembly,* such as inserting or placing components on printed circuit board, all setup must be eliminated: (1) universal fixtures and "pass-through" material handling can eliminate loading setup; (2) programs can be automatically generated or downloaded instantly; and (3) no setup is needed to change parts when all parts reside in the machine. This underscores the importance for part standardization for parts that will be assembled by programmable automation.

Using this programmability of printed circuit board assembly equipment, a cleverly designed bare board could be assembled in a variety of ways with different parts to produce a wide variety of assembled printed circuit boards, as shown in the electronic example of Figure 8-2. The main advantage of this programmable customization is that the number of bare board variations can be minimized; this is important because of long procurement lead time and large batch sizes inherent in bare board manufacture. Without this versatility, all versions of bare boards would have to be stocked with all the cost and risk of inventory and obsolescence. On the other hand, a standard bare board could be "ordered" simply by arranging a steady flow of boards coming into the plant, as advocated in Chapter 7.

For materials that are cut-to-fit, raw material should be standardized on the same size bar stock or tubing to avoid setup delays to change the stock size.

Ross Controls mass-customizes pneumatic control valves and manifolds through dimensional customization on CNC machining centers. These *Ross/Flex* products replace modular assemblies made up of pneumatic valves, pipe, and various fittings. The Ross/Flex process uses *virtual* modules for valve geometries and connecting holes, many of which are then machined automatically by CNC machine tools. Ross customers order prototype valves or enhanced performance valves for special production machinery. This creates learning relationships which makes the customer happier and more loyal with each order.[13]

Dimensional customization can also be used to circumvent problems cause by *tolerance stacks,* which happen when many stacked parts must achieve a specified tolerance for the stack or must mate with another part or another stack. Instead of the expensive approach of trying to hold tight tolerances on all parts on both sides of the stack, dimensional customization can be utilized to drill, pin, or spot weld "at assembly" *any part* in either stack, which *eliminates the need to hold tight tolerances on all the parts.*

Postponement

Postponement is a mass customization technique that is applicable for certain products that can have their variety postponed until just before shipping. The factory builds basic “vanilla” platforms and adds “flavors” upon receipt of the orders. Postponement is most suitable for product architecture that has a major platform part can be built without variation and then customized by various adjustments, configurations, or bolt-on modules.

Hewlett-Packard implemented a postponement strategy to handle different input voltages for worldwide sales of DeskJet printers. Instead of an integrated power supply, whose production would have been based on regional forecasts, HP went with external country-specific power supply that the customer plugs into the printer. This postponement increased manufacturing costs slightly but when considering the cost of manufacturing, shipping, and inventory, the total cost dropped by 25%.[14]

Postponement works best when there is only one version of the vanilla platform or else forecasting will be needed to determine which version(s) to build ahead of time.

Postponement is useful in companies who want to mass produce as much as possible either internally or as part of an outsourcing strategy. They feel they are getting economies-of-scale and still have some capability to do some product differentiation before shipping to customers.

When applicable, postponement can reduce the amount of finished goods inventory, but options may have to be forecasted unless the options can be built spontaneously.

WHAT TO MASS-CUSTOMIZE

Mass customization can efficiently and quickly build vast families of synergistic products, *but mass customization may not be able to build all your existing products.* In fact, it probably won't, especially if the product line has accumulated products for decades without ever having been rationalized to eliminate the most unusual products with the most unusual parts, materials, and processing (Chapter 3).

The determination of which products to mass-customize must be based on a combination of market needs *and* organizational capabilities. The common temptation is to only consider the market need and offer to "customize anything," which may be so hard to do that operations revert to the inefficiencies and delays of craft production or frequent setup changes in a mass production environment.

It is important to remember that mass customization operations can customize a certain range of products very efficiently, but just outside that range it may be difficult and beyond that it may not be feasible at all. And not all customization offering will be received equally in the marketplace. Some customers may only want just so many choices and may not go to the trouble of submitting too much information, for instance, if they have to make several body measurements or have their bodies scanned for custom clothing. The mass-customizer needs to find out if customers are more interested in standard sizes with many available options or products actually tailored specifically for them.

> MC can customize efficiently within limits, but not beyond those limits

A key element of mass customization strategy is the optimal grouping of products into product families that can all be built efficiently on-demand without setup delays with all the parts and materials always available. Product family grouping will be discussed in the implementation section in Chapter 13.

During the transition to mass customization, product variations that *look like* they will be part of the implemented system should continue to be offered, even if this means using current operations, until more efficient BTO&MC capabilities come on line. New offerings would probably be best deferred until that time, unless there is a pressing market opportunity and such introductions would not be a distraction to BTO&MC implementation.

EXTRA VALUE-ADDED OPPORTUNITIES

Mass-customizers can use their versatile design and manufacturing capabilities to offer new, related, value-added products/services that expand the scope of their markets. These offerings may cost little extra to add, especially if they can be done in existing CNC operations. And yet, they may save customers so much time and money that they would gladly pay for the options, especially if these features are difficult for customers to make.

> Extra operations that are easy for you to do and valuable = *high profit opportunities*

For instance, one of the author's clients, Hoffman Engineering, built a plant in Lexington, Kentucky, which can spontaneously build-to-order a wide variety of standard or mass-customized electrical enclosures.[15] It's high-profit value-added opportunity is to use the very same CNC laser cutters that make the boxes to make the holes and cut-outs that cost their customers much time and money to do on their own.

ORDER ENTRY

The main difference between build-to-order and mass customization is the order entry. Whereas build-to-order order entry is based on *standard* product model numbers or sizes, mass-customized products are at least partially specified by some customization data or dimensions.

Although spontaneous BTO does not require elaborate information systems for material procurement and production management, mass customization ventures may need effective web-sites and information systems that can rapidly handle a wide variety of orders (including mass-customized orders) through configurators and convert order entry data into parametric CAD models, CNC programs, electronic manual instructions, supplier pull signals, and shipping instructions.

For simple customizations, customization data can be obtained from answers to salesperson queries, filling in forms on web-sites, or simply filling out paper forms. More complex customization can use software *configurators,* which keep track of all the options and features and all the rules that apply to their selection. In essence, configurators are *expert systems* that "know" all the things that are usually determined in the lengthy scenario where customer queries must be sent back and forth between the plant and the sales force. The *knowledge* of the company "experts" is captured in the software and is readily available through a user-friendly interface. Configurators can be quickly updated, even over internet connections to units in the field.

A 2008 Aberdeen Group survey or 300 enterprises found that the best-in-class companies who used configurators coupled with modular/flexible design were able to offer 3.5 times more customizable features, got quotes out three times faster, and incurred lost revenue due to quote or order errors to only four percent of the losses of the average company.[16]

All the features and options along with their prices and delivery estimates, are embedded in the system. Configurators also contain all the rules that govern dependencies, conflicts, constraints, and resources. The configuration "engine" can do the calculations to resolve complex relationships quickly. The customer is able to request many "what-if" scenarios.

Window manufacturer Andersen Corporation developed its own configurator, the Window of Knowledge™, to let distributers work with customers to design their windows on a computer and interactively view prospective designs immediately.[17]

Some advanced configurators can display solid models and advanced graphics to show the customer what the contemplated product will actually look like. For example, the configurator could show a side view of an automobile in the color of choice with the wheels selected. It could also show the interior view with the selected steering wheel and all the dashboard options, exactly as the ordered car will appear. These images, along with pricing, can be printed out on a color printer and given to the customer — in essence, a custom brochure for a custom product. For very new products, this visualization capability can help early customers understand and comprehend new concepts and approaches. This can help get early feedback from customers. In the Aberdeen Group survey, three-fourths of the best-in-class companies used visual imaging in their configurators.[18]

Web-based configurators offer the tantalizing prospect of customers not only placing their own orders but also "designing" their own products, all through automatic systems that are carefully constructed to allow ease of use and permit only valid product configurations. For years, Cisco Systems has been the leader in internet revenue. Cisco customers configure routers and other products through an on-line configurator made by Calico, based in San Jose California.

A thorough discussion on configurators and selection criteria is included in the book, *Growing Modular; Mass Customization of Complex Products, Services, and Software.*[19]

MARKETING BTO & MASS CUSTOMIZATION

The most important principle for the marketing of mass customization is that *sales are not all equal.* Product families that are structured according to the principles of this book will have the ability to satisfy customers with exactly what they want then they want it at the lowest cost on the market. By contrast, products outside these families would be a nightmare for a BTO&MC company to produce and, further, would distract it from making the profitable product with the highest profit and growth potential. Thus, as said many times, the sales force must resist all temptation to "take all orders." Mass customization offers great growth potential (Chapter 14), but only if done right.

> All sales are not equal.
> Some are *dreams;*
> Others *nightmares.*

It is possible to *instantly* shift the sales force focus to profitable products by changing their compensation from *revenue or unit volumes* to *profitability.* Of course, to compensate based on profit, the company must be able to *measure* profitability by individual product or order. This would be one of the beneficial results of implementing total cost measurements (Ch. 12). Until the cost accounting system can *predict* profitability of an order, the sales force compensation could be based on a base salary with commissions or bonuses based on *computing* profitability by adding

> Base compensation on
> *real profit,* not revenue
> or unit volumes

up all the relevant costs after the product is delivered and all the overhead costs have been attributed to that order.

Expand the focus on MC products: Phase out the losers.

When shifting to mass customization, Marketing will expand some markets (the BTO&MC products) while eliminating losers and outsourcing products that need to be in the catalog, but do not fit into the new operations (Ch. 3). BTO&MC products will have lower cost and much faster delivery, while outsourced products will have the original slow delivery and high cost.

Build-to-order and mass customization products will be built on-demand without forecasts or inventory, whereas any products still built by mass production may still need forecasts, inventory, and all the associated challenges. Without the uncertainty of forecasts, BTO&MC products will never be overbuild so they will not have to be discounted to "move the metal." Discounting will most wisely be used to help level production by offering discounts for off-season purchases of seasonal products (Ch. 8).

BTO&MC products will not be out-of-stock or over-stocked; thus, no more missed sales or discounting

On the other hand, Sales and Marketing will not have to deal with the frustration of not having "hot" products available because scarce parts and production capacity were wasted on building products that are not selling, which, ironically, will now have to be discounted.

There will be price adjustments apart from the expected lower cost from BTO&MC efficiencies. Merely implementing total cost accounting (Ch. 12) will adjust all prices in proportion to their *real* costs. This will end cross-subsidies where high-volume products typically subsidize low-volume products and standard products subsidize custom products.

Many marketing costs will be reduced as thick catalogs listing hundreds of "standard" products are replaced by a few customizable platforms featured in simpler printed catalogs or on web-sites. Further web-site orders will be automated and eliminate order-entry errors typically encountered with too many manual data transfers. Product documentation will be presented and easily updated on web-sites rather than expensive print media that is difficult and expensive to update. Configurators automatically compute prices that can be updated quickly and easily; this replaces price sheets that may not readily provide the bottom line price and are slow and expensive to update.

Mass customization can yield great marketing benefits by developing *learning relationships* with customers that allow mass-customizers to *keep customers forever.* This was the theme of the Pine, Peppers, and Rogers Harvard Business Review article,[20] that concludes:

> "In learning relationships, individual customers teach the company more and more about their preferences and needs, giving the company an immense competitive advantage. The more customers teach the company, the better it becomes at providing exactly what they want -- exactly how they want it -- and the more difficult it will be for a competitor to entice them away."

Ross Controls[21] has used mass customization to benefit from learning relationships. One of its customers, Knight Industries, who makes material handling equipment, buys 70% of its standard valves and 100% of its custom valves from Ross. When approached by a competitor, Knight's president said, "Why would I switch to you? You're already five product generations behind where I am with Ross." [22]

Mass Customization can capture new niche customers that were previously not "in the market" or trying to do the customization themselves – the slow and expensive way. A Fortune article about Build-to-Order described how Hoffman's Mass Customization plant generated new business: *"Catering to customers' individualized needs, the plant has won new business by inducing companies to give up doing it themselves. Some of Hoffman's customers used to make their own enclosures so they'd have the right box at the right time."*[23]

Marketing efforts should be able to greatly capitalize on the competitive advantages of never being out-of-stock, having the lowest cost, offering customized products, opening up niche markets, creating learning relationships, and being the fastest to market with new technologies.

COMBINED MASS CUSTOMIZATION AND BTO

There is a natural synergy between build-to-order and mass customization, hence the title of this book. They share the same batch-size-of-one operations and spontaneous supply chain.

There is a natural synergy between build-to-order and mass customization

Build-to-order and mass customization operations are equally efficient and very complementary, unlike situations where a mass customization experiment must be run separately from the "batch-and-queue" operations of mass production. Building BTO and MC products on the same lines will often push the combined volume over the " mass" threshold necessary to justify these implementations. Higher total volume makes it more likely to split production into more lines or cells, each of which would be more specialized with a smaller range and thus would be easier to establish. Thus, combined operations are more like to be approved and succeed than either attempted independently.

Once MC is in place, *customization is free*

Once a viable mass customization operation is established, customization within the system will be free.

REDUCING COST AND WEIGHT WITH MC

In addition to cost-effectively building custom products, Mass Customization principles can also be applied to significantly lower the cost and weight of assemblies by enabling the cost-effective manufacture of "efficient structures."[24]

For instance, the most efficient structures are based on the catenary arch. A catenary shape forms naturally when a chain is suspended between to anchors. The most common examples of *tension* catenaries are suspension bridges. The most common examples of *compression* catenaries is masonry arches, like the Roman aqueducts and buildings based on domes and arches, which can support large loads without steel reinforcement and have been standing for centuries.

However, shapes based on catenaries can not be built with *standard* blocks, struts, or cables, so each part must be customized to its particular place on the catenary curve. In ancient times, these unique shapes were cut by masons using *craft* customization. The different vertical cables of suspension bridges are usually fabricated in different batches with setup changes in between (mass production).

Similarly, the arched truss is the most efficient truss, meaning, for a given loading, it has the least weight, least cost, and uses the least amount of material. The arched truss is also based on the catenary shape, or a close approximation, which means that the vertical struts have different lengths.

The most efficient *domed* truss is the geodesic dome, invented by R. Buckminster Fuller in 1951 (Patent # 2,682,235).[25] Although, to the casual observer, these domes look like a bunch of identical equilateral triangles between a few pentagons, the struts are many different lengths, with larger domes (and those with higher "frequencies") needing more struts of different length.

Mass Customization techniques can be used to cost-effectively fabricate all these different struts on programmable CNC machine tools without setup changes, as described in Chapter 8, thus resulting in stronger, more efficient structures at the cost of mass produced standard struts.

Similarly, the joints or "nodes" for geodesic, octet, or catenary trusses will have to accommodate struts coming in at various angles. This problem can be solved by two mass customization techniques: *adjustable customization*, where the joints adjust to varying angles or struts adjust to various lengths, or *dimensional customization*, where CNC machine tools are programmed to fabricate families of different joint parts at the speed and cost of standard joints.

By contrast, it would be difficult, and therefore expensive, for a mass production operation to make all these different struts and joints. So conventional designers would be discouraged to attempt geodesic domes or catenary arch trusses and, instead, design "zig-zag" trusses (e.g. like Warren, Pratt, or Howe trusses) which would be assembled from standard struts and joints (like an Erector set). However, these are less efficient (weaker or heavier) and incur more material cost than can be achieved by the mass customization scenario described above.

The same logic could be applied to many other design challenges where mass customization can cost-effectively produce an optimal *set of shapes* that are function better or are more efficient structurally than those using many of the same "standard" components.

For truss structures that need to hold precise shapes, the above mentioned "drill at assembly" approach can eliminate the tolerance challenges where several truss struts come together. This can save thousands of dollars on complex structures like telescope and solar power reflectors. After the precise surface parts (such as mirrors or reflecting surfaces) are positioned in a precise fixture, the remaining truss members would be assembled with key members drilled and bolted or pinned together at assembly to ensure the truss retains the shape set by the fixture.

KEEPING ORDERS WITHIN CAPABILITIES

The biggest caution about mass customization is to make sure orders don't exceed the capability of the system. Some companies become victims of their own publicity if they oversell mass customization capabilities to the point that customers develop unrealistic expectation and salespeople think they can customize anything.

It is important to keep emphasizing that a mass customization plant can only build products within its capabilities, which can be expanded to efficiently handle more variety, but it is imperative that current orders are *always* within the capabilities of the system. It may be necessary to find a way to handle oddball requests such as outsourcing to a versatile vendor (Chapters 3 and 6).

If there seem to be too many opportunities passing by, then the breadth of capabilities can be methodically expanded using all the principles of this book.

ENDNOTES/REFERENCES

1. The phrase *mass customization* was first coined in the book by Stan Davis, *Future Perfect,* (1987, Addison-Wesley; Chapter 5, "Mass Customization"). Inspired by that book B. Joseph Pine II then wrote the book, *Mass Customization* (1993, Harvard Business School Press), which makes an eloquent case for the paradigm shift to mass customization. After reading that book, this author approached Joe Pine to work together to write the "how to" book, *Agile Product Development for Mass Customization,* (by David M. Anderson, with an introduction by B. Joseph Pine II; 1997, McGraw-Hill).

2. Jim Collins, *Good to Great; Why Some Companies Make the Leap . . . and Others Don't,* (2001, Harper Business), Chapter 8, "The Flywheel and the Doom Loop."

3. B. Joseph Pine II, Bart Victor, and Andrew C. Boynton, "Making Mass Customization Work," *Harvard Business Review,* September-October, 1993.

4. Susan Mo, "Japan's New Personalized Production," *Fortune,* October 22, 1990, p. 132.

5. Carliss Y. Baldwin and Kim Clark, "Managing in an Age of Modularity," *Harvard Business Review,* September-October, 1997.

6. Mechanical interconnections are major source of reliability problems in electronics, especially for low-voltage connections. Think of the automatic reaction when a flashlight does not work: shake it until it comes on. Although metal contacts were "touching," the electricity was still not flowing.

7. Jean-Phillippe Deschamps and P. Ranganath Nayak, *Product Juggernauts, How Companies Mobilize to Generate a Stream of Market Winners,* (1995, Harvard Business School Press), pp. 35-36.

8. B. Joseph Pine II, *Mass Customization, The New Frontier in Business Competition,* (1993, Boston, Harvard Business School Press), p. 180.

9. David M. Anderson, with an introduction by B. Joseph Pine II, *Agile Product Development for Mass Customization,* (1997, McGraw-Hill), Chapter 1, Introduction.

10. Michael W. Pessina and James R. Renner, "Mass Customization at Lutron Electronics - A Total Company Process," *Agility & Global Competition,* Vol. 2, No. 2, Spring 1998; Guest Editor, David M. Anderson; pages 50 - 57.

11. William G. Phillips, editor, "Best of What's New," cover story, *Popular Science,* December 2000, p. 60. Also, see article by Gautam Naik, Wall Street Journal, October 14, 1998. More information on the adjustable eyeglasses can be found at www.adaptive-eyecare.com.

12. Pine, *Mass Customization*, p. 203-204.

13. Steven W. Demster, President, and Henry F. Duignan, COO, Ross Controls, "Subjective Value Manufacturing at Ross Controls," *Agility and Global Competition,* Vol. 2, No. 2, Spring 1998; Guest Editor, David M. Anderson; pp. 58 - 65.

14. Edward Fietzinger and Hau L. Lee, "Mass Customization at Hewlett-Packard: the Power of Postponement," *Harvard Business Review,* January-February, 1997, pages 116-121.

15. Chris L. Conway and Jerome M. Tiry, "Mass Customization Comes to Hoffman," *Agility & Global Competition,* Vol.2, No. 2, Spring 1998; Guest Editor, David M. Anderson; pages 16 - 23.

16. "Tailoring Products to Customer Preferences; Configuring Profits to Order," by the Aberdeen Group, March 2008, available at http://www.aberdeen.com/summary/report/benchmark/4666-RA-tailoring-products-preferences.asp.

17. Anderson and Pine, *Agile Product Development for Mass Customization,* Chapter 1.

18. "Tailoring Products to Customer Preferences; Configuring Profits to Order," by the Aberdeen Group, March 2008.

19. Milan Kratochvil and Charles Carson, *Growing Modular; Mass Customization of Complex Products, Services, and Software;* Section 6.5, written by Cincom Systems, pages 106-118.

20. B. Joseph Pine, II, Don Peppers, and Martha Rogers, "Do You Want to Keep Your Customers Forever," *Harvard Business Review,* (March-April, 1995), p. 103.

21. Steven W. Demster, President, and Henry F. Duignan, COO, Ross Controls, "Subjective Value Manufacturing at Ross Controls," *Agility and Global Competition,* Vol. 2, No. 2, Spring 1998, pp. 58 - 65.

22. Pine, Peppers, Rogers, "Do You Want to Keep Your Customers Forever?," p. 103.

23. Philip Siekman, "Where Build to Order Works Best," *Fortune,* April 26, 1999 (Industrial Management & Technology edition), p. 169F.

24. James Ambrose, Design of Building Trusses, (1994, John Wiley), Chapter 16, "Finding Efficient Forms for Trusses," by Edward Allen. This chapter presents various efficient (least weight) truss shapes with constant stress in either in the top plate (floor) of a truss or the curved chord of an arched truss. These techniques can also generate the most efficient shape for a cantilevered truss, which is the shape of the Eiffel Tower.

25. *Inventions, The Patented Works of R. Buckminster Fuller* (1983, St. Martin's Press).

10

PRODUCT DEVELOPMENT for BTO&MC

CHALLENGES WITH EXISTING PRODUCTS

Existing product designs may not be suitable for spontaneous BTO or mass customization. Products not concurrently engineered for flexible environments may impede implementations, diminish the payback, or even thwart success entirely. The product portfolio may have too many unrelated products that lack any synergy and, thus, too many different parts and processes. Even within a focused product portfolio, there may be a needless and crippling proliferation of parts and materials. The specified parts may be hard to get quickly. The products and processes may have too many setups designed in. Quality may not be designed into the product/process which results in disruptions in the flow when failures loop back for correction. The product/process design may not make optimal use of CNC as most CNC equipment is used in a batch mode, not flexibly.

Existing designs may lack synergy and have too many parts and processing steps

---

DEVELOPING PRODUCTS FOR BTO&MC

To be successful at designing products for BTO&MC, companies must proactively plan product portfolios for compatibility with BTO&MC, design products in synergistic product platforms, design around *aggressively* standardized parts and raw materials, make sure specified parts are quickly available, consolidate inflexible parts into *very versatile* standardized parts, assure quality by design with concurrently designed process controls, and *concurrently engineer* product platforms *and* flexible flow-based processes.

Further, product development teams need to eliminate setup *by design* by specifying readily available standard parts and standard tools (cutting tools, bending mandrels, punches, etc.), designing versatile fixtures at each workstation that eliminate setup to locate parts or change fixtures, and making sure part count does not exceed available tool capacity or space at each work station.

Finally, products must be designed to maximize the use of available programmable CNC fabrication and assembly tools, *without expensive and time-consuming setup delays.*

PRODUCT PORTFOLIO PLANNING

The very first step in product development is deciding what to develop. Product portfolio planning is the *proactive* determination of what products to develop.

Prioritize product development efforts to focus resources on the *most profitable products* to maximize the ratio of total gain over total cost. Total gain includes the potential gains for all variations and derivatives over time. Total cost includes all overhead cost to be incurred by the candidates. Toyota uses *multiproject management to optimize the sharing of resources across multiple, concurrent projects.*[1]

Prioritize customers and customer segments and focus on the ones with the best current and future opportunities for profit and growth.[2]

Allocate resources proportional to the challenge (count and mix) for new technology, risk, cost, time, and so forth.

Know the true profitability of all product variations to deliver the highest return from given resources.

Develop profiles that describe the characteristics of the most profitable customers and customer segments. Use these profiles to help prioritize product development opportunities, scrutinize unusual or low-volume sales, and rationalize away high-overhead, low-profit products.

Focus on product *families* that can benefit from synergies in product development, operations, and supply chain management. Identify opportunities where products could benefit from Build-to-Order of standard products and the Mass Customization of specials. Toyota produces an average of seven different vehicles on each platform, which maximizes reliability across vehicle types. Platforms are designed to be a basis for these vehicles for up to 15 years.[3]

Structure product families so all products share standard parts and flexible equipment

Plan the portfolio around versatile products and flexible processes that can (a) easily satisfy the anticipated range of product breadth and customization, (b) have the potential to satisfy even broader ranges of customer needs, and (c) easily adapt to evolving trends and upgrade possibilities

Plan the portfolio to avoid the commodity trap, with its inherently low profits, and evolve to innovative products designed for low cost and sold at high profit. Set aside resources and budget for *breakthroughs*[4] in *blue oceans* where innovative products would create uncontested markets. Based on analysis of 150 strategic moves spanning a hundred years and thirty industries, the *Blue Ocean Strategy* recommends using *value innovation* to pursue differentiation and low cost simultaneously,[5] a goal that is also supported by Mass Customization, which can achieve premium prices at lower cost.

Use all the above techniques to prioritize the portfolio to maximize the return from available resources. Make sure *all* approved projects will have enough resources *with the right mix of talent,* for thorough up-front work and consistently methodical work throughout the project.

Make all product portfolio decisions rationally and objectively. Avoid temptations to base portfolio decisions on automatic upgrades, enticing market opportunities, exciting technology, whims, pet projects, competitive precedents, and so forth. Do not allow portfolio decisions to be influenced by politics, powerful backers, attachments, inertia, fears, unrealistic expectations, folklore (e.g., "I heard somewhere that . . . "), and so forth.

Don't limit "the portfolio" to only *products.* Invest in research and module development that can benefit a broad stream of future products. At Toyota, *"managers might spend over half their time on a portfolio of ideas and projects."*[6]

Make total cost numbers the cost basis of all product portfolio decisions, as discussed in Chapter 12.

Relieve resource demands caused by *existing* products by planning versatile new products that can also replace hard-to-build/low-profit products that the company may not be able to rationalize away.

DESIGNING PRODUCTS FOR BTO&MC

Product Definition for BTO&MC

Once the product portfolio has been planned and the product families structured, those products need systematic *product definitions* so that they will satisfy the "voice of the customer."[7] For BTO&MC, *families* of products need to be defined to meet the "*voices* of the customers."

QFD methodically defined products to satisfy "the voice of the customer"

QFD (Quality Function Deployment) is a methodical procedure for *defining* products and systematically translating the voice of the customer into product design specifications and resource prioritizations.[8] Its strength is to translate objective and subjective customer wants and needs into objective specifications that engineers can use to design products.

The inputs to QFD are objective targets and subjective customer preferences. The outputs are product specifications and resource prioritization; see Figure 10-1.[9] The values in the "design specs" rows are the actual values and ranges that engineers will use to design products; having ranges for design specifications is a unique aspect of QFD for mass customization. The "resource prioritization" row is the percent of the design team's effort that should be spent on each aspect of the design. Using this prioritization will ensure that the design team devotes its efforts toward features that customers want the most. And, thus, they will not waste effort "polishing the ruby" more than the customer wants it polished. This can be a temptation when engineers are personally excited about certain new technologies that are being used in the product.

QFD inputs are specs and subjective preferences. Outputs are design specs and resource prioritization

"Customer preferences" are listed, one per row, in words that are meaningful to the customer, both objective *and* subjective. One of the most valuable aspects of QFD is also the easiest to obtain: simply asking customers for the "relative importance" of their preferences, from a low of 1 to a high of 10 (there may be more than one "10"). The result is a prioritized ranking of what is most important and what is least important to customers. This ranking is very important to help neutralize any internal biases toward exciting technology, fads, "pet" features, or misconceptions about customer preferences

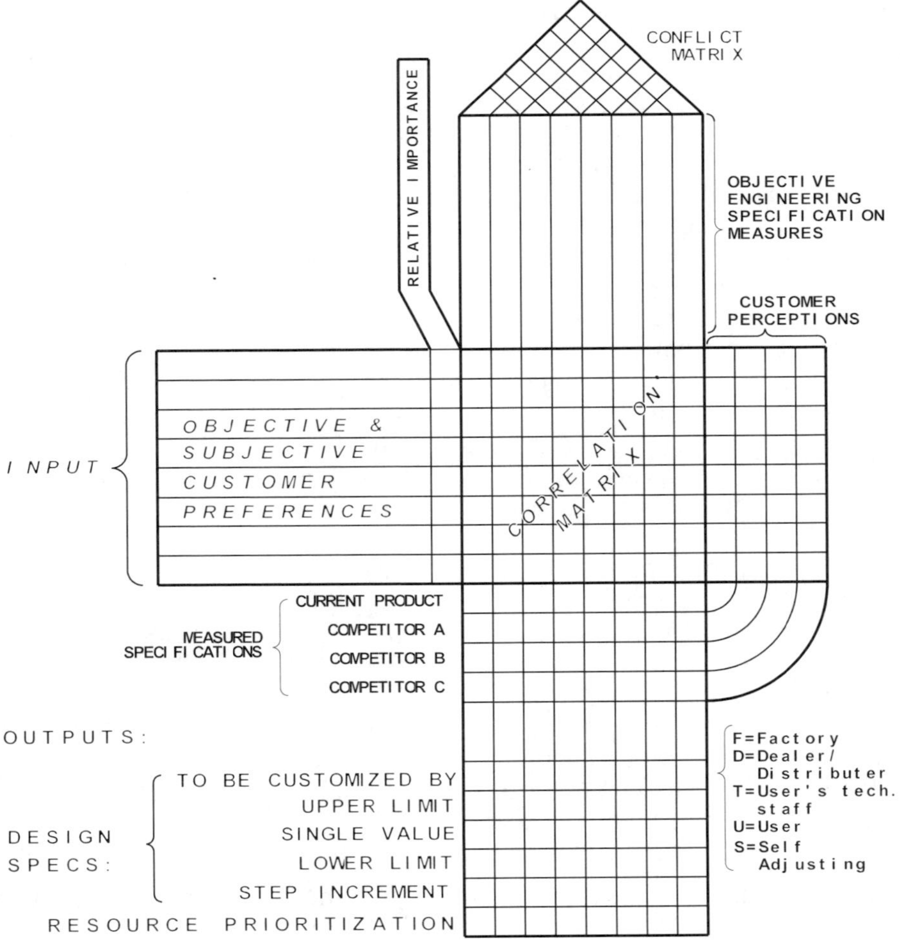

Figure 10-1 QFD for Mass Customization

In the "customer perceptions" region in the chart, customers are then asked how they would rank the current product against the competitors' products. Comparing this competitive ranking with the customers' "relative importance" ranking can give valuable insight into resource allocations. The full QFD process uses these inputs to calculate an optimal percentage of the engineering budget to devote to various tasks, but even a subjective comparison of customer importance to competitive ranking would be very insightful.

The most important analysis to determine for resource prioritization is to look for important customer preferences that the competition does better. But glancing at two uncorrelated lists will not provide much insight into resource prioritization. The author developed to following method to help clients comprehend the relationship to all preferences and competitive ratings to prioritize plans of action: plot all preferences (on the vertical axis) vs. competitive position; diagonal lines will then segregate the data into *zones.* The most critical data point will be the one that has the *highest* customer importance and your *weakest* competitive ranking. The remaining data point zones are given priorities proportional to how close they are to that point.[10]

The "objective engineering specifications" are the quantitative specifications that will eventually have target numbers for discrete products or ranges for mass-customized products. The objective measures would be in engineering units that quantify dimensions: force, torque, energy, decibels, and so forth. All of those objective measurements are actually measured for the current product and several competitors and entered in the chart under "measured specifications."

In the center of the chart, the "correlation matrix" correlates which customer preferences are affected by which engineering specifications. Symbols are placed in the square to indicate the type of correlation. Usually two to four rankings are listed: "positive" or "negative" correlation; "strong," "some" or "possible" correlation; or "strong positive," "medium positive," "medium negative," "strong negative" correlation. Roughly half the boxes should be checked. Results become less valuable as one approaches either extreme of all boxes checked or no boxes checked.

The "conflict matrix" tabulates any specifications that might be inherently in conflict with others, to aid in making tradeoffs of one feature versus another. An example would be a more powerful car engine that may adversely affect handling and mileage because of its extra weight.

The various cells in a QFD chart are used to make calculations and normalize them to useful percentages. One of the "bottom lines" is the design target row (labeled "design specifications"), which would be single values for discrete (non-customized) products and ranges for mass-customized products. The other is the resource prioritization which can be given in a percentage of the engineering budget or hours spent achieving the various design targets.

In the "design specifications" area the first row indicates who will be doing the customization: the Factory (F), at the Dealer or Distributer (D), at the point of sale by the user (P), by the users technical staff (T), by the actual User (U), or Self adjusting (S). Other symbol representing other customization scenarios can be used as appropriate. A blank or a question mark could mean "to be determined."

These entries will identify where and by whom the customization will be done. The design team will have to design the product for compatibility with this customization strategy. Factory customization would rely on flexible manufacturing to be able to quickly build any configuration that is called for. Dealer/distributer customization would have to take into account the tools, parts, patience, and skill level present at *all* dealers and distributers. User customization would have to be designed for the anticipated range of users skill level, patience, and tools needed, if any. A variation of user customization would be a situation common in large companies where technical staffs or departments would perform the customization, for instance, configuring computers for all office workers or customizing production tools for factories. Self customizing products literally adjust automatically to "fit" the user.

In the next three rows, the "single value" row which would be used for those engineering specifications that can be expressed by a single value and not a range of values. The upper and lower limits specify the range of dimensions for applicable specifications. The "step increment" would specify the steps or increments that could be appropriate for values within the range. A "zero" entry would indicate infinitely variable dimensions between the upper and lower limits.

Concurrently Engineer Products and Processes

In general, *concurrent engineering* is the practice of concurrently developing products *and* their manufacturing processes. If *existing* processes are to be utilized, then the product must be design for these processes. If *new* processes are to be utilized, then the product *and* the process must be developed concurrently.[11]

> In BTO&MC, product development teams concurrently engineer families of products and flexible processes

Product development for BTO&MC expands this definition to concurrently engineering *families* of products and flexible processes, including the following: developing products in synergistic families; specifying the absolute minimum variety of parts and materials; choosing readily available parts and materials; designing products for no setup; and designing for programmable CNC equipment.

In order to do this, product designers must not only thoroughly understand all the individual manufacturing processes, but also understand the philosophy, principles, and implementation of build-to-order and mass customization.

Products must be developed in synergistic product families, which will all be built on the same flexible line. Each family needs to be designed so that:

(1) All the parts can be distributed at all points of use, which is encouraged by aggressive standardization of parts (Ch. 4) and material variety minimization (Ch. 5). Too many parts would either clutter and confuse the work areas or force the parts to be *kitted* for a batch of products to be made, which is contrary to the one-piece flow of all the flexible paradigms. An important aspect of part standardization is fastener standardization, which is usually the easiest to do and can provide significant benefits. If screws can be standardized to one type at each work station, then *autofeed* screwdrivers can be utilized, which automatically orient screws and feed them through a flexible hose to a powered screwdriver. Of course, in order to arrange to have a single type of fastener at each station, the design team will have to be concurrently designing the entire production flow.

(2) Products and processes need to be designed so that there will be no setup for *any* part or product in the family (Ch. 8). This includes any setup to (a) find, kit, load, or replace parts or materials; (b) change and position dies, molds, or fixtures; (c) change tools; (d) load, position, clamp, or calibrate workpieces; or (e) adjust machine settings or calibrate machinery.

(3) Equipment programs can be located and downloaded instantaneously and manual instructions can be found, displayed, and understood quickly.

(4) Parts and materials have been chosen so that they can be resupplied spontaneously (Ch. 7). This includes specifying readily available parts and materials, designing around aggressively standardized parts, specifying raw materials to be cut on-demand from the same standard input stock sizes.

(5) Parts can be pulled quickly into assembly on-demand (Ch. 6). This includes finding suppliers who can build parts on-demand or working with in-house manufacturing to establish in-house capabilities.

The rest of this chapter will discuss in more detail, the two main aspects of this: designing for on-demand production and designing for a spontaneous supply chain.

DESIGNING PRODUCTS FOR ON-DEMAND PRODUCTION

Problems with "Un-lean" Product Designs

Existing product designs may not be suitable for BTO&MC if the following conditions exist.

There may be too many different products and, thus, too many different parts to distribute at all the points of use. Even within the ideal group of products, there may be a needless and crippling proliferation of parts and materials. The specified parts may be hard to get quickly.

Quality may not be designed into the product/process which results in disruptions when failures loop back for correction.

The product/process may have too many setups designed in. Product/process design may not use CNC flexibly, which is common since most CNC equipment is used in a batch mode, not flexibly.

Supply chains may be too geographically dispersed for just-in-time flows into factories. This can be even more challenging for BTO and mass customization which pull parts from suppliers through frequent and rapid deliveries.

Strategies to Develop Products for BTO&MC

Plan new product portfolios and structure existing products into synergistic product families and compatible production lines/cells.

Rationalize existing products to eliminate or outsource products and product variations that are not compatible with BTO&MC strategies, being careful to retain low-volume products that are synergistic with others in their family.

Structure product families so all products share standard parts and flexible equipment. All the products within a flexible cell should be compatible with respect to part standardization, raw material standardization, part/material availability, spontaneous supply chains, modularity, setup elimination, and flexible processing, including CNC machine tool operation.

Look for part/product families with the following characteristics:

- High *output* variety and potentially low *input* variety. Even if the input variety is currently too high, it could be reduced by standardization.
- Trends indicating increasing product variety over time.
- High labor cost; increasing labor shortages; significant or erratic overtime; excessive firefighting.
- High setup costs, low equipment utilization, and delays.
- Unreliable forecasts, which will get exponentially worse as product variety and market volatility increase.
- Inventory problems: excessive carrying costs, obsolescence, write-offs, out-of-stocks, and so forth.
- Delivery response pressures or opportunities
- The most difficult customizations.
- High visibility, the success of which will help spread implementations to other product lines.

Designing Products/Processes for BTO&MC

For lean production, build-to-order, and mass customization, products need to be designed so that:

All the parts can be distributed at all points of use, which is accomplished by (a) *aggressive* standardization because too many parts would either clutter and confuse the work areas and (b) dock-to-line part distribution to avoid *kitting*, which is a setup that is contrary to the one-piece flow of all the flexible paradigms.

Screws are standardized to one per work station, to avoid confusion and permit the use of autofeed screwdrivers., which automatically orient screws and feed them through a flexible hose to a powered screwdriver. To accomplish this, the design team will have to be concurrently designing the entire production flow.

Products and processes need to be designed so that there will be no significant setup for *any* part or product in the family. This includes any setup to: find, kit, load, or replace parts or materials; change and position dies, molds, or fixtures; change tools; load, position, clamp, or calibrate workpieces; or adjust machine settings or calibrate machinery.

Equipment programs can be located and downloaded instantaneously and manual instructions can be found, displayed, and understood almost instantaneously.

Parts and materials have been chosen so that they can be resupplied spontaneously by specifying readily available parts and materials, designing around aggressively standardized parts, and specifying raw materials to be cut on-demand from the same standard input stock sizes.

All parts can be pulled quickly into assembly on-demand. This includes specifying suppliers who can build parts on-demand or working with in-house manufacturing to establish in-house capabilities.

Sheet metal, plate, and bar stock mounts should all be on the *same side* of the metal, to support *future* standardization programs and weight reduction directives. Thus the metal thickness can be changed in the future without change orders.

Designing for No Setup

Part Setup. Product design has a profound effect on part setup. Excess proliferation of parts complicates internal part distribution and may make it impossible to have all parts available at all points of use. Even a moderate excess of part types will cause setup delays to distribute, find, and load parts into manual or machine bins. A serious excess of part types may make it necessary to kit parts for every batch, which is a significant setup. Part "prep" is another setup that can be avoided by specifying parts that are already prepared properly for the product and the processing equipment.

> Product design has a profound effect on setup

Fixturing Setup. Designers can eliminate fixturing setup by designing parts for versatile fixturing (Ch. 8) which, if not already in place, may have to be *concurrently* designed with the parts. Tool plate or guide rail setup for printed circuit boards can be eliminated by laying out all circuit boards on the same size panel (one large or several small) and "de-paneling" them later, as done at Tektronix and Nokia.[12]

Tool Setup. Designers can eliminate tool change setup by designing parts around common tools (cutting tools, bending mandrels, punches, etc.), ideally one tool that never has to be changed. If multiple tools are required, designers must keep tool variety well within tool changing capacity for the whole line.

Instructions. Design for Manufacturability (DFM) principles can greatly simplify assembly.[13] Designers need to work with manufacturing engineers to concurrently develop simple assembly procedures that can be understood in a few seconds either on a computer screen or on paper instructions that can be quickly located and understood.

Designing for CNC

Computer numerically controlled machine tools (hereafter referred to as "CNC") offer vast opportunities to eliminate machining setup, as discussed in the section, "CNC to Eliminate Machining Setup," in Chapter 8. CNC machine tools include metal cutting equipment (mills, lathes, etc.), laser cutters, punch presses, press brakes, printed circuit board assemblers, and basically any production machine controlled by a computer. Designers need to understand enough about CNC operation to use the versatility of CNC to eliminate setup.

> Designers need to understand how CNC can make parts flexibly and eliminate setup

Grouping parts. The first step in designing for CNC is to structure compatible groups of parts to be processed in each CNC machine – this was originally labeled *group technology.*[14] Of course, this must be based on the overall manufacturing strategy (Ch. 13) and flow of parts and products (Ch. 8). This evolves from a serious *concurrent engineering* activity in which the grouping and flow of parts are a key element of the design team's responsibilities. The grouping may determine the type of CNC needed, or existing machine tools may influence the grouping.

Understanding CNC. After thoroughly understanding of the range of parts to be made by each CNC, the designers will need to understand the capabilities and limitations of the equipment. This can be accomplished by studying the equipment specifications *and* talking to CNC operators. In fact, CNC operators should actually be on the design team to optimize the design and processing plans. Of course, the ultimate understanding would come from actual CNC operational experience, either through prior jobs or a job rotation program.

Eliminating CNC setup. The versatility of CNC provides unique opportunities for eliminating setup if parts are designed properly. Ideally, all operations should be able to be performed on one machine in a single fixturing. Multiple machinery require extra fixturing setups. Even if multiple specialized machines have a higher speed rating, the total flow time through all operations *including setup* is what is important. The value of eliminating these setups may justify a more sophisticated CNC, compared to more setups on many cheaper machines.

In order to process parts in a single fixturing, designers need to specify a suitable *datum,*[15] from which all dimensions are referenced *and* which is suitable for clamping to a milling center table or lathe chuck. The part also must be designed so that all the operations can be done in this fixturing.[16]

DESIGNING FOR SPONTANEOUS SUPPLY

Designing around Standard Parts

Standardization of parts (Ch. 4) and material variety reduction (Ch. 5) are the most important design contributions to the feasibility of spontaneous supply chains. *Aggressive* standardization can enable the easiest technique of the spontaneous supply chain: steady flows of very standardized parts and materials (Ch. 7). If there are too many different parts and material types, steady flows cannot be arranged because of the variety and unpredictability of demand. The total cost value of standardization and its contribution to the BTO&MC business model should motivate engineers and materials organizations to implement such aggressive standardization.

> Product designers must design around *aggressively* standardized parts

Most part proliferation happens because engineers do not understand the importance to supply chains and operations and, even if they did appreciate this, do not know which parts are best to choose. The effective procedures presented in Chapter 4 show how to generate standard parts and materials lists.

The usual operational mode in a part-proliferated environment, is to order the parts ahead based on forecasts, either for scheduled batch production or to have forecasted parts available for on-demand assembly (the Dell model). If part variety is truly unavoidable or parts are to be mass-customized, then they should be built on-demand using all the BTO&MC presented in this book.

Designing to Reduce Material Variety

An important task for design engineers is to specify not only the functional specifications of materials but also the order size and how they get cut into various parts. As Hoffman Engineering found out, ordering 600 different shapes of sheet metal was a logistics nightmare, but ordering became much easier when the company shifted to just a handful of basic types which were then cut on-demand as they were needed.

Another way engineers can reduce material variety is to specify a single tolerance or grade, instead of multiple grades. As pointed out in Chapter 4, any perceived cost increase of shifting all materials to the higher grade will be more than compensated by the total cost savings and value of the BTO&MC business model.

Families of parts should be designed so that every CNC machine tool uses the absolute minimum of raw material types, ideally just one per machine.

Multiple lengths of linear materials should be obtained by ordering reels or long lengths of materials, which are then cut to length on-demand, as shown by the "CNC cut-off" machines in Figure 8-1.

Inflexible raw parts, like castings, forgings, extrusions, custom silicon, and bare printed circuit boards, should be *consolidated* (Ch. 5) into very versatile parts that can be used for many functions on a wide range of products. Programmable chips should be programmed using the same standard "blank."

Designing Around Readily Available Parts/Materials

A spontaneous supply chain depends on readily available parts and materials. Therefore, it is an important aspect of engineers' jobs to specify parts and materials that are readily available. Usually, design engineers choose parts based on functionality and maybe quality. But for BTO&MC, availability is equally important.

An important aspect of engineers' jobs is to specify parts and materials that are readily available

Parts and materials that can be obtained from multiple suppliers tend to have better available, in general, and are also more likely to be standard parts. On the other end of the spectrum, parts and materials available from only a single source may some day be hard to procure at all. On the lecture circuit, the author often finds companies that have had parts that became obsolete before the product was released!

Rapid delivery is important for BTO&MC, so design teams should specify local suppliers who will be able to supply materials on-demand. Many companies preclude the ability to obtain parts and materials spontaneously by buying from supposedly "low cost" suppliers from another continent. Similarly, low bidder suppliers probably are not going to be able to deliver on-demand and will most likely not have adequate quality either (Ch. 7).

The guiding principle is to select parts and material to be readily available, delivered on-demand, and the lowest *total* cost, which includes material overhead, ordering, expediting, routing shipping, expedited shipping, incoming inspections, kitting, internal distribution, and all the costs of locating alternate sources of supply to counter availability problems.

OUTSOURCING ENGINEERING

As with the outsourcing of manufacturing (Ch. 6), many managers are intrigued with the prospect of outsourcing engineering as a way to "save cost" on product development with foreign engineers whose wages are a fraction of domestic engineers. Further there is the allure of speeding up product development by keeping engineering working in three shifts a day with work passed off around the world every day.

This might work with independent activities such as call centers or insurance or loan processing, but product development is a highly interactive and integrated team activity. Product development is most effective and efficient when products are designed by complete multifunctional teams that are "co-located" with manufacturing operations and close to customers and vendors.[17]

Any proposals to save cost must be based on minimizing *total cost.* Chapter 11 emphasizes many the many costs that make up the total, of which labor cost is just one element. Chapter 6 argues that some of the consequences chasing cheap manufacturing labor – lower labor efficiency and poorer quality – might cancel out the labor rate savings; adding in other costs

> Engineering expenses are best minimized by making the *whole process* more efficient, not going abroad for cheap engineers

could push results into a net loss. Similarly, the best way to lower product development expenses is to *maximize the efficiency of the whole process* through concurrent engineering (with co-located teams) and design for manufacturability (with continuous interactions with manufacturing people).[18]

A more efficient product development process consumes fewer labor hours to (a) develop new products and (b) to solve problems with the product launch, ramping up to volume targets, achieving quality and productivity targets, implementing engineering change orders, and all the firefighting associated with these activities, *which are really avoidable* with an efficient product development process.[19]

Optimal product development practices not only minimize *engineering* costs but also can substantially reduce *product* costs, which is one of engineering's primary goals. Spreading teams out geographically – with people who may not even be working at the same time – compromises that teamwork that is necessary to optimize product architecture, which determines 60% of the product's cost (Figure 11-5). And outsourcing subsystem and part design to remote engineers can produce products that are hard to integrate, have part interaction problems, or miss out on synergistic system opportunities.[20]

> Spreading engineers all around the world compromises the teamwork that develops the best products

Physically integrated teamwork is even more important when products need to be designed for lean production and build-to-order. Design teams need to work closely with manufacturing people to design for no setup, design for CNC, and concurrently designing versatile product families and flexible processes. For mass customization, engineers need to work closely with manufacturing, marketing, and the sales force to establish CAD templates that can automatically generate CNC programs (Ch. 9).

The transition to outsourced engineering incurs a certain cost and may cause significant consequences. There will be a certain cost to finding engineers, arranging to hire and pay them, and preparing and transferring drawings and CAD files. The effort to implement these changes will probably consume the available "change energy" and preclude or compromise other more valuable improvement efforts, like DFM, robust design, six sigma, build-to-order, and so forth.

The consequences can be even more severe if the "cost savings" from outsourcing engineering are to come from layoffs. In addition to the general consequences of layoffs (Ch. 13), there are special consequences for product development. Once word spreads, morale and efficiency will drop and the best engineers may leave, thus weakening engineering efforts and making product development even less efficient. Engineers who quit or are laid off will probably take with them valuable skills and knowledge, some of which may be unique and thus weaken teamwork. Continuity may suffer for ongoing projects. The remaining engineering capability will be less efficient with such losses and teamwork will be less effective because of missing talent, knowledge, and diversity.

> If the outsourcing is intended to "save" cost with layoffs, the *real* cost could be severe

Which Engineering Could be Outsourced

Certain engineering tasks *could* be outsourced, as long as they help rather than hurt the overall product development effort. Research may be more suitable than Development. Outsourcing certain tasks could help new product development efforts. Outsourcing certain other tasks could relieve domestic engineers from distracting tasks and thus indirectly benefit new product development efforts.

Tasks that support domestic new product development include: computational intensive research tasks, as long as they do not specify or imply product architecture or part design; computational intensive analysis of performance, strength, dynamics, heat flow, reliability, design of experiments, robust tolerances, etc.; materials research; structuring and adapting parametric CAD templates and CAD/CAM programs; and literature or web searches in general.

Tasks that usually distract new product development efforts include the following activities on current products: engineering change order processing; changing drawings; analyses of problems on current products; cleaning up or convert documentation, drawings, part numbers on acquired products; maintenance engineering and conversions from obsolete materials for older legacy products and spare parts; cleaning up documentation for outsourcing the manufacturing of legacy, spares, and oddball products and parts (as recommended in Chapter 6); conversion of drawings to Geometric Dimensioning and Tolerancing (GD&T); or conversion of drawings to metric or dual dimensioning.

DESIGN FOR MANUFACTURABILITY

The following are some DFM guidelines that support BTO&MC by simplifying the design and minimizing the incoming variety of parts and materials.[21]

Avoid Arbitrary Decisions

Successful product development requires that designers proactively design products for a BTO&MC environment using the principles presented herein. But this may be difficult or impossible if designers make *arbitrary decisions* that preclude implementation of these principles. *All* design considerations – functionality, cost, quality, *and* flexibility – must be taken into account *early*. If they are not, then designers will probably make many arbitrary decisions that will make it much harder to satisfy omitted design considerations later.

> Design motto:
> *No arbitrary decisions!*

Arbitrary decisions that complicate product development, in general, involve product definition, concept/architecture optimization, styling, outsourcing strategy, milestone deadlines, order of design, tolerances, and overhead allocation algorithms.

Arbitrary decisions that complicate BTO&MC product development include: not concurrently designing products and flexible processes; not designing products in families or platforms; product architecture not supportive of BTO&MC; not designing around standardization; not selecting parts and materials that can be pulled into assembly from a spontaneous supply chain; not designing for zero setup; and an outsourcing strategy that arbitrarily outsources all parts or all sub-assemblies.

Avoid Left/Right Hand Parts

Avoid designing mirror image, or "right and left hand" parts. Design the product so the same part can function in both right or left hand modes. Another version of this principle is to use identical parts instead of different ones for top and bottom, front and back, and so forth. If identical parts appear unable to perform both functions, add features to both right and left hand parts to make them the same.

This provides BTO&MC with four important benefits that should far outweigh any *perceived* increase in cost that might occur to make the left hand and right hand parts the same.

1) Part variety is cut in half for each pair of parts that can be made the same.

2) Purchasing quantities are doubled, thus resulting in twice the purchasing leverage which can result in better cost and pull-based delivery.

3) Tooling costs are cut in half, which can be a very important consideration for complex permanent dies and molds for casting and plastics.

4) Setup can be avoided to change dies and molds from left to right versions.

Specify Prefinished Material

Many manufacturing steps can be eliminated by ordering *standard* prefinished material that is prepainted, preplated, embossed, expanded, anodized, or clad with a different surface coating. Painting operations for sheet metal can be eliminated by switching to stainless sheet metal. This may be justifiable if the *total* cost of painting sheet metal is considered. Prefinished material can be ordered with the finished side protected by adhesive backed paper that can be peeled off after assembly.

Design Machined Parts to be Made in One Setup

Machined parts should be designed so that all the operations can be done in a single fixturing, which will routinely achieve the best tolerance of the machine tool, usually +/- .001" or better.[22] If all operations cannot, it is important that the *most* critical dimensions are cut in the same fixturing. However, removing a part to reposition for a subsequent cutting lowers accuracy of these critical dimensions because the tolerance will then depend on the accuracy of the machine tool *and* the accuracy of the subsequent positioning, which is usually much worse than machine tool accuracy.

Design for Easy, Consistent, and Accurate Alignment

For parts that need to be assembled, design the interface for easy, consistent, and accurate alignment with pairs of inexpensive, but tight-tolerance, dowel pins. Matching tight-tolerance hole diameters can be made with reamers. To eliminate the tolerance match problem between holes, use one *round* pin to locate in "x" and "y" dimensions and a *diamond* pin to locate the angle from the round pin.[23] The diamond is precision ground to locate in the angle direction, but is relieved in the direction of the hole spacing. Although this technique was developed to locate tooling, it can be useful for aligning parts for assembly. This technique allows for accurate alignment at low cost since readily available dowel pins cost as low as $10 for .0002" accuracy and reamed holes can also produce "sub-mil" tolerances (less than .001") inexpensively. The quick and precise assembly provided by this technique supports modular design (next).

MODULAR DESIGN

The previous chapter discussed three ways to mass-customize products: modular, adjustable, and dimensional customization. It is important to keep in mind that modularity is only *one* way to customize products, which works best for inherently modular products, the most obvious example being personal computers.

Modular design is a design technique in which functions are designed in modules that can be combined in subsequent designs. A related concept is "reusable engineering" where portions of previous designs become the basis of new designs.

The benefits of reuse and modular design are (1) less engineering effort, (2) more commonality amongst models for simpler supply chains and more flexible operations, and (3) better reliability from using proven parts and designs.[24] Reuse can be facilitated when designs are done on CAD because it is easy to bring up the previous drawing and copy or export the desired details.

When this concept is extended to manufacturing, products may actually be assembled from "building block" modules. Care should be taken in engineering to design standard interfaces for optimal flexibility. Catalog hardware often has standard interfaces, for example, motor mounts, shaft couplings, fluid power and electrical connections, and so forth.

An effective modular design strategy would be to intentionally create portions of designs that have general usefulness for whole families of future products. For instance, machinery could have modular bases, frames, drives, gearboxes, controls, and cabinetry which could be combined with other modules.

Pros and Cons of Modular Design

Plug-together modularity may simplify some assembly and accept third-party plug-in modules, such as in personal computers. Modularity can be a key element in a postponement strategy (Ch. 9).

Time-to-market for "new" products can be quicker if existing modules have already been designed, documented, debugged, and certified. Delivery may be quicker if products can be assembled from standard modules. Product assembly and test can be streamlined when modules can be assembled and tested separately. Modularity can lead to broader product lines and be the foundation of a proactive upgrade strategy. Further, modularity can simplify maintenance and field service when defective modules can be replace and then repaired off-line.

Modularity can lower inventory levels and overhead cost compared to several different versions of the integrated part or product. Widely-used modules have less vulnerability to lead time delays.

The cost of modularity can only be determined on a total cost basis. Engineering costs will be lower if existing modules can be utilized, but may be higher if new modules and interfaces have to be designed. The cost of manufacturing the interfaces may be more than an integrated product, but using versatile modules used on many products may lower costs through economies-of-scale. Selecting existing modules would save the costs and delays of testing and debugging, compared to an entirely new integrated design.

Similarly, product testing may be easier if existing modules have already been tested, debugged, and certified. Diagnosis efforts may be easier with less functionality on each module. However, if all the modules in the product are new, then tests will have to be devised and all the modules will have to be tested.

Modularity has some drawbacks too. Product development expenses may be increased by the necessity to design modular interfaces. Modularity may not be feasible for inherently integrated products. Modular interfaces may compromise functionality by adding weight, weakening structures, or causing signal interruptions at low voltage. In electronics, the extra connectors needed for modularity may cause reliability problems and slow down electronic signals, thus degrading performance. Module interfaces may result in undesirable visual aesthetics (such as seams, joints, etc.) or undesirable acoustics (such as squeaks and rattles). A subtle drawback of modularity is that since modules are such an obvious means to create product variety that it absorbs the entire focus and exclude adjustability and CNC machine tools cutting dimensions on-demand (Ch. 9). Modularity alone may not be able to cover a wide enough range of customer needs.

Modular Design Principles

Module investment. Companies should *invest* in the design of modules that are versatile enough for *many* products. Individual projects may not have the budget or resources to develop modules for many products. Modules should be versatile enough to have general usefulness for many current and future products.

Total cost basis. Do not look at module cost at the level of one project; look only at the *total* cost of all the products that will utilize the modules.

Reusable engineering. Modular design is not limited to physical modules, but also could be reusable engineering or software code.

Interfaces and protocols must be optimized at the system architecture level.

Standardization. For maximum usefulness, modules must have standard interfaces. Use industry-standard interfaces if available; if not, develop very versatile interfaces. Toyota engineers have a strong sense of the vehicle as a system and consequently focus a great deal of skill and energy at the design interfaces.[25]

Documentation. Document for modularity, with optimal CAD layer segregation, bills-of-materials, software objects, etc.

Debugging. Minimizing debugging cost by using existing modules that have already been debugged. Some leading companies, like Hewlett-Packard, feel that this is the best way to produce bug-fee software. With a high percentage of reuse, debugging efforts can focus on *new* aspects.

Consistency. Resist the temptation to "improve" modules unless the improvement is substantial and justifies the cost and time of design changes, production changes, evaluation, debugging, and re-certification/qualification. Do not attempt "cost reduction" on modules unless the total cost saving for all applications justifies the development of a new-generation module.

OFF-THE-SHELF PARTS

For *standard* parts with little variety, designers can greatly reduce development time and cost in addition to reducing product cost by never designing a part that is available out of a catalog. Rarely can anyone save any money or time by designing and building parts that are available from catalogs.

Many designers are often misled by their own accounting systems into thinking that internally produced parts can be designed and built less expensively than catalog hardware. Accounting systems rarely report the *total* cost of parts (see Chapters 11 and 12). Usually when designers look up part "cost" in an accounting database, they are only provided the material cost and labor cost with very little "burden" (overhead). Considering the enormous overhead in some companies, "part" costs may be understated substantially.

> Paradoxically, designers may have to choose off-the-shelf parts *first* and design *around* them

Using catalog parts has an inherent standardizing effect on a company's parts list because many catalog parts are used broadly enough to become *de facto* standards.

Paradoxically, designers may have to choose the off-the-shelf parts *first* and design the product *around* them, or else they may make arbitrary decisions that may preclude their use. But incorporating catalog parts early into the design will greatly simplify the design and the design effort.

Off-the-shelf parts are less expensive to design considering the cost of design, documentation, prototyping, testing, debugging, in addition to the overhead cost of purchasing all the constituent parts and the cost of non-core-competency manufacturing.

Off-the-shelf parts save time considering the time to design, document, administer, and build, test, and fix prototype designs.

Suppliers of off-the-shelf parts are more efficient at their specialty, because they are more experienced on their products, continuously improve quality, have proven track records on reliability, design parts better for DFM, have dedicate production facilities, offer standardized parts, and sometimes pick up warrantee/service costs.

Off-the-shelf part utilization helps internal resources focus on their *real* missions which are designing products and building products.

When to Use Off-the-Shelf Parts

Designers should specify off-the-shelf parts when the following conditions exist:

- **Parts are standard.** Off-the-shelf parts are especially applicable when they are *de facto* standards, always available, and can be purchased from multiple suppliers.

- **Volumes are high.** Standard parts can be manufactured in high volumes. Even companies with small order quantities can benefit from suppliers' economies of scale. Very high volume parts may be mass produced on dedicated lines, which further lowers cost and improves quality.

- **Quality/reliability are important.** Good suppliers will be able to focus more on their products and thus achieve higher quality levels than occasional production at an assembler's plant, especially when suppliers have been making them for a long time and are already "up the learning curve." Parts that have been produced for a while will have track records that can be used to predict and assure reliability targets will be met.

- **Specialized skill, expertise, or costly equipment.** Suppliers may have specialized skill and expertise that may be hard to equal in-house. High-volume suppliers may have costly equipment that can make parts faster, at lower cost, and at higher quality.

- **Different processing.** If certain parts require different processing than can be efficiently done in-house, it will be better to get those parts from off-the-shelf suppliers.

- **Growth or constrained capacity.** If in-house capacity is limited or growth is anticipated, off-the-shelf parts can relieve capacity constraints and make that capacity available for product growth.

- **Lowest total cost.** Off-the-shelf standard parts will usually have a lower total cost, even though (a) in-house cost systems may be misleading if they do not include all the overhead costs of in-house production or (b) in some applications, standard parts may *appear* to cost more than the "just right" sizing in Figure 5-3. But when the effects of purchasing leverage and material overhead are included, the standard parts will be the lowest cost solution for the company.

When *Not* to Use Off-the-Shelf Parts

In-house production of parts would be warranted under the following conditions:

- If the off-the-shelf parts are used in significantly different ways than intended.

- If they were designed and used predominantly in other industries with different demands and lifespans.

- If you are a very small part of the suppliers business which may make it hard to get help with the application or develop modified or custom versions. This situation also may result in undependable delivery when bigger customers "pull rank."

- If the supplier acts like a monopolist, charges high profits, won't help with the application, gives poor service, or may not be counted on for on-time delivery.

- If the supplier has not designed the product well for manufacturability and missed opportunities to lower cost and improve performance and quality. This is more likely for complex parts or subassemblies made in low volume.

- If you or your vendor/partners can manufacture it better: more consistently, at higher quality, and at a lower *total* cost

- When you want propriety control over manufacture, trade secrets, or intellectual property.

Finding Off-the-shelf Parts

The first task in finding off-the-shelf parts is to search of all the potential sources for the type of part needed. A common shortcoming here is to make a superficial search and then conclude that no one makes the needed parts. Then the designer feels justified in designing and building the part.

There are many directories and other sources that can show engineers where to find off-the-shelf parts including the Thomas Register, trade journal annual "directory issues," the yellow pages, internet searches, and trade shows.

Catalogs of off-the-shelf parts can be obtained from company headquarters, local representatives, reader service cards ("bingo cards") in trade journals, trade shows, company libraries, and peers private collections. Many web-sites now contain complete catalogs on-line.

There is a wealth of information in catalogs including specifications, selection guidelines, design guidelines, cross-references to other brands, listings of local reps, and policies on specials or custom parts.

Prices have generally not been published in catalogs for years. Usually prices are obtained on specific parts by asking the headquarters or local rep. Price lists for the entire line are more useful for doing "what if" analyses. They *can* be obtained with enough persistence. One method is to inform the supplier that in the "pricing stage" of the design, you use the catalog that has a price list and that these tentatively chosen parts often end up in the final design.

Designers can get some indications of relative availability by cross referencing other brands to reveal which sizes or models are supplied by the most suppliers. These are likely to be the standard parts with greater availability than parts only available from one or two sources. The best indication of availability is to find "quantity in stock" data. The size or model with the highest quantity in stock will probably be the most standard and have the best availability, not to mention the best pricing1.

After the needed part is located in a catalog, the designer will want to evaluate the actual part. The quickest way to get a part for evaluation is to ask for a *sample.* Samples are usually sent out immediately without charge unless the parts are very expensive or semi-custom in nature. This is quicker than purchasing and actually saves administrative expense at both ends. To maximize success in obtaining samples, be sure to mention yearly or lifetime projected consumption of the parts being considered.

ENDNOTES/REFERENCES

1. Morgan & Liker, *The Toyota Product Development System,* Chapter 4: "Front-Load the PD Process to Explore Alternatives Thoroughly."

2.Richard Koch, *The 80/20 Principle; The Secret of Achieving More with Less,* (1998, Currency/Doubleday), Chapter 4, "Why your Strategy is Wrong."

3. Morgan & Liker, *The Toyota Product Development System,* Chapter 4: "Front-Load the PD Process to Explore Alternatives Thoroughly."

4. Mark and Barbara Stefik, *Breakthrough; Stories and Strategies for Radical Innovation;* (2004, MIT Press);

5. W. Chan Kim and Renee Mauborgne, *Blue Ocean Strategy, How to Create Uncontested Market Space and Make the Competition Irrelevant,* (2005, Harvard Business School Press).

6. Matthew E. May, *The Elegant Solution*, (2007, Free Press), p. 41.

7. For a broader discussion of product definition including understanding customer needs and writing product requirements, see Section 2.11 (pages 60-67) in "Design for Manufacturability & Concurrent Engineering," by David M. Anderson, (2008, CIM Press).

8. John Hauser and Don Clausing, "House of Quality," *Harvard Business Review,* May-June, 1988; Reprint number 88307.

9. This figure was created for the author's seminars and first published in copyrighted class handouts. It was then published in *Agile Product Development for Mass Customization,* by David M. Anderson (1997, McGraw-Hill) and then in the Chapter 6, "QFD and Designing for Manufacturability and Customization," by David M. Anderson in *QFD Handbook,* edited by Jack B. Revelle, John W. Moran, and Charles A. Cox (1998, John Wiley & Sons).

10. David M. Anderson, *Design for Manufacturability & Concurrent Engineering; How to Design for Low Cost, Design in High Quality, Design for Lean Manufacture, and Design Quickly for Fast Production,* (2008, CIM Press, 448 pages); pages 62-63 show graphical plots of customer importances vs. competitive grades with importance zones.

11. Ibid., Ch. 2, "Concurrent Engineering,"

12. David Pringle, "How Nokia Thrives by Breaking the Rules," *The Wall Street Journal,* Jan. 3, 2003.

13. Ibid.

14. Charles S. Snead, *Group Technology; Foundation for Competitive Manufacturing,* (1989, Van Nostrand Reinhold).

15. The *datum* concept is a key element of Geometric Dimensioning and Tolerancing (GD&T), from the ANSI Y14.5 standard.

16. Anderson, *Design for Manufacturability & Concurrent Engineering,* Chapter 9, Guideline P14, "Design machined parts to be made in one setup."

17. Ibid,. Chapter 2, Concurrent Engineering.

18. Ibid.

19. Ibid. Sections 2.8 and 3.7

20. Ibid., Chapter 2, Concurrent Engineering, and Chapter 3, Designing the Product.

21. Ibid., Chapter 9, "Guidelines for Part Design."

22. Ibid., Chapter 9, Guideline P14, "Design machined parts to be made in one setup."

23. Ibid., Chapter 8, "Guidelines for Product Design," Guideline A3, "For critical alignment of parts use round/diamond pins." Also see Figure 8-1.

24. Ibid., Chapter 10, "Design for Quality, Guideline Q20: "Use proven parts and design features."

25. Morgan & Liker, , *The Toyota Product Development System,* Chapter 4, "Front-Load the PD Process to Explore Alternatives Thoroughly."

11

MINIMIZING TOTAL COST

The key to achieving the lowest cost product is to base all thinking and decisions on a *total cost* perspective. Unfortunately, the typical company cost system reports only material and labor costs. All other costs are called *overhead,* which is spread over corporate activities according to some arbitrary allocation algorithm (averaging), for instance, proportional to material, labor, or processing cost. And yet, all products do *not* have the same *overhead demands*. In fact, much can be done to lower overhead costs, which will be discussed below.

Many of the recommendations of this book, when viewed narrowly, may *appear* to be more expensive, and therefore may be resisted. Rationalization may appear to lower sales (Ch. 3); Standardization and consolidation may appear to require most products to use better materials than they need (Ch. 4, 5); Reintegrating unfamiliar processes may appear to be less efficient than using specialists (Ch. 6); The lack of competitive bidding (inherent in vendor-partnerships) may appear to raise procurement costs (Ch. 7); Small batches, approaching one, and dedicated flow lines using simple equipment may appear to be inefficient from a mass production standpoint (Ch. 8, 9). However, these programs will really lower the *total cost.*

This chapter presents several ways to substantially reduce cost in terms of total cost. The next chapter will discuss how to *measure* total cost to encourage optimal decision making and attribute the proper cost to all activities.

HOW *NOT* TO ACHIEVE LOW COST

First, it is important to dispel some myths about cost. Low-cost products do *not* come from volume alone; that is the Mass Production paradigm. With build-to-order and mass customization, products can be made at low cost virtually *independent* of volume.

Focusing only on parts and labor can lead to seriously counterproductive measures. Low-cost products do *not* come from cheap parts, which are often chosen because they *appear* to lower the *reported* material costs. And now on-line part auctions can effectively steer manufacturers to the lowest bidder. However, cheap parts will usually raise other costs: for quality, service, operations, and other overhead costs.

> Low cost does *not* come from cheap parts, low-bidding, cutting corners, or retroactive "cost reduction"

Moving production to "low labor rate" countries is another cost reduction fallacy (Ch. 6). Labor efficiency alone might cancel out the labor rate savings, for instance if labor cost is one fifth but labor productivity is also one fifth. Many decisions to move to production to low-labor-rate are based on labor-intensive designs. However, Design for Manufacturability can reduce labor content to the point where moving to low-labor-rate areas can no longer be justified.[1] Finally, cheap labor rarely stays that way and chasing it around the world is an expensive way to save money.

> Low cost does not come from chasing cheap labor.

Remote manufacturing compromises quality and product development

Greater distances between headquarters and engineers compromise teamwork and concurrent engineering. Quality may suffer if the new plant has not established an effective quality culture. Quality may also suffer if recurring defects are produced overseas and not detected until hundreds of defective products are discovered at the end of the long transoceanic "pipeline." Overseas production's slows delivery makes it hard to implement lean production, build-to-order, and mass customization.

Similarly, "cutting corners" in any manner will probably end up costing much more later. An unwise strategy to reduce cost may be to omit features. Although this may, indeed, reduce cost, it also may reduce revenue, diminish the reputation of the product and company, and hamper product evolution.

In their haste to rush early production units to market, many companies defer cost concerns until later with "cost reduction" efforts. The first problem with this strategy is that it probably will not happen because of competing priorities, and thus, costs remain high for the life of the product. The second problem is that cost reduction simply cannot be very effective at all!

For 120 companies identified as "cost cutters," *2/3 failed to achieve profitable revenue* over the next 5 years
– Mercer study

Mercer Management Consulting analyzed 800 companies over five years. They identified 120 of these companies as "cost cutters." Of those cost-cutting companies, "68% did not go on to achieve profitable revenue during the next five years." [2]

Cost is very difficult to remove after a product is designed

Cost is very difficult to remove after the product is designed. Eighty percent of a product's lifetime cumulative cost is designed into the product and is very difficult to *remove* later.[3] Shaving cost off reported costs (like parts) might raise other costs (like quality). If cost reduction is accomplished by changing the design, there is always the common possibility that one change may force other changes which, in Toyota culture, "are expensive, suboptional, and always degrade both product and process performance."[4]

Trying to significantly lower cost after production release is usually futile because of many early "cast-in-concrete" decisions, which limit opportunities.

The *total cost* of doing the change may not be paid back within the expected life of the product. Few companies really keep track of the total costs of changing designs.

There are also intangible impacts of an excessive focus on cost reduction: it absorbs effort and talent that could be applied to more productive activities, like developing better *new* products and implementing BTO&MC. One division of a large international company did not have time for the author's training on low-cost product development because they were too busy with 31 "cost reduction" efforts!

The real cost of performing the "cost reduction" may never be paid off during the life of the product

Low-cost products do not result from "saving" cost by cutting product development and continuous improvement efforts. This may not be a stated policy *per se* but improvement budgets can be impacted by corporate directives like, "all departments will reduce their budgets by 15%."

Low-cost products do not result from manufacturing misconceptions like purchasing policies based on low-bidders[5] or off-shore manufacturing[6] to lower labor costs, which too often raise many other overhead costs and lengthen delivery times.

Total cost accounting can quantify all costs (see below), but until that arrives, product designers must rely on total cost *thinking*.

COST MEASUREMENTS

Usual definition of cost. Traditional cost systems typically only measure parts cost and labor cost, as illustrated in Figure 11-1.

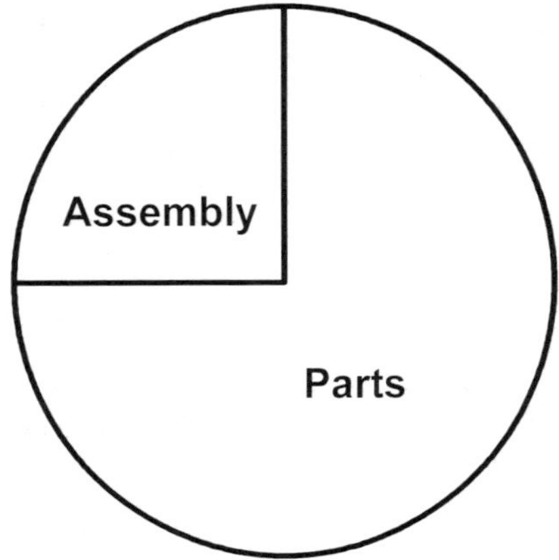

Figure 11-1: Usual Cost Breakdown

This leads to focusing all thinking about cost and cost reduction efforts on only these two costs. The rest of the costs are lumped together in several categories, collectively called *overhead,* which is then averaged (allocated) over all products. This practice results in the distortion of product costing, which leads to:

- Distorted pricing, since overhead costs vary[7]
- Cross-subsidies:
    - good products subsidize bad ones[8]
    - standard products subsidize customs & specials
- Overpricing of good products = less competitive
- Underpricing of bad products = lose money[9]
- Distorted profitability = poor product planning
- Bad cost decisions when the focus is only on parts/labor[10]
- Parts/labor "cost reductions" often raise other costs[11]

> Customers don't care about *your* cost. They only care about *their* cost which is *your price.*

However, the most important thing to realize about "cost" is that *you don't compete on cost; you compete on price.* Customers don't care about *your cost.* They only care about *their* cost which is *your price.*

Selling price breakdown. Total cost measurements provide the selling price breakdown as shown in Figure 11-2.

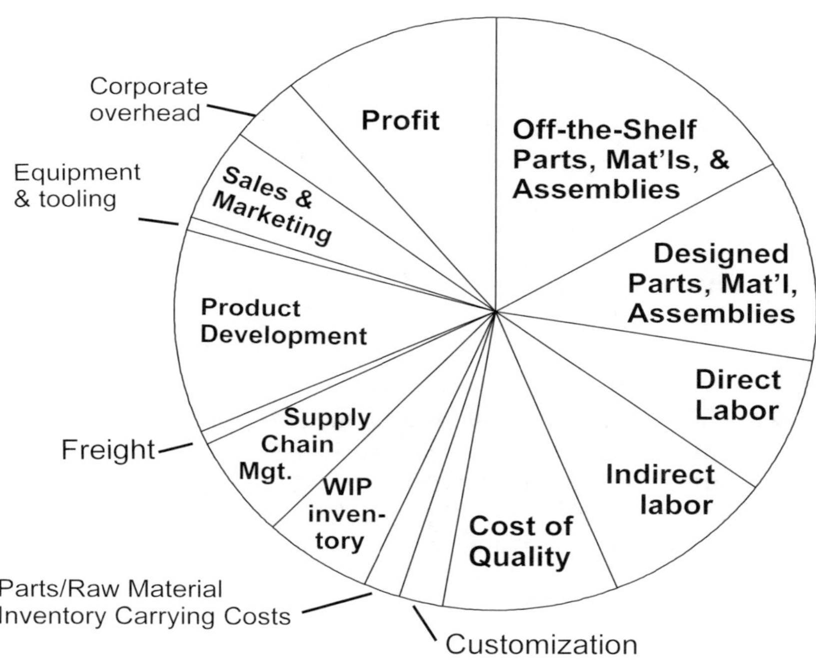

Figure 11-2: Selling Price Breakdown

This data makes it possible for strategies and decisions to be based on total cost. Decision makers must acknowledge that overhead costs are significant (usually more than labor and materials) and that companies really do have influence over overhead costs, whose cost saving opportunities will be discussed in detail below.

Part cost percentage throughout outsourced supply chain. The total cost pie chart shown in Figure 11-2 is easiest to comprehend for an *integrated* manufacturing operation. When companies purchase most of their parts and subassemblies, then they *perceive* that purchasing cost is the majority of product cost and therefore ignore the whole total cost concept.

However, every purchased part and subassembly has its *own* total cost pie chart, each with much opportunity to reduce overhead costs. Figure 11-3 shows total cost pie charts throughout an outsourced supply chain, starting with an OEM assembler with, say, two-thirds of cost spent on parts. But, each *subassembly* breaks out to another pie chart showing a higher proportion of overhead (twice, in this example). Similarly, each *part* breaks out to its own pie chart showing about the same percentage of parts and materials as the total cost pie chart of Figure 11-2. Therefore, the total cost pie chart ultimately represents all the costs – at one place or another – for integrated *or* outsourced business models.

> Every link in the supply chain has its own total cost pie chart. Ultimately, the majority of the pie charts consist of overhead.

In order to minimize total cost throughout an outsourced supply chain, suppliers would have to use all the principles of this book to minimize *their* total cost, which is the price you pay. If they do not do this spontaneously, then the assembler who understands these principles should ask for cost breakdowns and help the supplier minimize all the elements of total cost.

In fact, some overhead costs can be reduced best by joint efforts, for instance, eliminating parts inventory at both the suppliers *and* the assembler, by building parts on-demand, assuring quality at the source, and shipping them directly to assembly production lines just-in-time.

If the supplier will not share cost data and does not try to minimize overhead costs, then the assembler may be justified to bring in-house (integrate) key subassembly and part manufacture to have visibility and control over *all* the costs – and, of course, delivery time – as discussed in Chapter 6.

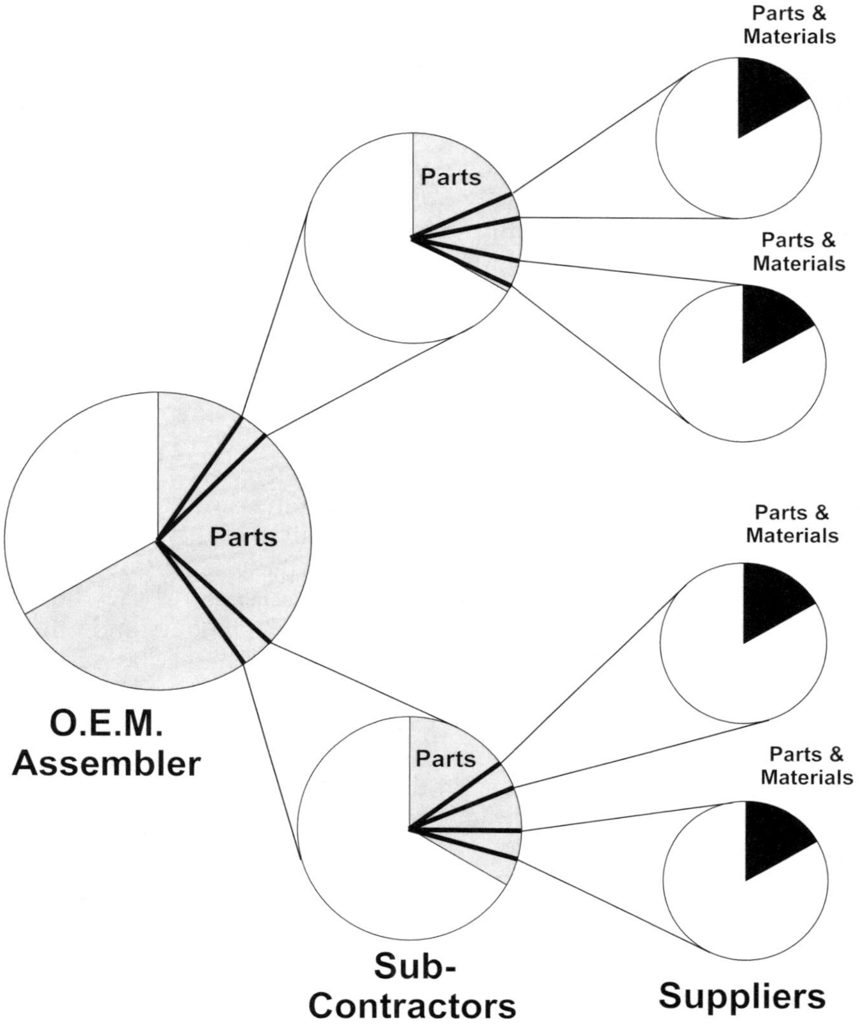

Figure 11-3: Part Cost Percentage
Throughout an Outsourced Supply Chain

TOTAL COST MINIMIZATION STRATEGY

Total cost can be minimized by addressing all the costs that contribute to the selling price. Figure 11-4 shows the overall strategy with design for manufacturability and lean production efficiencies minimizing direct labor, indirect labor, and quality costs. Standardization and product family synergies minimize material overhead, product development, shipping, and sales and marketing costs. Build-to-order and mass customization techniques, working with on-demand lean production (Ch. 8), can virtually eliminate setup changeover costs, customization costs, raw material inventory, work-in-process (WIP) inventory, and finished goods (FG) inventory both at the factory and in the distribution channels.

The remainder of this chapter discusses specific cost minimization opportunities for the various elements of total cost.

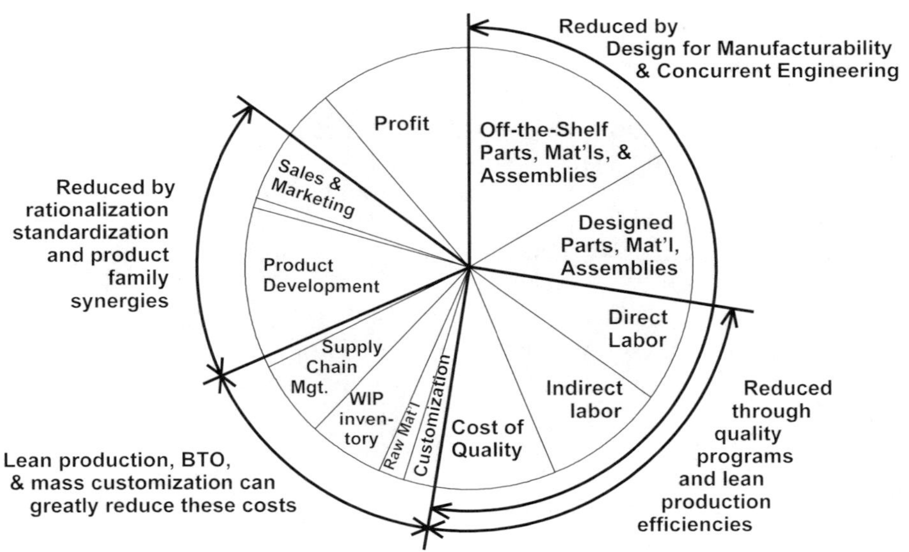

Figure 11-4: Total Cost Minimization Strategy

STRATEGY TO CUT TOTAL COST IN HALF

The author's book-length web-site, *www.HalfCostProducts.com*, presents a comprehensive cost reduction strategy consisting of the following eight strategies, each of which can offer significant returns as a stand-alone program. When combined, these coordinated cost saving strategies support each other synergistically for a whole-is-better-than-its-parts effect, instead of the more common, but ineffective, "cost reduction" attempts that can raise other costs, compete for resources, and compromise longer-term cost reductions. The site has the equivalent of a 250 page book and 700 hyperlinks. Here is a summary of the eight half-cost strategies:

1. Cost Reduction by Design. Product development determines 80% of product cost (Chapter 10). The concept/architecture phase alone determines 60% of cost! Other strategies are enhanced by designing for quality, designing for Lean production and Build-to-Order, Concurrent Engineering with vendors, which saves more than bidding, and designing around standard parts, which simplifies supply chain management.

2. Lean Production Cost Reduction. Lean production benefits include eliminating many forms of waste and the ability to double labor productivity, cut production throughput times by 90 percent, reduce inventories by 90 percent, cut errors and scrap in half, thus lowering other costs of quality, procurement, and inventory.[12]

3. Overhead Cost Reduction. Build-to-Order can build standard products to-order without forecasts of inventory, which costs 25% of its value to carry per year. Specials can be quickly and easily built by Mass Customization.

4. Standardization Cost Reduction. Standard part lists can be 50 times less than proliferated lists, thus generating purchasing leverage, lowering material overhead and inventory costs, and improving quality (Chapters 4 and 5).

5. Product Line Rationalization. Eliminating or outsourcing unusual, high-overhead, low-volume products and options can lower total cost immediately and free up valuable resources for other cost strategies. It eliminates the "loser tax" on cash-cows to subsidize products that have low-margins or are really losing money (Ch. 3).

6. Supply Chain Management Cost Reduction. Supply chain resources can be most effective in reducing cost by supporting product development teams, establishing vendor/partnerships, buying high-quality parts, driving and encouraging standardization, rationalizing products (since the most unusual products have the most unusual parts and materials), supporting lean efforts and automatic resupply of parts (that eliminates forecasts, purchase orders, inventory, and expediting costs), and keeping control of local in-house manufacturing that supports product development, lean production, build-to-order, inventory reduction, and quality programs.

7. Quality Cost Reduction. The *Cost of Quality* can be a significant proportion of revenue or selling price. Improvements are supported by designing for quality, lean production, and rationalization because rationalizing away unusual products raises net factory quality and avoids wasting quality resources on inherently lower quality products (see Chapter 3 on how rationalization improves quality).

8. Total Cost Measurement to support, justify, and quantify the savings of all cost reduction strategies. Until total cost can be quantified, everyone should make decisions based on *total cost thinking* (See Chapter 12).

At the bottom of the home page of www.HalfCostProducts.com, there is a section titled, "How Not to Lower Cost," which summarizes the problems of attempting cost reduction after design, low-bidding, and offshoring, with links to corresponding articles. The beginning of the offshoring article, shows how *offshoring thwarts six of the above eight cost reduction strategies!* These effects are presented in Chapter 6.

MINIMIZING COST THROUGH DESIGN

As mentioned earlier, 80% of the lifetime cumulative cost of a product is determined by the product's design, as shown in Figure 11-5. An even more important fact is that 60% of a product's cost is determined by product *architecture*. Low-cost product design is based on the premise that cost is *designed into* the product, especially by early concept decisions.

Tools like Design for Manufacturability[13] can help design products that are easier, and thus less costly, to build. Concurrent Engineering can ensure the lowest cost processing when the processes are concurrently designed with the product (see Chapter 10). Quality can be designed into the product with *robust* design techniques (Taguchi Methods™ based on design of experiments) and then built into the product with process controls instead of the more expensive inspection paradigm. Maximum utilization of catalog hardware can minimize part cost. Involving vendors early can result in lower cost outsourced parts. Total cost accounting data can lead to decisions that result in the lowest total cost and "reward" good products with appropriately low overhead charges.

Optimizing Architecture. The highest leverage opportunities for minimizing cost are in the architecture stage, which generally determines 60% (or more) of a product's lifetime cumulative cost, as shown in Figure 11-5. And yet, this high leverage opportunity is virtually ignored in many product development projects, when designers make snap decisions or just assume that the product will have the same architecture as previous or competitive products. The architecture phase of product development abounds with opportunities to greatly lower cost through creative concept simplifications and product architecture optimization.

80% of a product's lifetime cost is committed by the design;
60% by the concept!

Figure 11-5 shows that by the time a product is designed, 80% of the cost has been determined.[14] And by the time a product goes into production, 95% of its cost is determined, so it will be very difficult to remove cost later. The most profound implication for product development is that *60% of a product's cumulative lifetime cost is committed by the concept/architecture phase!* This emphasizes the importance of optimize product architecture.[15]

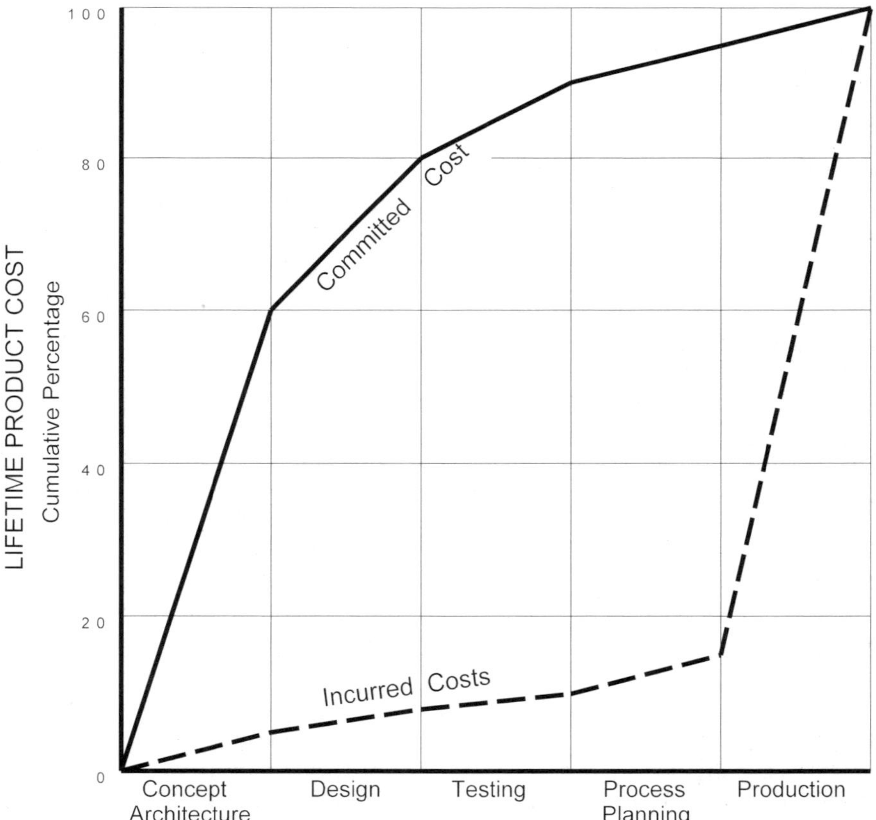

Figure 11-5: When Costs Are Committed

from *Design for Manufacturability & Concurrent Engineering,* © 2008 by Dr. David M. Anderson

MINIMIZING OVERHEAD COSTS

Minimizing Product Development Expenses

Advanced product development can significantly reduce product development and related expenses with:

Product portfolio planning. The *very* first step in product development is deciding what to develop. For development efficiency and operational flexibility, this selection process should define *families* of products that share parts, modules, software, processes, engineering, and logistics support. Decisions should be made in terms of product families, or platforms, and their evolution over time.

Before anyone can decide in which market segments to compete, they should understand the *true* profitability of their existing products. The *reported profitability* of existing products is only as meaningful as the *reported costs.* As will be discussed later, typical cost reporting systems are too "aggregated" to distinguish the real cost differences between products, thus distorting product costing. Distorted product costing leads to distorted perceptions of profitability and, thus, to distorted decision making on which products to develop. Developing products for *truly* profitable market segments will result in the most efficient utilization of product development resources. Developing products for market segments that are *thought to be* making money, but are really marginal or losing money, is a waste of product development resources.

> Before deciding what to develop, it is imperative to know the *true* profitability of existing product lines and markets

Decisions on which product families to developed must be based on *a balance* of customer/market needs and opportunities *and* corporate strengths, such as talent, technology, patents, manufacturing expertise, reputation, and so forth.

One final consideration for product development selection is to *make decisions rationally.* It is very easy to get "sidetracked" by exciting technology and enticing market opportunities. The rational approach would be to objectively assess potential technology, methodically define product specifications to satisfy the "voice of the customer," and base all decisions based on *total cost* numbers (as will be discussed later).

Multi-functional design teams. To minimize product development expense, product development teams must be efficient. They must "do it right the first time" because engineering change orders are expensive and redesigns are even more expensive. Designing products right the first time required good product development methodologies.

Using multi-functional teams to raise and resolve issues early will save the considerable expense of trying to do this later, after things are "cast under several layers of concrete."

> Thorough design *does it right the first time,* instead of expensive change orders later

Multi-functional teams can concurrently design products and optimal processes for the lowest total cost. With manufacturing and vendors involved the team will be better able to make rational decisions regarding tooling and automation.

Product development expenses can be minimized by utilizing the most efficient designers, who are not always in-house engineering working alone. The multi-functional team *with active manufacturing and vendor assistance* would be more efficient than an isolated group of engineers working alone. The vendors who make the parts may be the most efficient at designing them, and will be able to design the lowest cost parts.

And remember, the most efficient part design effort is none at all! This is accomplished with optimal use of off-the-shelf hardware.

Methodical product definition. Similarly, product development expenses can also be minimized by *defining* it right the first time because it is very expensive to make product definition iterations at the prototype stage when the customer says, "that's not what I wanted." Designing in unwanted features would have wasted product development resources and cause the product to be overpriced.

Total cost decision making. All decisions must consciously be based on a total cost *focus,* even if total cost cannot be quantified. The product development culture must support decisions that "just make sense" from a total cost perspective, even if they cannot be justified quantitatively by the current cost system.

Arbitrary decisions must be avoided. Making arbitrary decisions assumes that all choices have the same cost impact, which is rarely the case.

On the other hand, "cost" concerns can have a counterproductive effect if decision makers are giving too much focus on *reported costs,* like labor and materials, and not enough focus on *total costs.* Sometimes product development teams limit their opportunities by making major decisions based on rough estimates of reported costs, instead of using good cost models based on *total* costs for *multiple approaches, ideas, and scenarios.*

"Reinventing the wheel" can be avoided in product development with maximum use of reusable engineering and versatile modules. There is a common tendency for product development projects to ignore previous work and completely start over, when they could be more efficient by leveraging previous work. This can be facilitated by proper CAD practices: a good *layer* convention will segregate designs into well-defined layers, so that subsequent development projects can easily find and reuse previous engineering. Modular design can allow new products to be derived from standard modules.

Reuse engineering, parts, and modules instead of "reinventing the wheel"

Simplify designs with optimal product architecture, simpler concepts and part combinations.[16] As the title of a classic *Business Week* article put it, *the best engineered part is no part at all.*[17]

> The best engineered part is none at all

Basing designs on off-the-shelf parts can eliminate the cost of designing those parts and, at the same time, lower part cost and quality costs for those parts.[18]

Extending product life can be accomplished through upgrades rather than redesigns. Products can be *designed* to be easy to extend the product life with upgrades. Modular design can facilitate upgrading if anticipated changes can be confined to the fewest number of modules. Many common functions of a product can be reused for many iterations of the product.

Minimizing debugging cost can be accomplished by using existing modules that have already been debugged. Some leading companies, like Hewlett-Packard, feel that this is the best way to produce bug-free software. With a high percentage of reuse, development *and* debugging efforts can focus on *new* aspects.

Diagnostic test development can be avoided by designing quality into the product and building it in with process controls. In such a quality environment, failures would be so low that diagnostic test development could be avoided, because the cost of discarding failed parts would be less than the cost of test development, test equipment, and repair efforts. At IBM, printed circuit boards that were expected to have higher than a 98.5% first-pass-accept rate could avoid diagnostic testing entirely. This eliminated time-consuming test development and the expensive ATE (Automatic Test Equipment) "bed-of-nails" equipment. Above this threshold, it was more cost-effective to discard defective printed circuit boards than to pay for the test labor, test equipment, test development, and the repairs. Avoiding diagnostic tests can have a significant effect on product cost since diagnostic test equipment for circuit boards can cost up to two million dollars and, for some products, diagnostic test development can exceed the cost of product development!

Development expenses would be paid off sooner because of quicker product development cycles. This lowers the interest or "opportunity" cost of the money invested in product development. Similar logic applies to research expenses for technology development and tooling expenses.

More efficient development costs less. Advanced product development methodologies are more efficient because the product is well defined and the architecture is simplified around optimal concepts. This thorough early development work means fewer false starts, "looping back" to do things over, fewer changes, and redesigns.

Faster developments have less obsolescence risk. Shorter product development cycles result in less chance of market shifts and technical obsolescence by the time the product reaches the market, thus resulting in fewer changes and redesigns.

Minimizing Engineering Change Order Cost

Advanced product development methodologies result in more products designed right the first time and, thus, fewer engineering change orders (ECOs). Better product definition results in fewer changes to satisfy customers. Early "changes" in the planning stage are less expensive than late changes at the hardware stage. Reused engineering and modules have fewer "bugs" because of their widespread usage over time and across many product lines.

Early "changes" are much less expensive that later change orders

Well-designed products can minimize a substantial overhead expense that is not always included in ECO cost reporting, which is "firefighting" or problem solving. Usually this represents a considerable effort to solve production problems, which is even more intense when a new product is launched into production.

The cost of changes rises drastically as the product progresses toward production. Figure 11-6 shows how the cost of changes escalates as product development progresses:

Time of Design Change	Cost
During design:	$1,000
During design testing:	10,000
During process planning:	100,000
During test production:	1,000,000
During final production:	10,000,000

Figure 11-6 Cost of Engineering Changes[19]

Minimizing the Cost of Quality

The "cost of quality" is really the cost of *poor* quality. Companies without a strong quality program can have quality costs equal to 15% to 40% of revenue.[20] Advanced product development can *design in* quality. Process controls can *build in* quality. This dual approach to quality can substantially reduce both the internal and external cost of quality, which are defined as follows:

Internal cost of quality includes the cost of non-value-added activities such as testing, diagnostics, rework, scrap, waste, etc. It also includes the cost of test development, which, for some products can exceed the cost of product development, and millions of dollars for test equipment. Designing in quality and designing well-controlled processes can produce products with such a low failure rate that diagnostic test development would not be required.

External cost of quality includes the cost of field failures, warranty expenses, legal liabilities, and hard to quantify costs, such as bad publicity, that damages a product's reputation.

Companies can have poor internal quality but, with a good "test screen" can keep defects within the factory, and, thus, have a "high" external quality. However, achieving external quality totally by testing and rework is very expensive, and such a company would have a high cost of quality. Many luxury products, such as luxury automobiles, enjoy a quality reputation, but are very expensive because quality is achieved at a high cost. Before TQM spread to this industry, some luxury automobiles require several times more labor effort to *fix* than a well-designed car requires to *build*.[21]

The cost of quality can be a substantial part of revenue. When it exceeds profits, those profits could be doubled by eliminating it.

Another cost of quality is the "defects by the batch" effect in which large batches of parts are made with recurring defects and this is not noticed until the batches have traveled through many more work stations. By the time the defects have been spotted in final inspection, hundreds or thousands of defective parts would have been made, which would then have to be reworked or scrapped. In lean production, parts are made in a *one-piece flow* with immediate feedback after each "handoff," so there should be very little chance of recurring defects being produced.

Flow manufacturing eliminates "defects by the batch"

Rationalization will *raise corporate quality*, and thus lower corporate quality costs, by eliminating the unusual, low-volume products which usually have the lowest quality. Rationalization will also make quality improvement programs, like six-sigma, more effective and easier to implement because quality improvement efforts can be focused better on the remaining products.

Standardization will lower quality costs when there are fewer part types in the plant so there will be less likelihood of using the wrong part. Further, standardization reduces the number of suppliers because fewer parts types will come from fewer sources. Dealing with fewer suppliers can strengthen ties with those suppliers, thus resulting in the very desirable *supplier partnership* relationships, which can help lower total cost in many ways.

Continuous improvement, or *kaizen* in Japanese, is an effective technique to keep driving costs down with incremental improvements that can have significant cumulative effects. Continuous improvement programs can be performed spontaneously by in-house workers,[22] as a part of TQM programs,[23] or in cooperative efforts with suppliers. [24, 25]

Maximizing Factory Efficiency

Rapid product development can more quickly phase out older, more costly products with new generation cost-effective products. The older, less efficient products, in addition to having higher direct costs for labor and materials, have higher overhead demands for ECOs and firefighting. More efficient production, from better designed products, can result in more output from existing plants and equipment. For growing companies, this extra output might spare the company, or at least defer, the expense of adding new equipment or expanding facilities.

Rational Selection of Lowest Cost Supplier

This decision encompasses both the make/buy decision and supplier selection. These decisions must be made rationally on a total cost basis. As discussed throughout this chapter, conventional cost systems can mislead decision makers if they only report labor and materials. Not quantifying and including all the overhead costs for in-house manufacture creates a bias toward that option, since purchases include all costs, by definition, and in-house production may not.

Regarding supplier selection, total costs can actually increase as a result of choosing the supposedly "low-bidder" if purchase cost is emphasized over such subtle, but important, characteristics such as quality, delivery, flexibility, and help with product development. Jordan Lewis, in his book about customer-supplier alliances, *The Connected Corporation,*[26] commented about the effects of General Motor's 1992 demands for double-digit price cuts from suppliers, which were instituted by J. Ignacio Lopez De Arriortua:

> For low-bidding of parts, its "save now – pay later." Actually it's *save a little now – pay a lot later*

"By emphasizing price alone with its suppliers GM won immediate savings – and ignored total cost. At GM's plant in Arlington, Texas, an ill-fitting ashtray from a new, substandard supplier caused a six-week shutdown of Buick Roadmaster production."

Another GM plant saved 5 percent by going with the low bidder on another part. When the parts were delivered, one-half failed quality tests. The other supplier, who lost the bid, had to gear up production in four days and fly parts to GM by chartered plane. The second supplier commented that "My guess is that their 5 percent savings turned into a 15 percent loss."

> Low-bidding alienates your best suppliers, thus missing out on the biggest cost reduction opportunity – *cooperation*

Peter Drucker, writing in *Managing in a Time of Great Change,*[27] encourages lowering total cost by minimizing "interstitial" costs between suppliers and manufacturers or between manufacturer and distributer:

"But the costs that matter are the costs of the entire economic process in which the individual manufacturer, bank, or hospital is only a link in a chain. The costs of the entire process are what the ultimate customer (or the taxpayer) pays and what determines whether a product, a service, an industry, or an economy is competitive.

"The cost advantage of the Japanese derives in considerable measure from their control of these costs within a *keiretsu,* the "family" of suppliers and distributors clustered around a manufacturer. Treating the *keiretsu* as one cost stream led, for instance, to "just-in-time" parts delivery. It also enabled the *keiretsu* to shift operations to where they are most cost-effective."

There is also a significant *cost of bidding*, which was discussed in Chapter 7.

Lowering Overhead Costs with Flexibility

Flexibility can reduce overhead costs significantly. There are some interesting parallels between flexibility and quality. Decades ago, it was commonly believed that quality cost more. Then Philip Crosby wrote the book, *Quality is Free,* and showed that the gains from lowering the cost of quality would pay for quality, thus making it free. Similarly, the financial gains derived from flexible operations can more than pay for the effort to make operations flexible. These gains will become a source of competitive advantage over competitors that do not embrace lean production, build-to-order, and mass customization.

Eliminating Inventory Costs

There is a lot of *working capital* tied up in various forms of inventory: raw materials and parts inventory; Work-in-Process (WIP) inventory; and finished goods inventory in factory warehouses, at distributers, and at the dealers. Fortune Magazine estimates that, for Fortune 500 companies, working capital averages an amount equal to 20% of sales.[28]

The US Department of Commerce tracks inventory/sales ratios for various industries.[29] For all manufacturing industries the ratio was 1.29 for data collected in 1999 and 2000. For all of "manufacturing and trade," the ratio was 1.31. This means that for manufacturing industries and their sales channels, the value of inventory equals about 1.3 times sales. Using an inventory carrying cost of 25% (Ch. 2), this means that the *cost* to carry inventory is 33% of the value of sales. *Thus, manufacturers, distributers, and stores together pay a third of sales dollars to pay for the cost of carrying inventory!* Regardless of who has to pay for it, it becomes part of the selling price and anyone who can lower inventory can either sell for less or reap more profits or both. If these numbers are applied to the price pie-chart, the conclusion is that all the inventory wedges should add up to 33% of the price. Any manufacturer that can achieve less inventory will have a strong cost advantage. BTO&MC can eliminate almost all of inventory carrying costs.

> On average, inventory carrying *cost* is 1/3 of sales dollars

Ironically, inventory shows up on the balance sheet as an *asset* when, in fact, inventory is really a *liability* to the operation of any manufacturing plant, especially those needing to be flexible. This point was one of the revelations presented in Eli Goldratt's *The Goal* when managers of the fictional plant, faced with extinction, realized that they had to focus on *the goal* – making money – instead of letting their behavior be dictated by irrelevant cost accounting metrics.[30]

One progressive materials manager of a processing equipment company told the author that after much successful work to reduce inventory, he got a call from the company Controller, who was having trouble preparing the annual report because the inventory had been reduced so much that it was "lowering company assets," according to their traditional accounting rules. Companies must make sure they are pursuing the real "goal" instead of irrelevant metrics.[31]

Inventory is
not an asset

The following discussion presents several opportunities to lower the costs of inventory and other overhead costs by designing flexibility into products and plants.

Minimizing Customization/Configuration Costs

Many companies offer customized goods, but do not do it cost-effectively. By *mass* customizing products, the customization process is built into the system – the product design and the manufacturing operation.

Thus, the mass-customizer has cost advantage over companies that are inefficient at customization and configuration. Two under-reported costs related to customization are custom engineering and changing or modifying standard designs and processes. Both of these activities usually cost much more than are indicated by current cost systems. Many companies do not even keep track of engineering costs by project. In addition, engineers often try to get a lot of "free" help from many support people who are "just on overhead."

For the mass customizer,
customization is free

The extra manufacturing cost to do *ad hoc* customization and configuration reactively is much more than it would be for mass-customized products. These customization costs often include extra tooling, lengthy setups for small runs, inefficient production control, low equipment utilization, special programming, slow and frequent "learning curves," special tests and inspections, and lots of "fire drills" to shove customized products through mass production factories. Further, these low-volume customized products may disrupt the manufacturer of the *standard* products, thus increasing *their* cost. Thus, it is often the case, in companies with traditional accounting systems, that custom products are really being *subsidized* by the standard products.

Fire-drill customization costs a lot more than companies realize

Minimizing the Cost of Variety

A large part of working capital is tied up in the cost of variety, which is defined in depth in Chapter 3 of *Agile Product Development for Mass Customization.*[32] The cost of variety rises exponentially as a function of market variety, as shown in Figure 2-3. Eliminating setup and reducing the batch size to one eliminates most of the cost of variety.

The key element of lean production is set-up elimination. If setup could be eliminated, then operations would be flexible, meaning that every product could be different and, yet, still reap "mass production" efficiencies. In low-volume operations, setup could be caused by any effort to do something "different," for instance, to get parts, change dies and fixtures, download programs, find instructions, or any kind of manual measurement, adjustment, or positioning of parts or fixtures.

The cost of variety rises *exponentially* as market variety rises

Work-in-Process inventory can be virtually eliminated by set-up reduction, JIT, and design commonality of parts and processes. WIP inventory costs rise proportionally to batch size, except when the batch size is one, in which case WIP inventory can be virtually eliminated Since WIP inventory carrying cost could be 25% of its value per year, eliminating this inventory could result in substantial yearly savings.

Floor space can be reduced because of reduced inventories, elimination of the fork lift aisles necessary to move large batches of parts, elimination of kitting, and higher utilization of machinery and people. Appreciation of the cost or value of floor space varies according to the need to expand manufacturing. But floor space reduction should be constantly pursued. The cost of expanding manufacturing space is a very large step-function that may force a company to move away from an area that is too crowded or expensive. Floor space reduction can provide an attractive alternative to expansion or relocation. Further, floor space requirements can be reduced faster than new facilities can be built. The lead time for physical plant expansion is so large that such plans must be started well ahead of the anticipated need, often based on inaccurate long-range marketing projections. Between 1991 and 1994, Compaq Computer quintupled production without increasing factory space by implementing programs like WIP inventory reduction.[33]

Internal transportation costs, such as fork lift activity, can be reduced. This can be eliminated when parts and products flow individually, between adjacent work stations, instead of in large, heavy bins between distant work stations.

Utilization is improved with less setup, thus reducing equipment cost, a very big cost savings potential for expensive equipment, like CNC machining centers, surface mounted printed circuit assembly equipment, or expensive testers. Machine tool utilization can be as low as 10%, which means the equipment is only producing parts 10% of the time the machine is available for work; the remainder is setup or waiting. It is important to realize that *doubling the utilization rate will double output.* If production equipment had a utilization of 30%, output could be *doubled* by raising the utilization to 60% and another 1.5 times by raising the utilization to 90%. Utilization improvement is a cost-effective way to increase production, *and* it is quicker, considering the lead time to procure and install new production equipment.

> Raising utilization from a common 30% to 60% *doubles output* from the same equipment

Set-up labor expenses can be eliminated including the labor cost to change machine set-ups and to retrieve parts, tools, and drawings.

Flexibility can improve the balance of labor and machinery utilization in sequential operations such as assembly lines. Products built in flexible lines can be optimally ordered to offset imbalances in the workloads of adjacent machinery or people, using a concept known as *product complementarity.*[34] Products with high demands on the A machine and low demands on the B machine can be interspersed with products with low A demands and high B demands.

Production can quickly adapt to changing market conditions by building all the products on the same flexible line. Inflexible operations are always faced with a dilemma when the demand for "model A" has exceeded capacity while "model B" is having a sales slump. Manufacturing may have adequate overall capacity, but the model A line or plant will not be able to satisfy demand while the model B line or plant is partly idle or laying off people. A slightly more flexible approach would be to have the ability to move people from the "B" line to the "A" line, but that assumes adequate equipment capacity on line "A." Flexible operations would simply pull more model A products, and fewer model B products, through the flexible line(s).

Operational flexibility can allow companies to transfer production from one flexible line or plant to another to respond to changing market conditions, rather than the more expensive alternatives of overtime, rapidly bringing contract labor up to speed, and expedient outsourcing to ease production bottlenecks for the product that is in demand. Similarly, by transferring production, companies can avoid layoffs at plants making products in low demand.

Kitting cost and space can be eliminated. Without flexibility, there will be labor costs and space requirements to gather all the parts for a batch, "kit" them together and deliver them to manufacturing.

Minimizing Materials Management Costs

Purchasing costs can be reduced if there are fewer purchasing actions for fewer part types. Standardized parts will cost less because of the greater purchasing leverage of higher volume parts. Further, purchasing costs can be reduced by automatic (non-purchasing) resupply techniques like steady flows, kanban, min/max, and breadtruck.

Vendor fabrication and assembly are more feasible, and thus quicker and less costly, if parts are well designed *and* documented, especially if the vendor was part of the design team. Sometimes it may be more cost-effective to have the vendor design the parts, as is common now days in the automobile industry.

Fewer part numbers mean less material overhead for raw materials and parts inventories, documentation, controls, and so forth. There will also be less expediting cost for seldom-used parts that are difficult to obtain. Pareto's law (the "80/20" rule) applied to inflexible plants would say that 80% of the material overhead costs would be consumed on low-usage parts that may only represent 20% of total part volume.[35]

Spare parts logistics and field service can be greatly simplified, and thus cost less, with part standardization and modular design. Products designed around common parts have smaller spare parts kits. This could lower the effective product "price" for customers who add the cost of spare parts kits to the product's list price. Part standardization can also result in less downtime due to part shortages. Service costs can be reduced if failed modules can be quickly replaced and repaired in more efficient facilities.

Minimizing Marketing Costs

When manufacturers keep listening to customers' wants and needs and keep designing and manufacturing products to satisfy these evolving needs, this results in *learning relationships* which result in the ability to *keep customers forever.*[36] Not only is this good for generating revenue, but it also saves the considerable cost of acquiring new customers to meet growth objectives. Studies, such as one done by the Technical Assistance Resource Project for the U.S. Office of Consumer Affairs, show that the price of acquiring new customers is five times greater than the cost of keeping existing ones.[37]

Minimizing Sales/Distribution Costs

There are considerable cost reduction opportunities in the warehousing and distribution of products. The physical distribution system accounts for 10 percent of the Gross National Product.[38]

Designing modular products and concurrently engineering products and production systems to build products on-demand can save a lot of money with respect to the way products are configured, packaged, shipped, distributed, and sold. In fact, build-to-order can eliminate most of the distribution chain as we know it. Being able to build-to-order and ship from the plant eliminates warehousing and associated distribution costs from the plant to the customer.

In industries where product variety is considerable, like blue-jeans and shoes, this cost can be enormous, considering the number of sizes and styles. Sung Park, founder of Custom Clothing Technology Corporation in Newton, Mass., which developed the technology that Levi Strauss used for its customized Personal Pair™ line, said, "You've got to look at the whole value chain. Zero inventory. No markdown. No distribution-centered costs. The product doesn't sit in the warehouse."[39]

Minimizing Supply Chain Costs

Supply chain management has become a strategic competitive advantage, especially for companies like Hewlett-Packard.[40]

Peter Drucker points out the opportunities of minimizing cost in the supply chain: "Process-costing from the machine in the supplier's plant to the checkout counter in the store also underlies the phenomenal rise of Wal-Mart. It resulted in the elimination of a whole slew of warehouses and reams of paperwork, which slashed costs by a third."[41]

Minimizing Life Cycle Costs

An often neglected part of total costs is *life cycle* costs, which are those costs that are incurred over time, such as service, repair, maintenance, field failures, warrantee claims, legal liabilities, changes over the life of the product, and subsequent product developments. Products can be designed to minimize life cycle costs. The cost of changes can be minimized by a methodical product definition and thorough product development. Change costs and subsequent product development costs can be minimized with modular product architecture, where many modules can remain unchanged as other modules are updated or redesigned.

Designing for reliability can minimize many costs related to product reliability. There are several techniques that can be used to maximize reliability.[42]

Saving Cost with Build-to-Order

If products and production processes can be designed to build products to-order, then many costs can be saved:

Factory finished goods inventory can be eliminated by building products to order, instead of building to forecast and then holding products in a warehouse until ordered by distributers or customers. Like WIP inventory, finished goods inventory may cost the same to "carry" except that finished goods are completed and therefore are more valuable. Using 25% of value per year, $10 million worth of finished goods in inventory would cost $2.5 million per year to carry. Build-to-order can eliminate factory finished goods inventory and, thus, save its yearly inventory carrying cost.

Dealer finished goods inventory can be almost eliminated if resupply orders can be quickly filled and delivered to the customer. As with factory inventory, dealer inventory has a carrying cost. Even if the dealer/distributer is separate from the manufacturer, the carrying cost will have to be paid, ultimately by the customer. For example, an automobile dealer with 200 vehicles "in stock" with an average value of $30,000 each would represent $6 million worth of inventory. Using a yearly carrying cost of 25%, this inventory will *cost* the dealer $1.5 million per year.

When new car prices exceed customers' ability to afford them, customers buy more used cars as a cost-effective alternative.[43] Built-to-order new automobiles could compete well against used cars, with lower prices, since the new cars could avoid dealer inventory expenses, whereas used cars must be stocked in inventory by definition.

Supply chain inventory can be minimized since build-to-order products do not need to be stocked at various warehouses along the supply chain: at distributers, consolidators, forwarders, and so forth. Regardless of who "pays" for this inventory, the customer ultimately must pay a higher price.

Similarly, a build-to-order system "pulls" parts from suppliers on a just-in-time basis, thus eliminating parts inventory along the supply chain. Eliminating excessive supply chain inventory costs will allow customers to pay less for equivalent products.

Companies known for rapid deliveries, like Federal Express, are providing companies with "inventory-less" direct deliveries of parts and products to and from factories. After National Semiconductor commissioned Federal Express to run National's storage, sorting, and shipping activities, delivery time was reduced from 45 days to four days with an ultimate goal of 72 hours. At the same time, distribution costs have been reduced from 2.6% of revenue to 1.9%.[44] Adding the value of increased sales from customer satisfaction would make inventory-less distribution even more attractive.

Less interest expense will be incurred for expensive components in products that sell sooner because of the quicker throughput of flexible plants and the elimination of finished goods inventory.

Inventory write-offs can be eliminated because there would be no products in inventory that could deteriorate, or incur damage, or become obsolete. If there is a substantial finished goods inventory of an expensive product at the end of that product's life, the obsolescence write-off can be enormous.

Quicker transitions to new technology are possible if there are no older technology products waiting in inventory that must be sold first. Rosendo G. Parra, Group Vice President of Dell Computer Corporation, summarized the advantage of built-to-order for Dell: "We were probably the first vendor to transition into the new Pentium FPU processor, simply because we didn't have a hundred and some days of inventory out in distribution that we had to move first."[45]

BOM/MRP expenses could be minimized. Build-to-order could minimize overhead expenses with respect to translating forecasts into materials ordering requirements with MRP (Materials Requirement Planning) systems.

ECONOMIES-OF-SCALE

Build-to-order and mass customization will change the notion of "economies-of-scale," which evolved from the mass production paradigm of "the bigger (the batch), the better." Traditionally, there are two main sources for economies of scale: setup charges, which would be less for larger batches, and dedicated automation, which can only be justified on high-volume production. A temporary third is the amortization of product development and initial tooling.

> BTO&MC will change the notion of *economies-of-scale*

Setup can be eliminated using the techniques presented in this chapter, so manufacturers will no longer have to spread setup charges over large batches to minimize the setup charge per part or product.

Instead of fixed, dedicated "hard" automation, BTO&MC, utilizes flexible automation, which can be just as efficient, although it may appear to cost more. However, when analyzed on a total cost basis, flexible automation may be at parity or better because (1) it can be reused over many generations of products, whereas hard automation cannot be, (2) equipment development costs will be lower since CNC equipment is "automation off-the-shelf," and (3) overhead costs are lower in a BTO&MC environment with less cost for material overhead, setup changes, inventory, and floor space.

Any investments in automation will be much better focused after eliminating and outsourcing the unusual, low-volume products using the rationalization techniques of Chapter 3. The remaining cash cows can all then be refined using *kaizen* continuous improvement techniques.

Until such a change takes place industry-wide, BTO&MC will be able to benefit from economies of scale *at the part level* for standard parts, while mass producers keep trying to achieve economies of scale at the product level, the folly of which was summarized in Chapter 2.

ENDNOTES/REFERENCES

1. David M. Anderson, *Design for Manufacturability & Concurrent Engineering; How to Design for Low Cost, Design in High Quality, Design for Lean Manufacture, and Design Quickly for Fast Production,* (2008, CIM Press, 448 pages).

2. Robert G. Atkins and Adrian J. Slywotzky, "You Can Profit From a Recession," *Wall Street Journal,* February 5, 2001, p. A22.

3. Anderson, *Design for Manufacturability & Concurrent Engineering*, Figure 1-1, "Product Cost vs. Time."

4. James Morgan & Jeffrey K. Liker, *The Toyota Product Development System* (2006, Productivity Press); Chapter 4, "Front-Load the PD Process to Explore Alternatives Thoroughly."

5. For more on low-bidding for parts, see the section, "Cheap parts – Save Now, Pay Later," in Chapter 7.

6. For more on trying to save cost with overseas manufacturing, see the section, "Overseas Manufacturing "to Save Cost," in Chapter 6.

7. Robin Cooper and Robert S. Kaplan, "How Cost Accounting Distorts Product Costs," *Management Accounting* (April 1988).

8. *Relevance Lost, The Rise and Fall of Management Accounting,"* by H. Thomas Johnson & Robert Kaplan, Harvard Business School Press, 1991.

9. Douglas T. Hicks, *Activity-Based Costing for Small and Mid-Sized Businesses, An Implementation Guide,* (New York, John Wiley, 1992), p. 20.

10. Michael R. Ostrenga, Terrence R. Ozan, Robert D. McIlhattan, Marcus D. Harwood, *The Ernst & Young Guide to Total Cost Management,* (New York, John Wiley & Sons, 1992), p. 146.

11. Anderson, *Design for Manufacturability & Concurrent Engineering,* Chapter 6,"Minimizing Cost by Design."

12. James P. Womack and Daniel T. Jones, *Lean Thinking; Banish Waste and Create Wealth in Your Corporation,* (1996, Simon & Schuster), p. 27.

13. Ibid.

14. This data was generated by DataQuest and presented in the landmark article that started the Concurrent Engineering movement: "A Smarter Way to Manufacture; How `Concurrent Engineering' can invigorate American Industry," page 110, *Business Week,* April 30, 1990. In the author's in-house seminars, he presents similar data from Motorola, Ford, General Motors, Westinghouse, Rolls Royce, British Aerospace, the Allison Division of Detroit Diesel, Draper Labs, Rensselear Polytechnic Institute, and several other published sources.

15. Anderson, *Design for Manufacturability & Concurrent Engineering,* Chapter 3, "Designing the Product."

16. Anderson, *Design for Manufacturability & Concurrent Engineering,* Section 3.13, "The Importance of Good Product Architecture."

17. *"The Best Engineered Part is No Part at All,"* Business Week, May 8, 1989, p. 150.

18. Anderson, *Design for Manufacturability & Concurrent Engineering*, Chapter 5, Anderson's Law: "Never design a part you can find in a catalog."

19. "A Smarter Way to Manufacture; How 'Concurrent Engineering' Can Reinvigorate American Industry," *Business Week,* April 30, 1990.

20. Phillip Crosby, *Quality is Free* (Mentor Books, 1979).

21. James Womack, Daniel Jones, and Daniel Roos, *The Machine That Changed the World* (Rawson Associates, 1990; paperback: Harper Perennial, 1991)

22. Kiyoshi Suzaki, *The New Manufacturing Challenge, Techniques for Continuous Improvement*, (New York, Free Press, 1987).

23. James H. Saylor, *TQM Field Manual,* (New York, McGraw-Hill, 1992), Chapters 3 (Continuous Improvement) and Chapter 4 (Continuous Improvement System).

24. Jordan D. Lewis, *The Connected Corporation, How Leading Companies Win Through Customer-Supplier Alliances,* (New York, Free Press, 1995), Chapter 8.

25. James P. Womack, Daniel T. Jones, and Daniel Roos, *The Machine That Changed the World, The Story of Lean Production,* (New York, Harper Perennial, division of HarperCollins, 1990), Chapter 6.

26. Lewis, *The Connected Corporation,* p. 38.

27. Peter F. Drucker, *Managing in a Time of Great Change,* (New York, Truman Talley Books/Dutton, 1995), p. 117.

28. Shawn Tully, "Raiding a Company's Hidden Cash," *Fortune Magazine,* August 22, 1994, page 82.

29. Data was drawn from the US Department of Commerce web-site page "Real Inventory-Sales Rations for Manufacturing and Trade, Seasonally Adjusted," which is tabulated for various industries at www.bea.doc.gov/bea/an/0400niw/table3.htm.

30. Eliyahu M. Goldratt, *The Goal,* (New York, North River Press, 1984, second revised edition, 1992), Ch. 33, page 268.

31. Eli Goldratt's *The Goal* makes this point often about the need to do "what makes sense" to achieve "the goal" rather than basing decisions and actions on attempts to satisfy arbitrary performance metrics and cost measurements that only make a small part of the system *appear* to look productive or cost-effective.

32. David M. Anderson, *Agile Product Development for Mass Customization* (1997, McGraw-Hill), Ch. 3, Cost of Variety.

33. Ronald Henkoff, "Delivering the Goods," *Fortune Magazine,* November 28, 1994, page 62

34. Marshall Fisher, Anjani Jain, and John Paul MacDuffie, *"Strategies for Product Variety: Lessons From the Automobile Industry,"* Working paper from the Wharton School, University of Pennsylvania; page 26; January 16, 1994, page 26

35. See the article on "Rationalization" on the articles page of www.build-to-order-consulting.com

36. B. Joseph Pine, II, Don Peppers, and Martha Rogers, "Do You Want to Keep Your Customers Forever," *Harvard Business Review,* (March-April, 1995), p. 103.

37. Wilton Woods, "After All You've Done for Your Customers, Why Are They Still Not Happy," *Fortune,* (December 11, 1995), p. 180.

38. Robert V. Delaney, *Sixth Annual State of Logistics Report,* (St. Louis, MO, Cass Information Systems, June 5, 1995), Figure #8.

39. Niklas von Daehne, "Database Revolution," *Success,* v42, n4 (May, 1995), pp. 38-42.

40. Dr. Corey Billington, "Strategic Supply Chain Management," *OR/MS Today,* April 1994, (published jointly by the Operations Research Society of America and the Institute of Management Sciences), pp. 20-27.

41. Peter F. Drucker, *Managing in a Time of Great Change,* (New York, Truman Talley Books/Dutton, 1995), p. 117.

42. Anderson, *Design for Manufacturability & Concurrent Engineering,* Chapter 10, "Design for Quality."

43. Douglas Lavin, "Stiff Showroom Prices Drive More Americans to Purchase Used Cars," *Wall Street Journal,* November 1, 1994, p.1.

44. Ronald Henkoff, "Delivering the Goods," *Fortune,* November 28, 1994, p. 64.

45. Niklas von Daehne, "Database Revolution," *Success,* v42, n4 (May 1995), pp. 38-42.

12

MEASURING TOTAL COST

TOTAL COST MEASUREMENT

In order to appreciate all of the cost savings and revenue enhancements cited in this book, it will be necessary to *quantify all costs.* If costs were tracked on a *total cost* basis, then the cost saving potential of BTO&MC would be known and given an appropriate overhead charge and thus competitive prices. However, if total cost is not tracked, then well-designed BTO&MC products (which really have much less overhead demands) will have to subsidize the high-overhead products; this would be an unfair burden and may ultimately compromise the success of the BTO&MC products.

Inadequate cost systems hinder good management and lead to bad decisions

Most companies have such inadequate cost systems that it actually hinders good management and good decisions. Merely reporting labor and material costs encourages companies to move operations to "low labor rate" countries. Trying to minimize return on capital encourages companies to scrimp on capital investments and outsource production. But, offshoring slows down the supply chain and actually increase total cost.

Products with too much setup, inventory, "firefighting," engineering change orders, excessive part variety, low equipment utilization, and high quality costs *should* have a *higher* overhead rate. Products that are designed, using the methodologies presented herein, for quick and easy manufacture should have a much lower overhead rate. Overhead rates should be proportional to overhead demands, which vary by product.

Overhead cost allocation would be simple if factories only produced one product

Let us consider the proverbial "Model-T plant" – a plant with only one product and no variations. The *Ernst & Young Guide to Total Cost Management*[1] discusses the cost management implications of a single product plant:

> "If product variety were absent, the business environment would be simple . . . In a world like this, you could do product costing literally on the back of an envelop. You would simply divide the total production costs by the total production volume to calculate a unit cost."

Overhead costs vary with product diversity and complexity
– Cooper & Kaplan, *How Cost Accounting Distorts Product Costs*

However, as Cooper and Kaplan[2] pointed out in an article with the profound title, *How Cost Accounting Distorts Product Costs,* overhead costs "vary with the diversity and complexity of the product line." And, product diversity and complexity are increasing because of market pressures and perceived opportunities. Thus, it is important to quantify overhead costs, since they can be much greater than the typically reported costs of labor and materials.

QUANTIFYING OVERHEAD COSTS

There are three steps involved in quantifying overhead. The first step is to acknowledge deficiencies in current product costing practices. The second step is to estimate the degree of cost distortions. The third step is to understand the value of total cost measurements.

A. Acknowledge Deficiencies of Traditional Accounting

The first step in quantifying overhead is to acknowledge the deficiencies in the current cost system, which is the central theme of Johnson and Kaplan's pivotal book, *Relevance Lot, The Rise and Fall of Management Accounting.*[3] Traditional cost accounting systems were designed to present operational results and the financial position of the organization *as a whole* for investors and for agencies that tax or regulate. On the other hand, managers and engineers need *relevant* cost information to make good decisions. Typical problems caused by conventional cost systems are:

Distortions in product costing, which is discussed at length in the Cooper and Kaplan reference cited above. Johnson and Kaplan concur:

> Accounting systems fail to provide accurate product costs
> – Johnson & Kaplan, *Relevance Lost*

> "The management accounting system fails to provide accurate product costs. Costs get distributed to products by simplistic measures, usually direct-labor based, that do not represent the demands made by each product on the firm's resources."

The *Guide to Total Cost Management* asserts that product costs "are distorted because each product typically includes an assignment of overhead that was allocated on some arbitrary basis such as direct labor, sales dollars, machine hours, material cost, units of production, or some other volume measure." [4]

Distorted product *costing* results in distorted *pricing* that can underprice some products so low that they actually lose money and overprice other products to the point where they are uncompetitive.[5] Distorted product costing results in a distorted perception of the profitability amongst the company's products. This distorted view of profitability can cause managers to "feed the problems and starve the opportunities" with detrimental effects on product development and operational improvement priorities. Understanding *the real* profitability will help companies to drop unprofitable products and focus on profitable ones.

Distorted *costing* causes distorted *pricing* and that causes distorted *profitability* & distorted *decisions*

Cross subsidies, where high-volume products subsidize low-volume products and standard products subsidize custom products. Johnson and Kaplan state unequivocally:

> "The standard product cost systems, which are typical of most organizations, usually lead to enormous cross-subsidies across products," [6]

Cooper, Kaplan, et. al., in their Institute of Management Accountants sponsored study, summarized what their eight case study manufacturing organizations discovered after they implemented Activity Based Costing:

Typical cost systems lead to *enormous cross-subsidies* across products – Johnson & Kaplan, *Relevance Lost*

> "The manufacturing companies generally found, as expected, that low-volume, complex products tended to be much more expensive than had been calculated by the existing standard cost system."[7]

One of the most dangerous consequences of cross-subsidies is penalizing new generation products and programs like build-to-order and mass customization by making them pay the same overhead charges that should have been paid only by the products that have the high overhead demands. Such unfair charges could ultimately thwart new generation products and programs.

Do not *induce* cross subsidies at the costing level to camouflage inefficiencies and intentionally make other products carry the burden. All activities should be costed based on actual total cost; if the companies chooses to *discount* certain products, all the cost/profit impacts should be understood.

Relevant decision making. Good decisions are the keys to success in any business venture, and this especially applies to BTO&MC. Unfortunately, many managers and engineers try to make decisions by the numbers, when the numbers are misleading or even irrelevant. Again quoting Johnson and Kaplan:

> Ironically, as numbers become less relevant, more people manage "by the numbers"
> – Johnson & Kaplan, *Relevance Lost*

> "Ironically, as management accounting systems became less relevant and less representative of the organization's operations and strategy, many companies became dominated by senior executives who believed they could run the firm 'by the numbers'." [8]

Again quoting the Ernst and Young Guide:

> "If the costs are wrong, then all decisions about pricing, product mix, and promotion could be undermining long-term profitability."[9]

Johnson and Kaplan state that accurate, relevant numbers, based on total cost, can lead to much better decision making:

"The management accounting system also needs to report accurate product costs so that pricing decisions, introductions of new products, abandonments of obsolete products, and responses to the appearance of rival products can be made with the best possible information on product resource demands." ". . . An ineffective management accounting system can undermine even the best efforts in product development, process improvement, and marketing policy." [10]

Typical costing is of little help to reduce cost & improve productivity
– Johnson & Kaplan, *Relevance Lost*

Cost Management. Since one of the major challenges today is to produce products at low cost despite the other challenges of speed and quality, cost management takes on a new level of importance. But conventional cost management systems are of little help here:

"Management accounting reports are of little help to operating managers attempting to reduce cost and improve productivity." [11]

Downward spirals. Cost accounting distortions can cause "reinforcing" behavior (reinforcing loops) that can cause a business to "spiral down." The concept of reinforcing loops was presented by Peter Senge in his book, *The Fifth Discipline.*[12] Companies making both high-volume and low-volume products, as pointed out by the above cost management references, really do have different overhead demands. However, if overhead is averaged, then the following spiral can occur, as shown in Figure 12-1.

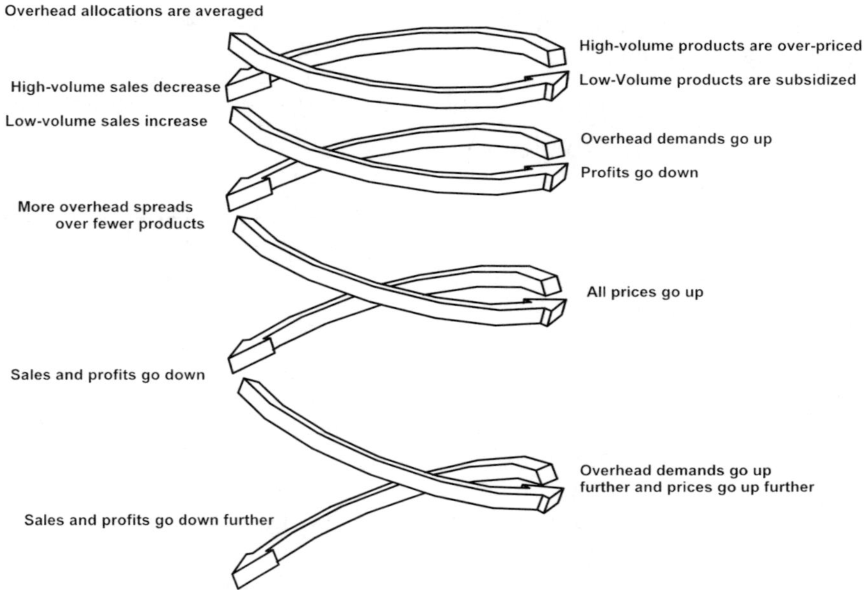

Figure 12-1: Cost Distortion Downward Spiral

Thus, the loop reinforces and the company continues to "spiral down." This may drag a company down to the point of unprofitability, or weaken an otherwise strong company.

B. Estimate the degree of cost distortions

- Subjectively draw conclusions about how much cost distortion is probably occurring when overhead is spread "like peanut butter."

- Conduct a pilot investigation to quantify one worst case product cost distortion by interviewing *everyone* involved and manually adding up all the time they spent. Choose a low volume or seldom built product suggested by the product line rationalization. Use polls and surveys to narrow the search by asking manufacturing people to *vote* on a candidate.

C. Understand the value of total cost measurements

Total cost measurements can be extremely valuable for the following decisions:

- It is essential to know the relative profitability of existing products in order to identify:
    - The most profitable products and market segments, to be the basis for product development prioritization and resource allocations
    - The least profitable products to rationalize away, when products can be sorted by profitability
- Quantifying the value of the BTO business model and developing metrics that support implementation.
- Justifying automation, tooling, and CNC machine tools
- Supporting and encouraging standardization, especially justifying the total cost savings of consolidating expensive parts
- Justifying design tools, training, and overhead reduction programs
- Sales goals and incentives based on *profitability* instead of volume of *sales*

RESISTANCE TO TOTAL COST MEASUREMENTS

Despite the deficiencies pointed out earlier about traditional cost systems and case study testimonials, like the one cited above, many managers still resist company-wide total cost measurements. Their resistance to change is usually based on the following misconceptions. Many managers:

1) **Do not accept the deficiencies** in the current system. This was the topic of the first section of this discussion on total cost accounting. Many of the books referenced in this section make thorough arguments about this point, especially Johnson and Kaplan's *Relevance Lost, The Rise and Fall of Management Accounting.* [13]

2) **Underestimate the benefit.** This chapter reinforces the value of relevant cost numbers as a basis for good decision making, in general, and, specifically for decisions governing product line rationalization, standardization efforts, implementing manufacturing flexibility, and many aspects of product development.

3) **Overestimate the effort** to make any improvements. The last half of this chapter presents easy ways to quantify total cost.

TOTAL COST THINKING

Even before any formal total cost accounting programs are implemented, companies can improve some decisions subjectively by using *total cost thinking.* The principles presented in this chapter can help individuals make better *subjective* decisions by correcting many misconceptions about cost and instilling the proper attitudes and beliefs. But, in order for this to happen, the corporate culture must encourage this. Management policies can either encourage or discourage this. If all proposals must meet strict criteria for payback and ROI (return on investment), this will govern the decision making process. If the criteria are based on traditional cost

> Before total cost is measured, use *total cost thinking*

accounting, then the decisions will tend to be governed by irrelevant numbers (mentioned above) and many truly good proposals will fail to win approval, because much of their benefit comes from benefits that are not quantifiable by the current system. If companies rigidly adhere to criteria based on incomplete costs, then attempts to inject subjective total cost decision making will fail, even if it is in the best interest of the company and its customers.

One subjective approach to this dilemma was proposed by Robert Kaplan in the article about justifying Computer Integrated Manufacture (CIM): *Must CIM be Justified by Faith Alone?*[14] Kaplan's technique to "work around" deficiencies in accounting systems was to:

1) compute how much the proposal fell short of the objective criteria: the *shortfall*
2) summarize the "intangible" benefits (all the benefits that could not be quantified)
3) pose the question, "Is it worth the shortfall to gain all these intangible benefits?"

One of the examples of part standardization, cited in Chapter 4, was the author's effort to standardize all resistors to 1% tolerance to replace the previous duplication of resistors in both 1% and 5% tolerance versions. There were no numbers available to justify the change. But it "just made sense" to cut in half the number of resistors in all three factories. Subsequently, the author has learned of people who did the same part consolidation and concluded quantitatively that the purchasing power of the combined orders offset the "cost" of the higher tolerance.

Sometimes, subjective decisions must be made *in spite of the numbers*. One of the author's clients, who makes water meters, consolidated seven raw castings into three by adding extra brass (for test ports) to every product whether or not they needed the optional tapped holes. This extra material *appeared* to add cost to some of the raw castings because the extra brass was not needed for test ports. In fact, the person who did the change felt like he would be "beat up" for raising the "standard cost." But, company management supported the change, knowing subjectively that it would lower the cost of variety enough to be a net gain and make operations more flexible.

Management policies can encourage total cost thinking by *empowering* people to make the best decisions, in their judgement, instead of trying to *limit bad decisions* by making them pass some predetermined threshold based on irrelevant numbers.

IMPLEMENTING TOTAL COST ACCOUNTING

Programs to implement total cost accounting are sometimes called Activity Based Costing (ABC), Activity Based Cost Management, Total Cost Management,[15] or the earlier "transaction-based" costing. Activity Based Costing is a system which focuses on the *activities* performed to produce products. Costs are either assigned directly to products or to activities, which are then assigned to products based on how much of these activity costs were incurred by each product.

ABC creates a separate decision making model

Cost systems. Total cost systems are *not* intended to replace the existing finance system. In most cases, implementations create independent decision making models. In the study that Cooper, Kaplan, et. al. did for the Institute of Management Accountants, this was the case:

> "No modifications to existing financial systems were required, and companies continued to run all their existing systems in parallel with their new ABC model." "The activity-based model was treated as a management information system, not as part of the accounting system." [16]

The numbers from this model were more useful than that available from the existing cost system:

> "Managers found the numbers generated from the activity-based analysis more credible and relevant than the numbers generated from the official costing system." [17]

COST DRIVERS

A cost driver is defined as the *root cause* of a cost – whatever "drives" the cost. Identifying cost drivers make the root causes visible and this has two important consequences.

1) Total cost can be measured
2) The behavior that actually lowers total cost can be encouraged

> Cost drivers identify the root causes of cost that should be quantified instead of averaged

The cost driver approach identifies key drivers of cost that should be quantified instead of lumped in with all other overhead. The cost driver approach is easy to implement and starts with the most important overhead costs that need to be quantified. New data collection efforts are focused on only a few key cost drivers. Cost drivers can be based on estimates, as long as there is universal consensus. Cost drivers can provide a more rational basis for performance measures.

For example, the *activities* that incur the following costs could be analyzed for significant ranges beyond the averages that usually are the basis for overhead allocation.

- Engineering Change Order costs
- Material overhead
- Quality costs, scrap, rework, and other non-value activities
- Inventory costs and inventory related costs
- Setup costs
- Equipment utilization
- Process yields, scrap, rework, etc.
- Costs of field service, repairs, warrantees, claims, litigation, etc.

The activities that cause these costs should be analyzed to find out what is causing the variation. Experienced managers will probably be able to identify the key cost drivers that are *driving* different costs in these activities, for instance:

- Volume: high volume or low volume
- Degree of customization: standard or custom
- Part standardization: approved or preferred
- Part destination: for production products or spare parts for products that are out of production
- Distribution costs: direct or through channels
- Product age: launching, stabilized, or aging (experiencing processing incompatibilities with newer products and/or availability challenges for parts and raw materials)
- Market niches: commercial, OEM, military, medical, and nuclear markets have varying demands for quality, paperwork, proposals, reports, certifications, traceability, etc.

Those activities that incur different costs from the variations in these cost drivers should be investigated. The costs of these activities should be charged accordingly. For instance, if low-volume products do incur more cost than high-volume products, then this should be reflected in the overhead allocation. If standard parts do incur less material overhead, then they should be charged a lower overhead.. If certain operations incur more overhead than others, then the cost drivers should reflect this, as will be shown in the following examples.

Intel's Systems Group. When the author implemented a part standardization effort, using the procedure described in Chapter 4, the result was that 500 "commonality" parts were identified as being preferred for new designs. These common parts really did deserve lower material overhead than the 13,000 remaining "approved" parts because they were purchased in higher quantities. And the standardization program wanted to encourage engineers to use these parts. To accomplish both these goals, the Accounting Department structured material overhead into a two tiered system: one rate for the 13,000 approved parts and a lower rate for the 500 commonality parts. This reflected greater "material world" efficiencies and encouraged their usage.

Tektronix Portable Instruments Division. To encourage part commonality and assign accurate material overhead, Tektronix assigned a material rate that was *inversely proportional* to volume. Thus, a high volume part had a very low overhead rate; conversely, a "low runner" was assigned a very high rate.[18]

HP Roseville Network Division (RND). HP RND formerly had only two cost drivers for its printed circuit board assembly: direct labor hours and the number of insertions. A special survey showed that resistors and capacitor insertions were about one-third the cost of integrated circuit insertions; manual insertion was three times as expensive as automation; and "low availability" parts had an additional cost of ten times their materials cost. So they implemented the following nine unit-based cost drivers.[19]

1. Axial insertions
2. Radial insertions
3. DIP insertions
4. Manual insertions
5. Test hours
6. Solder joints
7. Boards count
8. Part count
9. Number of slots

HP Boise Surface Mount Center (BSMC) implemented the following ten cost drivers for surface mount printed circuit board manufacture.[20] Note driver #7 which encourages part commonality.

Cost Pools	Drivers
1. Panel Operations	Percent of a whole panel; if one panel contains four individual boards, then each board is charged 25% of the panel rate
2. Small component placement	Number of "small" components placed
3. Medium component placement	Number of "medium" components placed
4. Large component placement	Number of "large"components placed
5. Thru-hole component insertion	Number of leaded components inserted
6. Hand load component placement	Minutes required to place all components that must be hand loaded rather than automatically placed on the board
7. Material procurement & handling	Number of unique parts in the board
8. Scheduling	Number of scheduling hours during a six-month period
9. Assembly setup	Number of minutes of setup time during a six-month period
10. Test & rework	Number of "yielded" minutes of test & rework time per board

Figure 12-2 shows the changes in product costing after implementing these cost drivers. Note that one-third of the products had their costs go down and two-thirds had their costs go up, with one product doubling in cost!

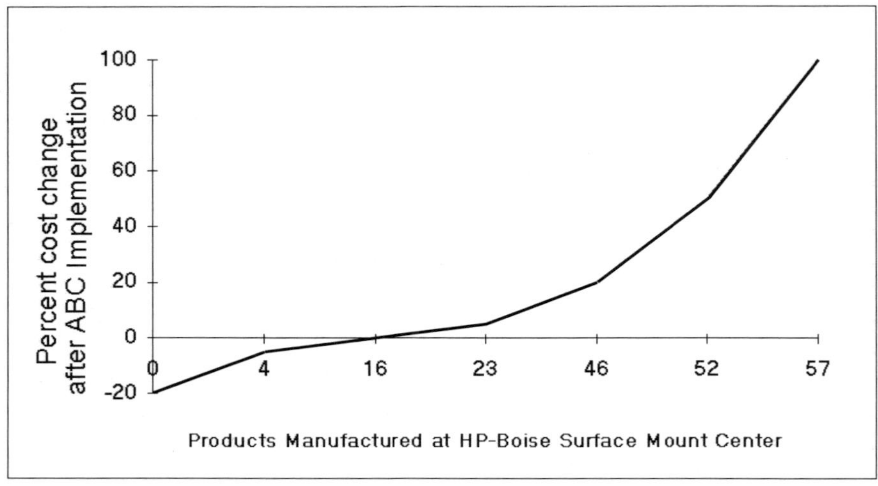

Figure 12-2: Changes in Cost after Implementing ABC

Results: "Accountants now provide important inputs into product design and development decisions. Under the prior cost system, all overhead was applied as a percent of direct material cost, and it was difficult to understand how changing a board's design would change manufacturing costs. Also, designers had little motivation to optimize the board for efficient production. With ABC, however, the cost system attempts to mirror the manufacturing process, so that engineers and production managers easily can see how design changes will affect cost."[21]

"abc" – The *Low Hanging Fruit* Approach

> It is better to be *approximately correct* than *precisely wrong*
> – Doug Hicks,
> *ABC for Small and Mid-Size Businesses*

An excellent "how-to" book oriented toward cost-effective ABC implementations is Douglas Hicks' *Activity-Based costing for Small and Mid-Sized Businesses.*[22] It is based on the valid premise that *it is better to be approximately correct than to be precisely wrong; accuracy is preferable to precision.*[23] Said another way, it would be better to be *approximately right* than *precisely wrong!* Knowing that a product has a negative profit margin between -55% and -65% is more valuable than *thinking* it has a positive profit margin of exactly 10.89%. Hicks' claims that the false pursuit of precision in product costing is unrealistic: "No cost accounting system provides an organization with precision. *All* product costing is approximate. *All* cost systems contain too many estimates and allocations to be precise." [24]

With this focus on relevancy over precision, it is easier to implement total cost measurements than the "full-blown" ABC, which strives to measure all activities. Hicks calls this "activity based costing" with the lower case acronym, "abc."

> "In *abc*, activities are defined as *groups of related processes or procedures that together meet a particular work need of the organization.* Under this definition, the activities of the Accounts Payable department would most likely be accounts payable. Period." [25]

However for material overhead, many purchasing costs are avoided by *kanban* and "bread truck" deliveries. In addition, fewer types of parts ordered in larger quantities reduces purchasing costs and increases purchasing leverage. Companies implementing JIT can expect to cut expediting efforts by a factor of *three*.[26]

One of these shortcuts that makes abc easy is to *estimate* the percentage of an activity's cost caused by a particular cost driver, for instance, high-volume products compared to low-volume products. Thus, instead of averaging the cost allocation where all products get the same charge, the low-volume products would be charged, say, 80% and the high-volume products 20% (for the typical Pareto effect), as shown in Figure 3-2.

For instance, to set material overhead that is relevant, show the purchasing manager the standard parts list generated by the techniques of Chapter 4. Ask what percentage of purchasing agents' efforts focus on these standard parts. The answer is usually that only 10% of material overhead is associated with standard parts since they are simply higher volume reorders of easy to find parts. Then adjust the material overhead from 50/50 (all parts get the same overhead) to 90/10 (overhead applied according to overhead effort).

Ask purchasing to estimate what % of effort goes to standard parts. If 10%, then allocate overhead 90/10 instead of 50/50

Hicks presents a simplified approach to implementing *abc*, with the emphasis on accuracy and relevance, rather than precision. He proposes an abc cost model in a format suitable for spreadsheets, like Excel.

An alternative to creating a model on a spreadsheet would be software specifically developed for ABC analysis. The eight case studies, cited in the Institute of Management Accountants ABC study, all used such software packages on PCs.[27]

Implementing "abc"

Understand the importance of total cost measurements for relevant costing, pricing, and decision making. For any "cost reduction" program, the measurement of cost is just as important than the steps to reduce cost because total cost measurements:

- helps the company make the right strategic decisions that will be most effective
- keep directing behavior that will continually reduce cost
- quantifies the real cost savings (or losses), which then affect subsequent decisions

The drive for implementing abc should come from the most *motivated* group while the implementation could be by the most *willing and able* group, which may or may not be the Finance Department.

Approach and label the program in a way that mitigates resistance and generates support. Some formal "Activity Based Costing" programs have been so cumbersome that they have died under their own weight, unlike the low-hanging-fruit approaches like "abc," which quickly generate benefits without consuming a lot of time and resources. If "Activity Based Costing" is not an appropriate label, it could be called something like, "Costing Model."

Identify the *cost drivers* of activities that should be quantified instead of lumped in with other overhead.

Roll the quantified cost driver data into the total cost model. Keep it up to date.

Make sure the total cost information is readily accessible, easily understood, used for all cost-based reporting, and the basis for decision-making.

IMPLEMENTATION EFFORT FOR TOTAL COST

Most companies do not need a system as complex as would be needed for the multinational mega-corporation, despite misconceptions to the contrary. Implementing some degree of total cost measurement can be achieved with modest resources. Of the eight companies that implemented ABC in the Institute of Management Accounts study, the companies that used "medium involvement" of outside consultants took an average 6.5 months by 2.1 FTEs (full-time equivalent workers) to implement the ABC model. Companies that used "active involvement" of consultants took an average of three months by 1.6 FTEs.[28]

> Implementation efforts can be measured in hours, not years

One practitioner reported that implementation efforts have "ranged from 80 hours for a small commercial printer to 500 hours for a large automotive supplier with very poor historical financial and operating records." [29]

TYPICAL RESULTS OF TOTAL COST IMPLEMENTATIONS

When total cost is implemented, companies start to see the real picture about product cost, which is often surprising. Cooper, Kaplan, et. al. refer to the "typical ABC pattern," where several offerings are shown to be highly profitable, most at or near breakeven profitability, and a few highly unprofitable.[30] A Schrader-Bellows case study[31] showed that, out of seven products originally thought to be profitable, three actually were profitable, one was barely breaking even, and three were unprofitable, with one highly unprofitable.

> Typical ABC discovery: several profitable, most break-even, some highly unprofitable
> – Cooper, Kaplan, et. al., *Implementing ABC Management*

Total cost analyses often adjust manufacturing costs up for most products, while lowering them only for a few "deserving" products. After HP implemented the nine cost drivers cited above, they found that 72% of the products were really costing more than assumed, as shown in Figure 12-2. This same percentage of products costing more is shown graphically in Figure 3-7. In the HP example, cost adjustments ranged from slightly lower to double! [32]

ENDNOTES/REFERENCES

1. Michael R. Ostrenga, Terrence R. Ozan, Robert D. McIlhattan, Marcus D. Harwood, *The Ernst & Young Guide to Total Cost Management,* (1992, John Wiley & Sons).

2. Robin Cooper and Robert S. Kaplan, "How Cost Accounting Distorts Product Costs," *Management Accounting* (April 1988).

3. H. Thomas Johnson and Robert Kaplan, *Relevance Lost, The Rise and Fall of Management Accounting,"* by (1991, Harvard Business School Press).

4. Michael R. Ostrenga, Terrence R. Ozan, Robert D. McIlhattan, Marcus D. Harwood, *The Ernst & Young Guide to Total Cost Management,* (1992, John Wiley & Sons), p. 27.

5.Douglas T. Hicks, *Activity-Based Costing for Small and Mid-Sized Businesses, An Implementation Guide,* (1992, John Wiley), p. 20

6. Johnson and Kaplan, *Relevance Lost, The Rise and Fall of Management Accounting."*

7. Robin Cooper, Robert S. Kaplan, Lawrence S. Maisel, Eileen Morrissey, Ronald M. Oehm, *Implementing Activity-Based Cost Management,* (1992, Institute of Management Accounts, Montvale, NJ)), page 4.

8. Johnson and Kaplan, *Relevance Lost, The Rise and Fall of Management Accounting."*

9. Ostrenga, Ozan, McIlhattan, and Harwood, *The Ernst & Young Guide to Total Cost Management,* p. 146.

10. Johnson and Kaplan, *Relevance Lost, The Rise and Fall of Management Accounting."*

11. Ibid.

12. Peter M. Senge, *The Fifth Discipline, The Art and Practice of The Learning Organization,* (1990, Doubleday/Currency).

13. Johnson and Kaplan, *Relevance Lost, The Rise and Fall of Management Accounting."*

14. Robert S. Kaplan, "Must CIM be justified by Faith Alone?" *Harvard Business Review,* (March-April, 1986), p. 87.

15. Ostrenga, Ozan, McIlhattan, Harwood, *The Ernst & Young Guide to Total Cost Management.*

16. Cooper, Kaplan, et. al.,*Implementing Activity-Based Cost Management,* p. 7.

17. Ibid.

18. Robin Cooper and Peter B. B. Turney, "Internally Focused Activity-Based Costing Systems," *Measures of Manufacturing Excellence,* edited by Robert S. Kaplan (1990, Harvard Business School Press) pages 292 - 293.

19. Cooper and Turney, "Internally Focused Activity-Based Costing Systems," from *Measures of Manufacturing Excellence.*

20. Mike Merz, Professor of Accounting at Boise State University, and Arlene Harding, Finance Supervisor at HP BSMC, "ABC Puts Accountants on Design Team at HP," *Management Accounting,* (September 1993), page 22 - 27.

21. *"ABC Puts Accountants on Design Team at HP,"* by Mike Merz, Professor of Accounting at Boise State University and Arlene Hardy, Finance Supervisor at HP BSMC; published in Management Accounting, September 1993, pp. 22 - 27.

22. Douglas T. Hicks, *Activity -Based Costing; Making it work at Small and Mid-Sized Companies,* Second Edition (1998, John Wiley & Sons).

23. Douglas T. Hicks, *Activity-Based Costing for Small and Mid-Sized Business, An Implementation Guide,* (1992, John Wiley & Sons), p. 7.

24. Ibid., p. 8

25. Ibid., p. 35.

26. A. Ansari and B. Modarress, *Just-In-Time Purchasing,* (New York, Free Press, 1990), p. 44.

27. Cooper, Kaplan, et. al., *Implementing Activity-Based Cost Management,* pp. 6, 25, and 256.

28. Ibid., p. 296.

29. Douglas T. Hicks, *Activity-Based Costing for Small and Mid-Sized Business, An Implementation Guide,* (1992, John Wiley), p. 9.

30. Cooper, Kaplan, et. al., *Implementing Activity-Based Cost Management,* p. 5.

31. The Schrader-Bellows case study is described in Harvard Business School Case Series 9-186-272; A summary of the findings appears in "How Cost Accounting Distorts Product Costs," by Robin Cooper and Robert S. Kaplan, *Management Accounting,* (April, 1988).

32. Merz and Hardy, *"ABC Puts Accountants on Design Team at HP,"*, pp. 22 - 27.

13

IMPLEMENTING BTO&MC

THE FLEXIBILITY SPECTRUM

The flexibility spectrum (Figure 13-1) shows how manufacturers evolve from inflexible mass production to the integrated business model proposed herein. This spectrum indicates some of the key steps along the journey to build-to-order and mass customization.

Many companies may have competed some of the steps as part of a "lean" effort, but some get stuck after a couple of steps. Others may jump ahead to try to tackle some worthwhile goals, but, because they skipped several important prerequisites, their efforts are more difficult than necessary and the results may achieve less than what would otherwise be possible.

Working our way through the journey from Mass Production in Figure 13-1, most efforts to improve flexibility start with some kind of lean production effort to reduce setup times, batch size, and work-in-process (WIP) inventory (Ch. 8). In addition to efforts to reduce delays to set up machines and eliminate inventory between them, the processing itself, whether manual or mechanized, needs to be made flexible to process the needed variety without delays (Ch. 8).

Least Flexible **Most Flexible**

Mass production
Setup/batch/WIP reduction
Flexible processing
Total cost measurements to justify subsequent steps
Compensation that supports business model
Rationalization; focus on strongest; drop losers
Standardization of parts and materials
Products/processes designed for BTO&MC
Setup/batch/WIP elimination
Computer controls, links, & automatic changes
Product family lines/cells
Batch-size-of-one for standard products
Build on-demand; forecasted materials
Aggressive standardization
Build on-demand; pulled materials
Std., custom products on-demand
Integrated business model

Figure 13-1 Flexibility Spectrum

The next several capabilities in Figure 13-1 are all-important prerequisites that will be needed before proceeding further. Total cost measurements (Ch. 12) quantify total cost and help rationalize products by true profitability, encourage aggressive standardization, and justify further commitments in resources and capital. Compensation based on profit, instead of sales units or dollars, supports the business model by focusing sales on the most profitable products that fit within BTO&MC operations.

Rationalization (Ch. 3) simplifies supply chains, operations, and standardization efforts and frees up people to do the subsequent changes. Standardization of parts (Ch. 4) and materials (Ch. 5) is the foundation for flexibility and greatly simplifies procurement and internal logistics.

Existing products that are not designed for BTO&MC may have a needless and crippling proliferation of parts and materials. Specified parts may be hard to get quickly. The products and processes may have too many setups designed in and the product/process design may not make optimal use of CNC capabilities. BTO&MC companies *concurrently engineer* product families and versatile processes which includes designing around standard parts and materials, designing for optimal use of CNC, and designing products and processes to eliminate setup (Ch. 10). Since designing products for BTO&MC, and the prerequisite standardization, will take some time develop and get into production, these should be started as soon as possible.

All of the above will allow the *elimination* of setup, batches, and WIP inventory (Ch. 8). With the addition of computer controlled process and production lines rearranged into product family lines or cells, companies will be able to achieve a batch-size-of-one for standard products.

With these capabilities in place, manufacturers will be able to build standard products on-demand using forecasted materials (the Dell Model). After implementing aggressive standardization and other supply chain initiatives (discussed in Chapters 3 through 7), companies will be able to build standard products on-demand with materials spontaneously pulled into assembly operations. Adding mass customization capabilities (Ch. 9) will allow manufacturers to build standard or customized products on-demand without forecasts or inventory.

Although many of these capabilities may be implemented by various groups or departments, they need to be integrated into the BTO&MC business model.

BTO&MC IMPLEMENTATION

Identify Goals and Drivers

Identify goals and challenges that can be improved by BTO&MC. Start with bottom line goals like profit improvement, growth, market share, stock performance, and so forth. Then identify the *drivers* for those goals, in other words, the activities and programs that would *drive* the achievement of those goals, for instance cost reduction, delivery time reduction, and better customer satisfaction through better cost, delivery, quality, and mass customizing specific needs.

Obtain Customer Inputs

Obtain customer input on the relative importance of their *preferences* and *competitive rankings,* which are two of the main inputs for the QFD process discussed in Chapter 10. Even if major new product developments are not forthcoming soon, obtaining this information will be helpful for BTO&MC planning now and can eventually be used in the full QFD process.

Plot customer preferences as a function of competitive rankings (Ch. 10) to identify the top *zones* of importance – things that are important to the customer where your company is viewed as weaker than the competition.[1]

Identify Where to Start

> Start with products with the most variety, labor, overhead, delays, inventory, and customization difficulty

First, rationalize the product lines (Ch. 3) to remove, at least tentatively for the future: products, options, and variations that will not be in the BTO&MC system. Then look for product families (or products that can be grouped into families) with the following characteristics:

- High *output* variety and potentially low *input* variety. Even if the input variety is currently too high, it can be reduced by standardization (Chapters 4 and 5).
- Trends indicating increasing product variety over time
- High labor cost; increasing labor shortages; significant or erratic overtime; excessive firefighting
- High setup costs and delays
- Unreliable forecasts, which will get worse as product variety and market volatility increase
- Inventory problems: excessive carrying costs, obsolescence, write-offs, out-of-stocks, and so forth.

- Delivery response pressures or opportunities
- The most difficult customizations
- High visibility, the success of which will help spread implementations to other product lines

For various potential activities, estimate relative benefits and cost in terms of human resources and capital. These can be prioritized by plotting benefit vs. effort and identifying and grouping activities that are:

- "Low hanging fruit" opportunities with a high payback for a low effort
- Related activities already supporting other programs and goals
- Activities that may be needed to complete the system, such as bringing in-house long-lead-time parts production (Ch. 6)
- Expansion of mass customization capabilities to expand the ranges of customization or branch out to adjacencies, niche markets, derivatives, and so forth.

Defer enhancements that would be nice to have, but not critical to the initial implementation, like networking data transfers, on-line manual instructions, and so forth. For the interim, it may be advantageous to defer challenging tasks that have near term expeditious alternatives, which can be improved or automated later.

Decide which product lines to implement first and which activities within those product lines to address first. Tentatively map out a progression of subsequent implementations.

Training

In general, training is a key enabler in successful manufacturing companies. For the 25 finalists in Industry Week's annual Best Plants rating, 64.2% schedule 40 or more hours of training per employee annually compared to the average of 7.3% for all 2,511 manufacturers surveyed.[2] *The best plants scheduled almost nine times more training than the average plant!*

For implementing something as broad as build-to-order and mass customization, training is an important early step. Some companies may be working on some elements, but they may be "islands of excellence" that may not be coordinated well with others or may not gel into coherent business model. Existing programs may not be going far enough: Lean programs may only be tackling the low-hanging-fruit and may only be *reducing* setup and batches, instead *eliminating* them where necessary for spontaneous build-to-order. Standardization efforts may not have been *aggressive* enough for the spontaneous resupply as outlined in Chapter 7.

> BTO&MC training can show how to integrate many initiatives into a coherent, viable business model

In other cases, programs involving lean production and supply chain management may already be bogged down because they didn't simplify operations and supply chains first with product line rationalization (Ch. 3), standardization (Chapters 4 and 5), automatic resupply (Ch. 7), and design for lean (Ch. 10).

Other companies may have neither implemented anything useful nor had any training, in which case training would help the most by showing them all that needs to be done in the context of a coherent business model.

The worst category would be companies that have done nothing except actions that are *counterproductive* to BTO&MC, like excessive outsourcing, manufacturing off-shore (Ch. 6), or low bidding to find cheap parts or pressure suppliers (Chapters 7 and 11).

The right training would need to be customized for the company's products, operations, and culture, taking into account how much progress has been made in all the important activities. Although any motivated group could sponsor and arrange the training, the training itself should be given to all relevant functions and departments *and* corporate leadership. When Freudenberg-NOK implemented lean production, Joseph Day, Chairman and CEO, said, "We required our executives to be participants in the first wave in training. In that first wave we also had all our union executives."[3]

A good facilitator would encourage discussions on all topics as they are presented to discuss how they would be applied and alleviate doubts and concerns. In addition to all the elements, such as rationalization, standardization, integration, spontaneous supply chains, on-demand lean production, training should show how these elements fit into an overall business strategy to achieve a viable business model.

Training should be presented by someone who is thoroughly familiar with all the relevant principles, has credibility and an open-minded perspective, and has enough experience to answer questions and engage the audience in discussions on how to apply these principles.

As an example, the author's baseline agenda for his in-house seminar is shown on page 469.

Free Up Resources

First free up time for key implementers by:

- Rationalizing product lines to eliminate high-overhead/low-profit products that waste a lot of resources throughout the organization (Ch. 3). This can free up people *immediately.*

- Outsourcing legacy products and spare parts production and management, unless synergistic with BTO&MC production processes *and* supply chains.

- Scrutinizing requests for unusual product variations to avoid fire-drill modifications and customizations.

- Using configurators to automate the processes of determining the feasibility, cost, and time to do customizations, instead of the time-consuming and less accurate manual estimating.

- Delegating firefighting and routine tasks to the lowest levels; to free up key implementers, make available more aids, helpers, interns, "gofers," and administrative help in general.

- Shifting parts and materials from MRP to automatic resupply techniques; refocus purchasing people from buying parts for every order to setting up spontaneous supply chains for all future orders (Ch. 7).

- Shifting to vendor/partnerships would free up many purchasing people from all the tasks associated with low-bidding , in addition to making product development teams more efficient when these pre-selected vendors help them design custom parts.

- Avoid all the resource drains that result from trying to lower cost with bidding (Ch. 7) and offshore manufacture (Ch. 6).

- Don't waste resources trying to remove cost from products after they have been designed.[4]

- Minimizing competing demands, especially for activities that may not be supportive of BTO&MC or be obsoleted, such as warehouses and many aspects of ERP systems.

- Minimize resource demands for new product development from Engineering, Manufacturing, and Purchasing. The author's product development seminars cover 26 ways to free up resources including: prioritizing projects to focus on the highest return efforts; understanding "lessons learned" to avoid repeating past mistakes; buying off-the-shelf parts instead of designing them; reusing parts, modules, and engineering; pre-selecting vendor/partners who will then help design the parts they will make; designing *parametric templates* for product variations; designing in quality; and avoiding later problems with thorough up-front work.[5]

Minimize Fears and Inhibitions

When implementing any change, it is very important to overcome any potential resistance to change and provide motivation to enthusiastically implement the changes. Chapter 14 presents the business case for BTO&MC and this should be clearly communicated to everyone, including how these changes will improve the company's performance and everyone's career prospects. All inhibitions, resistance, and fears should be addressed and nullified proactively. Part of overcoming inertia is to emphasize the risks of no change at all, which could be considerable in fast changing industries. The President of Olympus, Tsuyoshi Kikukawa, said *"If you don't take a risk on a new idea, that, in itself, becomes a risk."*

> Offer a *no layoff pledge* that no job will be lost to efficiency improvements

One of the greatest fears surrounding any changes that improve "efficiencies" is the fear that improving efficiencies may eliminate jobs. Unfortunately, there are enough horror stories circulating to support these fears. The most effective way to allay these fears is to issue a *no layoff pledge* that assures all employees that no one's career will be at risk in any way from implementing the improvements. In *Lean Thinking,* Womack and Jones advise that "you must guarantee that no one will lose their job in the future due to the introduction of lean techniques. And you must keep your promise."[6]

If this is not the case and there is *even a perception of job risk,* then workers will probably offer strong, if not always visible, resistance to the proposed changes. On the other hand, a no-layoff pledge could allow companies to offer more job security to union workers in exchange for relaxing work rules, which would then allow more flexibility to implement BTO&MC.

At the other end of the employee spectrum, senior managers and sales people must not perceive that the changes will bring about any risk to their salary, bonuses, stock options, or commissions.

Establish Criteria for "Within the BTO&MC System"

As mentioned throughout this book, BTO&MC operations can quickly and efficiently build any product "within the BTO&MC system." Therefore companies need to proactively establish these criteria, for instance:

- Profits are above a pre-established threshold *by total cost measures;* configurators can use total cost measures to predict each order's profitability before the order is accepted or rejected.

- Orders are within capabilities of Engineering, Manufacturing, and Procurement to deliver products within time, cost, and quality expectations *without compromising other profitable operations.*

- Parts and materials should be easy to procure spontaneously. Some unusual procurements could be accepted if they are easy to integrate and can be procured within product delivery expectations.

- Orders are expected to have minimal overhead demands, fire-drills, hand-holding, customer-induced changes, or time-consuming manual intervention. Remember that mass customization resources (Engineering, Information Technology, Manufacturing, Procurement) should focus on making operations fast and efficient; Fire-drill craft operations have a different focus and staffing, as will be discussed below.

- During the transition to BTO&MC, product variations that *look like* they will be part of the implemented system should *continue* to be offered, even if this means using current operations, until more efficient BTO&MC capabilities come on line. Each new product should be designed to fit into some BTO&MC line that exists or will be ready at the time of release.

- Replacement parts and spare parts, as a category, should *not* be part of the BTO&MC system, unless certain spare parts could satisfy all the criteria presented herein with respect to spontaneous supply and on-demand production.

- Companies with multiple plants can designate one plant for a pilot and then one or more to be exclusively devoted to BTO&MC. Other plants could continue to build, in the conventional way, the products that don't fit into BTO&MC operations. One plant should be designated to build older legacy products, spare parts, and "oddballs" that survive product line rationalization, as discussed below.

Phases of Business Idea Implementation

An excellent new that book investigates the process of implementing new business ideas is: *What's the Big Idea? Creating and Capitalizing on the Best Management Thinking.*[7] Successful "idea practitioners" went through several phases the author called Progenitor, Pilot, Project, Program, Perspective, and Pervasive.

Progenitors are the champions who propose the idea and advocate its implementation. A *pilot* can be a modest or small-scale implementation to prove feasibility in the organization and get attention for the idea. The idea becomes a *project* when senior management support the idea and direct resources and money to explore the idea further.

A successful project then migrates into a *program,* which would "generally involve many parts of the organization at once and would go on for several years. It is at the program stage that the idea has its maximum visibility and awareness levels throughout the organization."[8]

Regarding staffing and implementation, the authors say that "It is relatively easy to staff up for a pilot, or even a single project. But by the time an organization treats an idea as a program, it's difficult to work on the new idea as well as the existing business. At this point it's common for the firms to seek outside help from consultants or other advisers."[9] The book adds that consultants have credibility that insiders sometimes don't possess and that "consultants can also help by being a private sounding board for the formulation of the idea, and by gathering materials and evidence in support of it."[10]

The result of good idea implementation is when it becomes *pervasive* behavior

---

Successful long-term programs then become a *perspective* that has "penetrated the mind of the organization." Finally, the idea becomes *pervasive* when everyone automatically practices the idea as an ingrained part of the culture. One savvy Wall Street analysis said he actually *downgrades* the stock of any company whose executives tout quality programs! His attitude is that *true* quality happens when it is sufficiently *pervasive* throughout the company so that no one has to talk about it.[11]

CREATE IMPLEMENTATION ROAD MAP

Road maps for implementing BTO&MC should emphasize several parallel, coordinated tasks that can be implemented simultaneously to support the business model vision.

Each category of tasks should be structured so that it can generate some *early useful results* and a *progression of useful deliverables* throughout the time line. This is important to deliver early paybacks, keep generating support, and support other activities (Figure 13-2).

> *Revolutionize* your business model with *evolutionary* self-supporting steps

Implementation Activities with Early Deliverables

Overall Planning

Establish the vision
    Establish the overall implementation plan

Product Line Rationalization

Profiles to identify "red flags" for more scrutiny on custom orders
    Data gathered: sales histories for products and variations
    Profiles to scrutinize orders; prioritize by profits, criteria to reject
        Pareto plots & analyses
            Decisions to eliminate, outsource, or improve
                Refocus resources on remaining products

Standardization

    Data gathered: historical usage for each category
        Pareto plots, analysis, consolidation
            Standardization lists generated & implemented

Supply Chain

- Convert to automatic resupply
  - Supplier lead time reduction
    - Arrange to pull from suppliers on-demand
      - Develop partnerships
        - Bring in-house/acquire as necessary

Lean Production

- Setup/batch/WIP reduction
  - CNC utilization optimized
    - Expedient integration
      - Computer integration

Product Development

- Develop products for lean, BTO, & MC
- Pre-engineering of options, variations, & customizations
  - Parametric templates
    - CAD/CAM; on-demand CNC program generation

Configuration/Information Technology

- Compile knowledge used to evaluate sales, orders, and bids
  - Paper/spreadsheet "configurator"
    - Configuration software developed/purchased
      - Integrated information flow
        - Configuration software deployed

Total Cost Measurements

- Profiles for early identification of low-profit products/orders
- Cost drivers identified
  - Cost drivers quantified
    - Cost model developed
      - Compensation based on profit

Figure 13-2: Implementation Activities

Any implementation efforts that take too long without spinning off early useful results will be slow to generate paybacks and may lose support and backing before finishing "the whole package." Further, other activities will not benefit from early input and coordination. Worse, the entire implementation could be jeopardized if critical activities take too long or lose support.

One of the earliest deliverables are *profiles,* which are based on subjective judgements and therefore can be compiled before the more formal programs produce quantitative results. For instance, a profile can be generated to identify low-profit, money-losing, or time-wasting products from rudimentary historical data or even polls and surveys that ask workers in many job functions which types of products or variations "cost more than we think." Profiles also can identify fire-drill products, those with high overhead demands, those that would be hard to complete on-time, and those that are difficult to build or customize. These profiles can, at the minimum, be a trigger or "red flag" for further scrutiny on which products, options, or variations should be eliminated/outsourced or which customizations should be rejected for automated orders. Such scrutiny could trigger manual investigations to add up enough elements of time lines and total cost to make informal decisions. More detailed profiles could be used to make rationalization/rejection decisions without additional investigations.

Various activities, as suggested below, may be implemented by multi-functional task forces or by specific departments. However, any implementation driven by company departments must ensure coordination with other departments, all working to the common good.

Each activity should be led by a capable leader who understands "the big picture," has leadership abilities, understands design and manufacturing processes, and has the support and respect of the team and management. The overall BTO&MC effort should be led by a dedicated project leader who throughly understands the vision and can coordinate all the activities to achieve that vision.

Pilot implementations for product lines may be started as soon as enough supporting activities have reached "critical mass." Smaller implementations for individual products or modules could be started earlier.

Implementation Cautions

The acquisition/merger strategies must support the overall business model. *Supply chain* acquisitions can speed deliveries and give control over cost and quality. However, acquisitions or mergers with *competitors* may result in integration distractions and take resources away from internal programs. The worst case scenario for BTO&MC would be any situation that results in acquired products being built in the same plant as the original product lines, which could drastically increase internal processing complexity and easily double the variety of parts and materials. Acquiring or merging with competitors might be acceptable if the focus is on distribution and markets and incompatible products are *not* manufactured in the same operations.

Protect mass customization from counterproductive policies or actions by resisting:

- Sales goals that encourage the sales force to "take all orders"
- Outsourcing key manufacturing operations; instead, strategies should focus on outsourcing products that don't fit into BTO&MC operations.
- Return on capital investment measures and goals that discourage equipment investments
- Low-bidding for parts or services; develop partnerships instead
- Mergers or acquisitions incompatible with the BTO&MC business model
- Temptations to build other company's products as a contract manufacturer to "fill the factory"
- Competition for resources from programs that do not complement BTO&MC or even conflict with it (like all previous points)

Supporting, Self-justifying Programs

The following programs support BTO&MC and also generate so much benefit on their own that they can be self-justifying:

Flexibility efforts that reduce setup time and batch size could be justified alone based on:

- Improved response time
- Reduced setup costs
- Reduced inventory carrying cost
- Reduced spoilage, obsolescence, & overproduction discounting costs
- Reduced floor space needs
- Higher people productivity and equipment utilization
- Best utilization of scarce labor, especially during peak seasons, by not wasting resources building anything that may not sell.
- Best utilization of scarce parts, by not building them to products that may not sell.
- Contribution to related programs

Product line rationalization would be justified alone based on:

- Eliminate low-profit high-overhead products
- Eliminate "loser tax" on cash-cow products = lower price or more profit
- Eliminate or outsource products incompatible with flexible plants
- Free valuable resources
- Simplify supply chains
- Contribution to related programs

Total cost quantification to:

- Identify low-profit products for rationalization
- Better decisions based on total cost
- Quantifying the value of the BTO&MC business model and developing metrics that support implementation.
- Justifying automation, tooling, and CNC machine tools
- Supporting and encouraging standardization, especially justifying the total cost savings of consolidating expensive parts
- Justifying design tools, training, and overhead reduction programs
- Sales goals and incentives based on *profitability* instead of sales volume or sales dollars
- Help justify other programs

Product development for flexibility

- Support BTO&MC efforts with products designed (or redesigned) for BTO&MC, designed around standard parts, and designed for flexible manufacturing
- Develop boarder ranges of subsequent products within the capabilities of flexible production facilities
- Make test products in production environments for speed, breadth, and realism
- Concurrently develop broad product lines and versatile production lines, ultimately in *new* markets

Mass customization to allow fast and efficient production of products that are low-volume, test markets, niche markets, regional, for export, seasonal, or customized for stores, groups, events, or even individuals.

FOCUS AND STAFFING FOR BTO&MC

A BTO&MC site should focus on quick and efficient manufacture of a planned range of variation. The following activities should be focused and staffed as shown in Figure 13-3.

BTO&MC Focus & Staffing:	Not:
Sales/Marketing	
Take profitable orders within the BTO&MC system	Take all orders, regardless of profit or throughput implications
Compensated on profit	Compensated on units or revenue
Easy customizations	Difficult customization
Configurator quickly established validity and "what ifs"	Slow back-and-forth process to determine order validity
Orders complete and accurate	Order entry errors; surprises; changes
Few customer-induced changes	Many customer-induced changes
Engineering	
Parametric templates accept customization inputs	Manually modify or create unique drawings for every order
Automatic CNC program	CNC programs from separate efforts
Smooth predictable operations	Fire-drills
Most efforts improving system	Most efforts on each order
Manufacturing	
BTO and mass customization	Craft production or mass production
Automatic CNC fabrication	Skilled workers make parts manually
Many features in one setup	Many steps through many machines
No setup delays	Setup cost & delays
Smooth operations	Fire-drills
Smooth pre-planned procedures	Rough *ad hoc* procedures
Eliminate change orders	Many change orders
Supply Chain	
Simplify complexity	Try to *"manage"* complexity
Pulling parts and materials	Ordering parts & materials; waiting
Arranging automatic resupply once for each category	Purchasing parts and materials for each order

BTO&MC Focus & Staffing:	**Not:**
Information Technology	
Order entry, databases, & CNC	MRP & purchasing functions
Acquisitions	
Suppliers, distributors Help MC model at both ends	Competitors More parts, suppliers, & processes
Business Model	
Easily *earn* the numbers	Do what it takes to *make the numbers*
Grow sales with better cost, quality, time, satisfaction	Grow sales by mergers, acquisitions, or taking all orders
Grow profits with better focus, lower costs, & higher sales	Sacrifice profits for market share

Figure 13-3: Focus & Staffing for BTO&MC

Segregating BTO&MC and Oddball Products

To support the above focus and staffing, don't pollute a BTO&MC plant with products that are not compatible with BTO&MC, such as older legacy products, spare parts, craft production products, and "oddballs." Those should be eliminated, built at another plant, or outsourced to contract manufacturers. *Failing to designate such an site will result in either having to turn down work or forcing incompatible products into the BTO&MC facility.* Replacement parts, as a category, should be kept out of BTO&MC operations, but certain parts may be compatible if they "fit" in all respects.

> Don't pollute a BTO&MC plant with oddballs that don't fit

Ideally, craft and oddball products should be segregated from BTO&MC products by plant or building. The BTO&MC plant would be set up to build everything within predetermined ranges quickly and efficiently. Craft production plants would build unusual products using conventional practices.

Proactively designate a different site for oddball products

If this is not possible, then certain unusual processes could have parallel paths and unusual parts could be procured (at greater cost and slower delivery) *provided a smooth integration into the BTO&MC operation was assured.* Proactively establish interface standards so that easy integration is assured.

If the entire product falls outside the BTO&MC system, it would have roughly the same cost and delivery as conventional operations today or non-flexible competitors, except that converting to total cost accounting will probably "raise" the selling price to cover the *real* expenses. Oddball modules and subsystems that are integrated into the BTO&MC system will raise costs and cause delays if flows are disrupted.

Don't mix craft and BTO&MC activities in the same department. Don't expect a BTO&MC activities to change back and forth from mass customization to craft customization. Send unusual activities out of the BTO&MC system and integrate them back later.

Don't try to mass-customize everything right away. Remember, any products, options, or customization not included right away in the BTO&MC plan can still be done in the conventional way as long as they integrate well and don't compromise BTO&MC operations or cost, quality, or throughput goals.

Finally, for mixed operations, make sure that appropriate overhead changes are allocated, which is especially challenging if mixed operations operate on the same building or even the same site.

PRODUCT FAMILIES

A key step in implementing BTO&MC is segregating products into *families,* which are groups of products that can be manufactured on-demand efficiently. BTO&MC parts, subassemblies, and products (hereafter called "products") are built in *families* that (1) can all be built efficiently on-demand without setup delays with (2) have all the parts and materials always available without the cost, delays, risk, and uncertainties of forecasts and inventory. This can be accomplished by building product families together on the same line. Product families are defined accordingly by the these two criteria:

> Products are grouped to build families without setup delays

(1) Product Family Grouping for On-Demand Production.

For the on-demand production aspect, products are grouped together so all processes can build all product variations without any setup delays. This involves eliminating all setup (Ch. 8) and *designing* products to be built without setup (Ch. 10).

To accomplish this, a group of products (a *family*) should be identified for all the products can be built in all work stations without setup delays. Thus, product families are defined by grouping together products with similar equipment capability needs, fixturing, and CNC programming and manual instructions.

The breadth of the product family is determined by the versatility of all the equipment; actually, the *least* versatile work station in a line may determine this. Thus, the greater the versatility of the equipment, the broader can be the product family. Similarly, the greater the volume, the more narrowly the lines can be focused.

For efficient utilization of equipment, processing times on expensive machine tools should be fairly consistent for all products in the family. For less expensive equipment, this consideration is less important.

(2) Product Family Grouping for Part/Material Availability.

In order to have parts and materials (hereafter referred to as "parts") always available at all points of use, a spontaneous supply chain (Ch. 7) will have to be arranged for all the parts in the product family.

To accomplish this, the *same* family group of products (same as the above criteria) must *also* have all the parts available at all points of use without setup delays to order parts, wait for delivery, retrieve parts, load parts, or find and understand instructions. In some assemble-to-order operations, like Dell Computer, parts can be "pulled" from forecasted parts inventory (most of the time) but with a cost and risk penalty. In BTO&MC, the parts are made available *spontaneously* without forecasts or inventory using the techniques of Chapter 7.

The variety of parts in the family must be low enough for *all* parts for all products in the family to be distributed at *all* points of use. Thus, excessive part proliferation through the family tends to *narrow* the family, whereas aggressive standardization can *widen* the family. As mentioned above, greater volume will allow family product lines to be more narrowly focused.

Combining the build-to-order of standard products with the mass customization of custom products will increase the volume of the operation and thus allow the family to be more focused.

Grouping Products into Families

In order for BTO&MC to succeed, product families will need to be *grouped* into lines which can build any product in the group on-demand. Product families are probably not going to be based on which products appear on adjacent pages in catalogs or web-sites.

For true BTO&MC operation, product family groups *must satisfy both criteria* for on-demand production *and* spontaneous part availability. But before trying to group products, be sure to performed the prerequisite steps of eliminating setup (Ch. 8) and making material availability spontaneous (Chapters 3, 4, 5, 6, and 7).

One possible starting point would be to identify synergy or similarities among products with respect to processing, setup elimination, part variety, and part resupply strategies. Another possible starting point would be the most constrained criteria, which may be related to setup elimination, bottleneck machines, part variety, or part resupply.

Yet another factor would be the interrelation between line capacity and production volumes – one high-volume product and its variations could fill the line and be its own family. A few medium-volume products could be grouped into broader family. Products with lower volumes and higher variety are more challenging and may require (1) more flexible processing, which then becomes another criterion for product family grouping, or (2) more narrowly-focused lines with smaller capacities, which affects plant layout.

If an existing factory is being converted to BTO&MC operations, then minimizing the conversion effort becomes another criterion. Floor space availability may be a criterion, although inventory reduction efforts can drastically reduce floor space needs.

Various product lines will each have their unique product family strategies, but, with experience and a thorough grasp of these principles, products that are suitable for product families can be successfully grouped.

Products that *cannot* be grouped into any family may have to be eliminated or outsourced by product line rationalization (Ch. 3). If these "non-groupable" products are necessary for product line completeness, then they may have to be redesigned around standard parts (Ch. 4), standard materials (Ch. 5), optimal integration (Ch. 6), spontaneous supply (Ch. 7), and/or setup/batch/inventory elimination (Ch. 8). If neither of these options are possible, those products need to be moved to their own plant or area *with support demands and overhead allocations separated from BTO&MC operations.*

Special case product families could be created for *industry-specific* lines, which could build all or most of the needs of an industry or region, or *customer-specific* lines, which could build all the parts on-demand for a specific customer. These will be discussed later.

BTO&MC PRODUCTION STRATEGY

High-Volume/No-Variation Products

High-volume products that have no variations can continue to be mass-produced *on dedicated lines that will never have setup changes.* Avoid the temptation to introduce setups by adding different products to "fill the line" or maximize utilization for the following five reasons:

> Lines could be:
> - high volume/no variety
> - medium vol./no variety
> - low-vol./high variety

(1) The intervening products would be made in batches and inventoried, which is incompatible with the inventory-less BTO&MC philosophy.

(2) The cost savings of the BTO&MC model should exceed the cost of any perceived "loss" from underutilized equipment to keep the line dedicated.

(3) Adding something different to a predominant setup results *in two extra setups,* which is costly way to try to add capacity. Ironically BTO&MC itself increases capacity through *less* setup throughout the factory.

(4) The BTO&MC infrastructure should focus resources on tasks other than unnecessary setup changes and related firefighting.

(5) Dedicated lines can be optimized better for quality and throughput if they avoid product and setup changes.

There are more opportunities to apply continuous improvement (*kaizen*) to a dedicated line, thus allowing older equipment to be used longer.

Capacity needs for other products should be satisfied by medium-volume dedicated lines, flexible lines, or outsourcing for unusual products.

Medium-Volume/No-Variety Products

Medium-volume products that have no variations can also be mass-produced on smaller dedicated lines each of which produces one product without any setup changes. These lines could consist of equipment that is older, used, or somehow "obsolete," which can usually be procured at a fraction of their original cost on the used equipment market. Quality would be assured by kaizen continuous improvement, which can be more effective if settings never have to change for setups. Lower equipment costs and other overhead cost savings would allow these to remain dedicated and cost-effective, even if not always running at full capacity. The floor space needed for these lines would be made available through the reduction of inventory, which has been documented in many companies to cut floor space needs in half or more.[12]

Low-Volume/High-Variety and Custom Products

Low-volume products and high-variety or custom products can be built on flexible lines that can produce any product in a family without setup changes. These lines can be implemented according to the methodologies described in Chapter 8. Proven production processes and process controls could be utilized with innovation focusing on setup elimination. Custom products could also be made on these lines with the methodologies described in Chapter 9. Over time, these lines could be enhanced to include broader ranges of products.

Flexible and mass production lines could share certain equipment as long as all products could pass through without setup changes. For instance, flexible and dedicated lines could converge on a common heat treating furnace providing that there were no delays to change the temperature or other setup changes.

IMPLEMENTATION SCENARIOS

The following is a range of implementation choices for build-to-order and mass customization. These steps show the progression of build-to-order and mass customization from selective growth to industry domination. Some of these could be deferred to more appropriate times in the future, with the early steps as prerequisites, each of which will probably pay itself off as a freestanding program. Within each of these scenarios, opportunities can be further segmented by products, product lines, market segments, niches, and so forth.

Implementation can range from selective growth to industry domination

1) **Build *Standard* Products To-Order.** A logical starting point for implementation would be to develop the capability to build-to-order *standard products* without forecasts, batches, or inventory. For BTO of standard products, changes would be confined to developing a spontaneous supply chain (Chapters 3, 4, 5, 6, and 7) and on-demand lean production (Ch. 8). New markets would not have to be developed. Existing products would sell to the same markets and customers, with the reduced cost and rapid responsiveness expanding the markets and attracting new customers. Products could be ordered with their existing model numbers, although the order-entry system may need to be improved for speed and accuracy.

The mass-produced/no-variety products could be built on the dedicated lines described above. Demand over capacity could be satisfied on the flexible lines or through overtime on the dedicated lines. Demand at less than capacity would result in some underutilization, but some of this can be justified as contingency capacity for peak demand and growth. Any remaining "inefficiencies" would offset by the economic viability of the entire BTO&MC model, especially elimination of inventory (and all its costs, risk, obsolescence, deterioration, etc.) and the eliminating the costs of setup and changeovers. It will also decrease costs for indirect labor, quality, freight, material overhead, sales, and marketing. Further, building standard products to-order will improve customer satisfaction with quicker delivery, less "out-of-stock" delays, lower cost, and better quality (from the better kaizen continuous improvement possible without setup changes). This, in turn, would grow these businesses while lowering their costs.

2) **Build *Custom* Products On-Demand.** The same spontaneous supply chain and on-demand lean production developed for the build-to-order of standard products can be extended to include mass-customized products. The main difference would be the order entry system. Simple product lines may just have a few options or dimensions, which might be handled by the existing order entry system. More complex product lines may need configuration software as described in the illustrated examples at the end of Chapter 8.

3) **Regional/Industry-Specific Plants.** The BTO&MC expertise developed in a manufacturer's existing plants could be extended to regional or specialized plants to: (a) build certain retail products on-demand near market concentrations, distribution hubs, or individual customers; or (b) supply parts to specific industries or products to stores on-demand, especially if there are clusters of customers concentrated in certain geographical regions. These "mini-plants" would be able to build any mass-customized or standard products on-demand with delivery measured in minutes for customers in close proximity. This model could quickly dominate local markets and take away business from all competitors in each cluster because of rapid delivery, lower total cost, and learning relationships.

For individual retail customers, mini-plants can interact directly with customers and then build their customized products on-demand, maybe allowing the customer to observe the manufacture, and deliver them very quickly.

Some standard "vanilla" parts could be built in a central facility followed by modular, adjustable, or dimensional customization performed on-site (Ch. 9). Researchers from the Center for Mass Customization & Customer Integration at Technische Universitaet Muenchen (Technical University of Munich, Germany)[13] wrote about the economics of the business case for mini-plants and concluded this was particularly well suitable for "product-service bundles" and that such an attractive offering could command premium prices.[14]

For industrial suppliers, regional plants would encourage and support their customers' efforts to grow their businesses with their own programs to improve responsiveness and implement build-to-order and mass customization, which, in turn, will further increase business for the BTO&MC part supplier. Clusters of many suppliers in "supplier parks" have been encouraged around automobile assembly plants in the U.S., Europe and South America.

4) **Customer-Specific Lines.** A more focused version of industry-specific lines would be *customer-specific* lines inside, or next to, certain customer plants. Each customer-specific line would build on-demand any parts needed by each customer, for mass-customized or standard products. Customers would get near-instantaneous delivery in a seamless integration with their operations. The single-customer focus would benefit both customers and the manufacturer with learning relationships to continuously improve customer satisfaction. As compensation for such a commitment, customers would agree to long-term arrangements which would assure long-term business for the BTO&MC supplier and justify the investment. This, and the learning relationships, would result in mutually satisfying long-term relationships. Furthermore, the BTO&MC supplier could leverage its mass customization expertise to its customers to encourage and support them in *their* effort to grow their businesses with their own programs to improve customer responsiveness, build-to-order, and mass customization, which, in turn, will further increase business for the BTO&MC supplier.

The profile for customer-specific lines would be a customer who had a need to pull these parts on-demand to build *its* products to-order for web sales, lean production, inventory reduction, cost reduction, mass customization, and rapid resupply of product to stores and dealers. The customer should have taken steps to level production, using the techniques of Chapter 8, to ensure such a line did not swing from overtaxed to underutilized. And most important, the customer would need to work with BTO&MC suppliers to ensure product designs have adequate standardization of parts and processing so that all parts in the family can be built flexibly on the minimum number of versatile lines.

5) **Expand Downstream.** If the BTO&MC supplier is convinced that the customer-specific line concept presents significant opportunities and all potential customers decline such an opportunity, then the supplier could capitalize on these opportunities by expanding downstream, by setting up or buying existing plants to manufacture intermediate or end-products on-demand and offer them built to-order or mass-customized. In such a situation the supplier-turned-assembler would have unique competitive advantages and be able to grow and profit significantly in these new businesses.

This is one of Slywotzky and Morrison's "Profit Patterns:" [15]

> "When value shifts occur up and down the value chain, the challenge is to expand the scope of the business model to

include activities that were once viewed as upstream or downstream from your own position."

6) **Expand Upstream.** Chapter 6 addressed the issue of internal selective integration to manufacture parts in-house that cannot be pulled quickly enough from suppliers for BTO&MC operations. Such an analysis may reveal supply chain opportunities if your company and other potential customers would benefit from rapid part delivery at lower cost. If this looks promising, your company might consider acquiring (at least a controlling interest in) key suppliers and convert them to BTO&MC. Revitalizing these suppliers could increase *their* profit and growth while ensuring fast delivery and low cost for *your* BTO&MC operations.

As mentioned in Chapter 6, Solectron and Flextronics, leading electronic contract manufacturers, have been acquiring some of *their* part suppliers.

CAPITAL COSTS FOR IMPLEMENTATION

High-Volume Dedicated Production Lines

High-volume mass-production lines may not need any new capital; in fact, the capital budget for these may go down as production shifts to more flexible lines (below). New equipment purchases to enhance quality and output could be avoided as a result of product/process consistency and kaizen continuous improvement efforts that can maintain or improve quality and output. If these lines are dedicated, then time lost to setup changes will be avoided, thus maximizing equipment utilization and output.

The higher utilization and the dedication of the line may provide some extra capacity for peak demands and growth. This may postpone or eliminate expansion needs for those lines. In fact, if growth comes mainly from medium and low-volume/custom products, the need for expanding mass production lines may be eliminated indefinitely.

Medium-Volume Dedicated Production Lines

For medium-volume dedicated production lines, the capital expenses depend on growth. If growth is zero, then additional money might be needed to modify or procure equipment to implement smaller dedicated lines. However, if growth is envisioned, then a BTO&MC program would *save* money compared to a traditional expansion. Capacity could be expanded with less capital expense by procuring simple (perhaps used or "obsolete") equipment, as part of a setup elimination strategy. In fact, used equipment could be bought from less wise companies (even competitors) that are either buying megamachines or selling off equipment to focus on non-manufacturing "core competencies." Used equipment can achieve quality and throughput goals through *kaizen* continuous improvement, which can be more effective when all process variables are fixed.

Low-Volume and Customized Products

It may require some capital expenses to establish flexible lines that can build a high-variety of low-volume or mass-customized products. The processing machinery may be similar to off-the-shelf production equipment but may be smaller, and less expensive, so they could be deployed for several product families. It may be necessary for the equipment to have programmability features for flexibility, and this may or may not cost more than traditional equipment. If this is not readily available at a reasonable cost, then manual settings could be used until the programmable features could be added later.

There may be an additional capital cost to enhance the order entry system, if mass-customized orders cannot be handled in the existing system. Software configurators and their computer platforms may need to be procured and implemented. However, these systems may pay for themselves simply by eliminating order entry errors. Configuration capability may already reside in existing order-entry or ERP/MRP software, as was the case with Hoffman. Developing web-sites to accept orders on-line might be another capital expense, but this should be paid off quickly with savings in sales force costs expand sales, and screening out unprofitable orders.

Capital Cost Summary

The total net cost of implementing build-to-order and mass customization, in a growth mode, would be any cost beyond the cost of a traditional expansion for growth.

The bottom line for capital costs for BTO&MC, assuming growth is envisioned, would be that (a) mass-production lines might delay or eliminate expansion expenses, with higher utilization from existing equipment and demand shifting to lower-volume lines, (b) medium-volume production capacity expansion could save money by procuring less-expensive equipment for small dedicated lines, and (c) these savings may help offset the cost of implementing the flexible capabilities for the low-volume/customized products. Of course, having an economically viable means of producing low-volume and custom products could expand this market greatly, thus bringing in more income to pay for new equipment costs.

Another way of looking at capital expenditures would be to summarize the typical yearly capital improvement budgets. It may be the case that much of these budgets could go toward implementing build-to-order and mass customization instead of beefing up mass production equipment.

There is the potential to save a lot of capital expenses for operations software (ERP, MRP, etc.), associated computer hardware, and implementation consulting, since much of this software is unnecessary in BTO&MC operations, for instance, MRP parts ordering and shop floor tracking systems.

Ross Controls Investment. Over a 7-year period, Ross Controls has vigorously pursued advanced, quick-turn capability and invested $30 million in sophisticated CAD systems and automated product equipment. Regarding the justification, Ross COO, Henry Duignan, noted that traditional return-on-investment analysis would never have justified the strategy, adding "You have to make a leap of faith."[16]

Ross Controls results: The quick-turn *Ross/Flex* operation has grown from 5% to 20% of revenue in its first four years.[17]

This customization is quite valuable to Ross/Flex customers. A glass-products manufacturer determined that reducing a valve cycle time from 12 milliseconds, of the off-the-shelf valve, to 2 milliseconds would improve the productivity of its bottle-making equipment by 20%. Ross/Flex designed and produced the desired valve in two days. "How much is a product worth that represents less than 2% of [a customer's] machine investment, but allows the customer to increase his production by 20%?" -- Henry Duignan[18]

Ross supplied General Motors Metal Fabricating Division with 600 integrated-valve systems for stamping presses. The integrated systems perform better with the custom valve at one-third the price of the valves GM replaced.[19]

HUMAN RESOURCES FOR IMPLEMENTATION

View BTO&MC as an Investment. The foundation for implementing any successful program is to view the implementation as an investment, which will be paid off by the returns summarized in Chapter 14. Subsequently, implementation of BTO&MC should not be a "spare time" activity of people who have competing obligations and distractions. Dedicated resources should be applied by reassigning current employees or hiring new people with appropriate skills to implement, refine, and possibly manage BTO&MC operations. Hiring new people for specific tasks brings in new talent and an open-minded perspective not limited or constrained by past internal practices or inertia.

Bring in Outside Expertise. Much of the slow and costly "learning curve" can be bypassed by bringing in leading-edge experts early to optimizing planning and maximize program effectiveness and efficiencies. Opportunities include the critical stage of developing strategies, overall implementation approaches, and specific implementations like rationalization, standardization, setup elimination, and total cost accounting. Look for expertise that is focused on high-leverage ("big bang for the buck") results rather than endless studies and benchmarking reviews. Be wary of "big name" consulting firms that throw buzz-words around, but lack experience in build-to-order and mass customization, sell a standard conventional package, have hidden agendas, are biased toward their own or affiliates' products, and send in the high-powered partners to make the sale after which less experienced employees do the work. For a summary of the author's consulting services, see page 470.

Start with Pilot(s). If implementation resources and capital are limited, start with a pilot to demonstrate the value of BTO&MC with a pilot product line(s) that has: (a) significant variety, volume, and customization challenges that are not being solved well by current operations, (b) minimum risk, (c) little resistance to change, (d) sufficient visibility and credibility to be a good pilot, including a generally perceived applicability to other product lines, and (e) potential for growing that business quickly. Leverage lessons learned to other programs.

Combine Programs. BTO&MC implementation could be combined with (or absorb) related programs that already have resources, such as lean production, cost reduction, customer responsiveness, lead time reduction, order entry, sales force automation, and so forth.

Minimize Competing Programs. Minimize programs that compete for resources, especially programs that could be greatly simplified, changed, or obsoleted by BTO&MC, like enterprise software installations. The techniques presented in Chapters 3 through 10 do much to simplify processes which may reduce the scope of other programs or make them unnecessary. The BTO&MC program would benefit from *being* the principal program and getting the dedication of these resources. Also, be cautious of acquisitions or foreign ventures that compete for resources and complicate supply chain management with an increased variety of parts and suppliers.

Focus on Compatible Products & Processes. Emphasis should be on products and processes that are compatible with BTO&MC, such as products that could be part of a current or future family and that fit into current or future BTO&MC processing lines. Both capital and resources can be liberated by redirecting resources from products and processes that are not compatible with short and long range strategies.

Redirect Cost Reduction Efforts. Valuable resources can be freed up by redirecting "cost reduction" efforts with the reasoning that (a) BTO&MC could provide a greater cost reduction return for a given effort and (b) many cost reduction efforts do not even pay off the cost of the effort within the life of the product or equipment. See article at www.HalfCostProducts.com/how_not_to_lower_cost.htm.

Delegate Firefighting. Delegate as much firefighting down to the lowest level possible (under proper supervision) to make key implementers more available. Limit firefighting to quality and throughput, not cost reduction (as per above point).

Offloading Responsibilities. Hire or reassign more lower-level workers for routine tasks to free key experienced people for BTO&MC implementations.

Identify Money-Losing Products. Identify money-losing products for elimination since these are usually the products that are wasting valuable resources on firefighting, unusual setups, and other wasteful tasks. Use product line rationalization (Ch. 3) and total cost measurements (Ch. 12) to identify these. Until that is in place, use profiles to identify money-losing products and then investigate each (manually gathering total cost data if necessary) to estimate the degree of unprofitability.

Rationalization. A quick way to immediately free up resources would be to rationalize products to get rid of very low-volume, high-overhead, infrequently sold products and immediately put "extra" people to work on improving more profitable products and supporting strategic programs like BTO&MC. High-overhead products that are needed for completeness, but do not fit the new environment, could be outsourced to free up even more internal people, whose jobs will now be easier without these difficult and incompatible products (see Ch. 3).

DOWNTURN STRATEGIES

Companies that have implemented BTO&MC will be much less affected by downturns, since they will be generating new growth, both in their cash cows and in new products. Companies that want to implement BTO&MC can use the "calm between storms" to make strategic investments.

Companies that have any financial challenges should start by *immediately* eliminating money-losing products, using total cost measurements (Ch. 12) to ascertain real profitability and the product line rationalization techniques of Chapter 3. The strategy here would be that if revenue is headed down anyway, get rid of the money-losers to boost profits. Chapter 3 presents a scenario where eliminating or outsourcing the 20% worst revenue generators can raise profits threefold!

Unfortunately, the typical knee-jerk reaction to the first hint of a downturn is to "cut cost" by laying off people, cutting prices, keep building products for inventory, and halting any "optional" programs. Robert Atkins and Adrian Slywotzky, writing in the Wall Street

Journal, advise *against* the "Management 101 arsenal of defenses: cut compensation and discretionary spending, hoard cash, delay product development, exact fixed cuts across the board." They contend that "nothing could be more shortsighted than this circle-the-wagons approach." Their conclusion is that "great managers embrace a recession because of its opportunities." [20]

Here are several important "do's and don't's" about what to do during downturns, which will all benefit BTO&MC implementations.

Don't Lay Off People

> Great companies don't lay off people
> – Jim Collins, *Good to Great*

Great companies don't lay off people. Of the eleven "great" companies profiled in Jim Collins' book, *Good to Great,* six had no layoffs at all in the ten years before making the transition from good to great and four other great companies had only one or two layoffs. In contrast, layoffs were used five times more frequently in the comparison companies ("not great" companies in the same industries).[21]

The largest manufacturer of electric motors in the U.S., Baldor Electric, rode out one recession, even though sales shrunk by 12%. Baldor's CEO, John A. McFarland said: "We don't believe in layoffs. You're just sending all the investment and expertise out the door."

Savvy smaller companies also realize the same principle. U.S. Tool & Die's Robert Moscardini pointed out that they spent 3,000 hours training workers per year and figures that, at $60 to $100 per hour, "That's a big investment. You don't just let those people go."[22]

Unfortunately, layoffs are the most common knee-jerk reaction to "cut cost fast" and "downsize" operations to the downturn level, which is presumed to be permanent; but *it rarely is permanent* – that's why they call downturns "cycles!"

> There are several compelling reasons not to lay off people

Problems of Layoffs. However, there are several compelling arguments against laying off people (with the possible exception of *discrete* early retirements of non-critical people that are not perceived as part of any layoff program). First of all, layoffs for any reason are not only devastating to those laid off but also can have debilitating effects on *employees that remain* first with company-wide fear and anxiety, then organization disruption, and following increasing workloads and possibly new jobs to learn. Further, the best and most sought-after employees may "jump ship" at the first hint at the coming disruptions and their consequences. This all lowers company productivity and effectiveness – all for the *illusion* of saving cost.

Layoffs really don't save as much money as expected, considering severance pay, unemployment fees and rate increases, and other termination costs. This is especially true if the workforce is protected by employee agreements or union contracts, which, as in the automobile industry, gives laid-off workers 95% of their pay for the remainder of the contract, which could be years. So any cost savings would only happen after a few years when the market may have picked up, at which time the company could (a) stubbornly stick with the plan because they have paid so much into it, and hire new, less experienced people, or (b) admit that all this human and organization cost would have been for naught and bring the same people back.

On a total cost basis, layoffs don't really save cost in the long run

Even without such contracts, human resource experts argue that firing/rehiring cycle has a 1.5 to 2 year breakeven point, meaning that if the market rebounds in less than that time, the company would be ahead by saving on all the severance and rehiring costs and simply keep the people on salary, *even if they theoretically had nothing to do!* Savvy readers of business books can probably predict where this train of thought is going: Put those people to work improving the company with programs like BTO&MC and use "excess" human resources to relieve work loads so all employees can enhance their skills through training and implementing improvement programs.

Other reasons not to lay off people are labor continuity and skill and knowledge retention. Losing key skills is usually more disruptive than perceived when deleting names on the employee database. Similarly, few companies have perfect documentation and knowledge sharing, in which case valuable information and skills will be lost when people leave. Many operations depend on the current employees knowing what to do and how to do it. Thus, laying off people slows down operations, raises the cost per task, and maybe degrades quality, especially when the transferred people have to learn new tasks and search for information.

Professor Kim Cameron, of the University of Michigan Business School, says that "layoffs tend to break down communications and the informal knowledge in organizations about how to handle certain customers, suppliers or operations."[23]

Some companies will layoff "temps" (temporary employees) at the first sign of trouble, arguing that hiring temps in the first place was a hedge and they were not really expecting job security anyway. However, temps may be performing important, maybe critical, tasks and, unless the company has perfect documentation and widespread cross-training, organizational performance may suffer.

Trimming Payroll without Layoffs. One approach that avoids layoffs, but immediately trims payroll temporarily without severance costs, is to shorten the hours worked – and payroll paid – for all employees, with the possible exception of critical people. For instance, a 10% payroll savings could be generated by closing business every other Friday; the employees would see a temporary 10% income drop, but would greatly appreciate keeping their jobs and wouldn't mind a three-day weekend every other week, an extra week off over the holidays, or adding a week to their summer vacation.

During the 2008 recession, the New York Times observed that: "At many companies, management is hanging on to as many workers as it can, cutting hours to try to limit layoffs, while hoping that business improves."[24]

Charles Schwab, when needing to temporarily cut labor costs, asked almost half of its employees to take some Fridays off. A Schwab spokesman summarized their philosophy as follows: "You don't engender loyalty and a sense of shared mission by resorting to layoffs first."[25]

For more challenging situations, some companies still avoid layoffs with unpaid mandatory furloughs, for example, the plant closes for one week per month or so. The company temporarily saves 25% on payroll cost without the cost or disruptions of layoffs. The workers retain their jobs and benefits and have a few weeks for extra vacations, spending more time with their families, or painting the house instead of paying for it. And when the downturn has passed, the workforce and its morale are intact.

If there are truly excess people who are not critical to the current level of business, they could be offered extended time off without pay (but with benefits) to pursue an advanced degree, travel, spend time with a newborn, take care of elderly parents, or other personal reasons for such a sabbatical. Of course, they would have to be assured that they would have a good job when they came back.

Another option to keep from laying off people is to "loan" employees to other companies that are nearby so as not force a move or a lengthy commute. A mutual bidirectional arrangement might be possible if the two companies had out-of-phase business cycles.

In any event, make sure that enough people will be on duty to adequately deal with customers and suppliers.

Don't Halt Training and Improvement Programs

It is a mistake to think of training and company improvement programs as "optional" and cut them out to try to save cost. This would be sacrificing the future for a questionable temporary gain. On the capital side, there would not be much saved since programs like many improvement programs like BTO&MC are not that capital intensive anyway. "Extra" people can either implement these programs or relieve others who may be better suited. And all employees will benefit from training and other forms of employee development, such as job rotation, trade schools, continuing education programs, college courses, and so forth.

Don't Compromise Product Development

Another temptation in downturns is to cut back on product development, because of relatively high salaries and the long-term nature of results. However, product development determines the near future and innovative research projects determine the more distant future (which still could be only a couple of years away). Compromising those now will guarantee an even greater and longer downturn later. In addition, engineers may possess critical knowledge and may be hard to replace later, especially if the current staff has unique experience and specialized skills. Even in the short term, "continuing engineering" and support may suffer from compromised continuity.

Don't Keep Plants Busy Building Inventory

One of the biggest mistakes made by mass producers is to keep plants "busy" building inventory. This is a bad strategy that will have counterproductive effects because (1) it costs real money *now* to pay for the materials and pay for the inventory carrying costs, (2) it will be hard to make a good guess of which version to build, thus leading to obsolescence or expensive reconfigurations, (3) customer preferences may change before products can be sold, thus obsoleting products, and (4) the distribution channel may then be clogged with obsolete inventory which will either have to be sold first, thus delaying new product introduction, or heavily discounted or scrapped, which is expensive.

In the build-to-order age, the only valid reason for building products without orders would be for donations for altruistic reasons or to stimulate future demand, for instance, computers to schools and telecom infrastructure equipment to developing nations.[26]

Don't Cut Prices

Downturns usually result in excess capacity, so it may be tempting to cut prices to improve market share – at the cost of profits – but this book advises against this *even in good times!* Chapter 3 pointed out this fact, citing the opening chapter in Slywotzky & Morrison's *The Profit Zone*[27] which is titled: "Market Share is Dead." The only exception would be if your company already had a *real* cost advantage.

Do Rationalize Products to Get Rid of Money Losers

Rationalizing product lines can cut costs, raise profits, and support BTO&MC programs by reducing product and part variety. Some companies are reluctant to prune back on any products for fear of decreasing revenue. However, if the company is going to take a temporary "hit" on revenue anyway, a downturn may be a good time to do it, since revenue drops are already expected by investors and have probably been explained away to other causes. Further, during downturns, companies have people available to perform the rationalization. Doing this during a downturn will set up the company for a stronger recovery.

Do Expand Into Related Products/Services

Fortune magazine's cover story on "Managing for the Slowdown," recommends the following strategy:

> "A strategy we've seen work well, though it's still not prevalent, is reviewing the total needs of your existing customers and imagining how you can satisfy a larger proportion of them. You're no longer focused on taking business away from competitors; you're focused on getting a larger share of your customers' wallets."[28]

Build-to-order and mass customization can do this using all the growth strategies presented in Chapter 14, for instance broadening the product line with the ability to cost-effectively make low-volume, niche products and offering mass-customized goods that may be new or better solutions than current means of customization.

Do Pull in Outsourced Production, Selectively

One way to justify keeping employees in downturns, and improve operational responsiveness, is to selectively pull in outsourced part production, as recommended in Chapter 6, to support BTO&MC in general. Downturns are a good time to do this since (1) this can reduce some outsourcing expenses right away, (2) it allows the company to retain valuable employees, (3) employees have time available for the "learning curve" and even to learn new skills. Capital equipment can be rented instead of purchased to save cash in the short term.

However outsourcers may not tolerate the practice, to pull in work at every slump and then send the same work back in every boom. This book recommends a strategy of reducing lead times and gaining control over total cost through *permanent* selective integration. When boom times return, outsourcing can be used more productively for the most unusual products instead of critical parts (Ch. 3).

Do Improve Productivity and Invest in the Business

Another theme of Fortune's cover story on managing for slowdowns was that "improving productivity during a downturn puts a company in a stronger competitive position when things turn up."[29]

Harvard business professor Donald N. Sull was quoted in Business Week's article on investing during the current recession as saying, "Making a big bet is the thing that allows you to build sustainable advantage – in technology, manufacturing improvements, or relationships.[30] The article concluded that during recessions, rivals also feel compelled to invest to keep up. Cutting back while others are investing can threaten your future, which is what happened to the U.S. semiconductor industry:

> In the mid '70s, when the U.S. semiconductor industry cut back on capital expenditures and made layoffs, the Japanese invested. When we came out of recession in the 1980s, the Japanese owned the memory chip business.[31]

A study by McKinsey & Company of 1,200 companies over the last 20 years showed that the companies that increased spending during the 1990-1991 spending became today's leaders.[32]

Do Implement Improvement Programs

Adrian Slywotzky, author of *How Digital Is Your Business?,* contends that:

> "The best-run companies have always been relentless in taking advantage of business downturns and market transitions."[33]

Companies contemplating BTO&MC should use slowdowns as an opportunity to implement BTO&MC or at least start on the prerequisites presented in Chapters 3, 4, 5, and 6. Skill levels of all employees can be also enhanced through training as "excess" human resources relieve work loads so everyone can have the time to attending training classes.

ENDNOTES/REFERENCES

1. David M. Anderson, *Design for Manufacturability & Concurrent Engineering; How to Design for Low Cost, Design in High Quality, Design for Lean Manufacture, and Design Quickly for Fast Production,* (2008, CIM Press, 448 pages); pages 62-63 show graphical plots of customer importances vs. competitive grades with importance zones.

2. David Drickhamer, "Aim High; How Industry Week's Best Plants Measure Up," *Industry Week,* October 2002, pp. 67-70. This article

is available online at http://www.industryweek.com/CurrentArticles/asp/articles.asp?ArticleId=1321.

3. Russ Olexa, "Freudenberg-NOK's Lean Journey," *Manufacturing Engineering,* January 2002, pages 34 - 45.

4. See article, "How Not to Lower Cost," at the author's web site: http://www.halfcostproducts.com/how_not_to_lower_cost.htm.

5. Anderson, *Design for Manufacturability & Concurrent Engineering,* Section 2.2, "Ensuring Resource Availability."

6. James P. Womack and Daniel T. Jones, *Lean Thinking; Banish Waste and Create Wealth in Your Corporation,* (1996, Simon & Schuster), Ch. 11, "An Action Plan."

7. Thomas H. Davenport and Laurence Prusak with H. James Wilson, *What's the Big Idea? Creating and Capitalizing on the Best Management Thinking,* (2003, Harvard Business School Press).

8. Ibid., Chapter 3, "Ideas at Work," page 52 in the section, "The Internal Idea Cycle,"

9. Ibid., page 152.

10. Ibid., page 36.

11. Ibid., page 53.

12. Jones, Daniel J., "JIT & the EOQ Model: Odd Couples No More!," Management Accounting v72, n8, Feb 1991, pages 54 - 57.

13. The Center for Mass Customization & Customer Integration at the Technische Universitaet Muenchen (Technical University of Munich, Germany) is headed by Dr. Frank T. Piller. The Center's web-site is *www.mass-customization.de* which has much content in both English and German languages.

14. Ralf Reichwald, Frank T. Piller, Stephan Jaeger, and Stefan Zanner, "Economic Evaluation of Mini-Plants for Mass Customization; a Decentralized setting for Customer-Centric Production Units," published in *The Customer Centric Enterprise; Advanced in Mass Customization and Personalization,* edited by

Mitchell M. Tseng and Frank T. Piller (2003, Springer-Verlag, 547 pages); printed in English.

15. Adrian J. Slywotzky and David J. Morrison, *Profit Patterns, 30 Ways to Anticipate Profit from Strategic Forces Reshaping Your Business,* (1999, Times Business, Random House), Chapter 5, "Value Chain Patterns," p. 117.

16. "Creating a 21st Century Business," *Industry Week,* April 19, 1993, p. 38.

17. B. Joseph Pine II, Don Peppers, and Martha Rogers, "Do You Want to Keep Your Customers Forever?" *Harvard Business Review,* March-April, 1995, p. 103.

18. John H. Sheridan, "Reengineering Isn't Enough" *Industry Week,* January 17, 1994, p. 61.

19. Pine, Peppers, Rogers, "Do You Want to Keep Your Customers Forever?" *Harvard Business Review,* March-April, 1995, p. 103.

20. Robert G. Atkins and Adrian J. Slywotzky, "You Can Profit From a Recession," *Wall Street Journal,* February 5, 2001, p. A22.

21. Jim Collins, *Good to Great; Why Some Companies Make the Leap . . . and Others Don't,* (2001, Harper Business), Ch. 3.

22. Clare Ansberry, "By Resisting Layoffs, Small Manufacturers Help Protect the Economy; Many Opt not to Sacrifice Hefty Investments They Made to Find Workers," *Wall Street Journal,* July 6, 2001, page 1.

23. Jon E. Hilsenrath, "Many Say Layoffs Hurt Companies More Than They Help; 'The Evidence is Very Weak' That Downsizing Boosts Productivity," *Wall Street Journal,* February 21, 2001.

24. Peter S. Goodman, "Workers Get Fewer Hours, Deepening the Downturn;" *New York Times*, April 18, 2008.

25. Ibid.

26. In early 2001, Hewlett-Packard and its partners planned to create sustainable business models by selling, leasing, or donating $1 billion

over the next year in products and services to governments, agencies, and nonprofit groups in developing countries. Source: David Kirkpatrick, "Looking for Profits in Poverty," *Fortune,* February 5, 2001, p. 174; The article subtitle reads: "The Third World is a ripe market for HP, argues CEO Fiorina. Is she a crackpot? No, a visionary."

27. Slywotzky & Morrison, *The Profit Zone,* Chapter 1, "Market Share is Dead."

28. Ram Charan and Geoffrey Colvin, "Managing for the Slow Down," cover story, *Fortune,* February 5, 2001, pp. 79 - 88.

29. Ibid.

30. Michael Arndt, Steve Hamm, Steve Rosenbush, and Cliff Edwards, "Signs of Life; A few gutsy companies think now is the time to grow," *Business Week,* July 14, 2003, pages 32 - 34.

31. Amy Johns, "The Man Who Put the Valley on the Map; Marketing sage Regis McKenna on why smart companies are spending through the recession," *Business 2.0,* January 2002.

32. Arndt, Hamm, Rosenbush, and Edwards, "Signs of Life," *Business Week,* July 14, 2003, pages 32 - 34.

33. Adrian Slywotzky, "Standing Tall in the Tech Slump," *Fortune,* February 5, 2001, p. 176.

14

THE BUSINESS CASE FOR BTO&MC

Build-to-order and mass customization represents a business model that offers an unbeatable combination of superior responsiveness, cost, and what customers want when they want it. It enables companies to build any product on demand without forecasts, batches, inventory, or working capital.

> BTO&MC offers unbeatable responsiveness, cost, and customer satisfaction

BTO&MC companies can grow sales *and* profits by expanding sales of standard, customized, derivative, and niche market products, while avoiding the commodity trap. BTO&MC companies are the first to market with new technologies since distribution "pipelines" do not have to be emptied first.

BTO&MC substantially *simplifies* supply chains – not just "managing" them – to the point where parts and materials can be *pulled* into production without forecasts, MRP, purchasing, waiting, or warehousing.

Build-to-order is the best way to resupply stores and dealers who demand rapid replenishment, low cost, *and* high order fulfillment rates without the classic inventory dilemma: too little inventory saves cost but creates out-of-stocks, missed sales, expediting, and disappointed customers; too much inventory adds cost and costly obsolescence risks.

Mass customization can efficiently customize products for niche markets, countries, regions, industries, and individual customers.

There is a natural synergy between build-to-order and mass customization. They share the same batch-size-of-one operations and spontaneous supply chain. Build-to-order and mass customization operations are equally efficient and very compatible, unlike situations where a mass customization experiment must be run separately from the "batch-and-queue" operations of mass production. Manufacturing build-to-order and mass-customized products on the same lines will often push the combined volume over the "critical mass" threshold necessary to justify these implementations.

The following are several elements of the business case for build-to-order and mass customization. This last chapter summarizes the benefits of the principles of the previous chapters and groups them by benefit category.

COST ADVANTAGES OF BTO&MC

BTO&MC companies enjoy a substantial cost advantage, which they can use for competitive pricing, reinvestment, or enhancing profits. They understand that the only "cost" that customers care about is *their* cost which is the manufacturer's *price*. BTO&MC principles attack several categories of *total cost* to achieve the absolute lowest prices – or competitive enough prices at higher profits.

> BTO&MC companies save cost in many ways; Here are a couple dozen.

One of the greatest cost advantages of build-to-order is the elimination of all the costs caused by inventory: inventory carrying cost (usually 25% of value per year; see Ch. 2), warehousing costs, administrative expense, obsolescence write-offs, and discounting to sell obsolete inventory. Incoming just-in-time part deliveries minimize parts inventory costs. Flow manufacturing and setup/batch reduction efforts can virtually eliminate work-in-process (WIP) inventory costs. Building to-order and shipping direct can virtually eliminate finished goods inventory costs.

Build-to-order minimizes or eliminates many other overhead costs for forecasting, MRP, purchasing, expediting, scheduling, planning, setting up production, and the extra sequence of activities required when forecasted inventory cannot satisfy demand: re-forecasting, re-purchasing, re-expediting, re-scheduling, re-planning, and re-setting-up production.

Building without batches eliminates the costs of setup changes, kitting, and the loss of expensive machine time and valuable resources. The rapid feedback aspect of flow production eliminates the cost of recurring defects, which are more likely to happen with large batches.

Lean activities eliminate many categories of waste and inefficiencies. Writing in *Lean Thinking,* Womack and Jones summarize how much lean production can do:

> "Based on years of benchmarking and observations in organizations around the globe, we have developed the following simple rules of thumb: converting a classic batch-and-queue production system to continuous flow with effective pull by the customer will: double labor productivity all the way through the system (for direct, managerial, and technical workers, from raw materials to delivered product) while cutting production throughput times by 90 percent and reducing inventories in the system by 90 percent as well. Errors reaching the customer and scrap withing the production processes are typically cut in half, as are job-related injuries.
>
> "Firms having completed the radical realignment can typically double productivity *again* through incremental improvements within two to three years and halve again inventories, errors, and lead times during this period."[1]

All the mass production inefficiencies discussed in Chapter 2 – lower productivity, and the lack of the ability to shift production between lines – raises cost and causes the mass producer to need more overtime than the more efficient and flexible BTO&MC company.

Eliminating inventory, incoming inspection, and kitting saves much floor space cost, which can delay or eliminate the need to build more factory/warehouse space for growth.

Customization and configuration costs are less since they are proactively planned and executed efficiently, instead of the very inefficient craft production and fire-drill activities used in most companies. Learning relationships make repeat orders more efficient.

Concurrent engineering products for manufacturability generally results in significant cost savings, but these are more profound for BTO&MC companies because of the greater opportunities to *design out* several categories of overhead cost.

Optimal utilization of flexible CNC automation saves labor cost; flexible fixturing and setup reduction makes CNC even more efficient.

Distribution costs are less for BTO&MC goods because shipping is more direct, finished goods inventories are eliminated, and expediting is not needed to rectify shortages.

Finally, using product line rationalization and total cost measurements to eliminate high-overhead/low-profit products eliminates the "loser tax" on cash-cow products, thus letting them to sell for less or make more profit. This advanced business model eliminates the need to discount or take low-margin sales.

RESPONSIVE ADVANTAGES OF BTO&MC

BTO&MC companies build products on-demand, instead of having to forecast, order, wait, build, and stock. For phone order and web-sales, 100% of orders can promptly be shipped directly from the BTO&MC factory.

> BTO&MC delivers the goods fast to stores, OEM's, or directly to the ultimate customers

Poor availability is intrinsic in any system that sells from forecasted inventory, which depends on inherently unreliable forecasts. Delays will be common when shipping from inventory, trying to build to-order in a mass production environment, or waiting for parts to be delivered after receipt of the customer order.

If customers want to buy something right now off the retail shelf, BTO&MC suppliers can resupply those shelves better than anyone can from inventory, so a complete selection can always be available to customers. Stores and dealers that order frequent shelf replacements from BTO&MC suppliers will develop a good reputation for availability and eventually generate a loyal customer base that keeps coming back because they know that they will always find what they want; for example, for clothing, customers will always find the style they want in their size. Learning relationships make repeat order fulfillment quicker with each order.

BTO&MC companies are the fastest to adjust to changing market conditions. New product development and introduction can be faster when "new" products are just "variations-on-a-theme" that are easier to develop because of modularity, parametric CAD, and flexible processing. Production ramps can be faster on flexible lines that don't have to be "tooled up" for new products. A Wall Street Journal article reported that flexible automobile makers are able to release different versions of a car throughout the year instead of the traditional single release in the fall.[2]

BTO&MC companies are the first to introduce new technologies into the marketplace, since they don't have to first empty the "pipeline" of obsolete products, which usually have to be discounted to clear the pipeline. Further, shorter direct distribution channels can speed new products faster to customers who can't wait for the latest and greatest.

If new product introductions result in greater than expected growth, the BTO&MC company can meet upsurge demands by transferring production to other flexible lines. BTO&MC implementations free up floor space which can then be used for growth so growth will be less likely to be hampered by shortages of floor space. Standard parts are available from more sources with more total capacity, so standard parts will be more readily available in times of rapid growth. Optimal responsiveness is also assured by supply chains that allow assemblers to *pull* parts quickly without lengthy hand-offs or depending on forecasted parts inventory.

Distribution is more direct, and therefore faster, for built to-order goods without delays caused by shuffling inventory from factories to warehouses to distribution centers and then to customers or stores. Eliminating inventory and many nodes in the distribution chain is not only quicker but also eliminates order aberrations and demand swings caused by the order/response lag time inherent in slow, multi-node distribution chains; an MIT simulation called the "Beer Game" demonstrates these escalating effects for players in the roles of manufacturers, distributers, and stores.[3] The BTO&MC company may elect to capitalize on increased distribution efficiencies by expanding into distribution or acquiring and streamlining distributers.

For capital equipment companies, quicker product delivery itself may be a competitive advantage. In addition, responding to RFQs (requests for quotations) will be quicker using configurators and less susceptible to subsequent delays due to better data and fewer order-entry errors.

A final responsiveness benefit is that quicker delivery results in less vulnerability to changes in markets, supplies, and customers themselves. A Fortune article on build-to-order cited the following benefit at Herman Miller's office furniture operations: *"When the lag between order and delivery is reduced, customers have less time in which to change their minds about what they need. More important, say others, the product gets delivered before the market cycles downward."* The article also cited the same advantage at Applied Materials where quick responsiveness can result in *"few changes in customer's specifications after the order is placed and the unit is party built.*"[4]

CUSTOMER SATISFACTION FROM BTO&MC

BTO&MC can provide unmatched customer satisfaction for industrial clients or the ultimate consumers. Consumers will be satisfied by a complete selection of products available at the best prices. OEM and industrial clients will be satisfied by receiving parts on-demand to support *their* build-to-order efforts.

> BTO&MC delivers
> what customers want
> when the want it
> at the price they want

Mass customization will enable even higher levels of customer satisfaction for customers who can quickly receive high-quality/low-cost products specifically customized to their individual needs. For customized products, customers can make better choices and consider more "what if scenarios" with a good configurator. Even if individual customization is not applicable, customer satisfaction will be enhanced when products are customized for their culture, group, country, or region.

High quality and reliability will be another source customer satisfaction because of one-piece flows, rapid feedback that prevents recurring defects, greater utilization of standard parts, part quality assured at the source, vendor/partnerships for custom parts, learning relationships throughout the supply chain, continuous improvement (*kaizen*), a higher proportion of automated CNC operations resulting in more consistent tolerances, and rationalizing away the worst-quality products.

Learning relationships enable the BTO&MC company to learn and adapt from each order, thus satisfying customers better on the next order and progressively developing more committed customers.

COMPETITIVE ADVANTAGES OF BTO&MC

Competition now is *between business models* – the company with the best business model will be the best competitor. Ironically, the subject of several best-selling business books, *leadership* and *execution,* will only drive a company faster down the wrong path if they have the wrong business model. Dartmouth Business School Professor Sidney Finkelstein, writing in *Why Smart Executives Fail, and What You Can Learn from Their Mistakes,*[5] concluded that "The real causes of nearly every major business breakdown are the things that put a company on the wrong course and keep it there."

> Competition is now between business models

> The real cause of business breakdowns is being on the wrong course
> – Professor Finkelstein, author of book on business failures

As a business model, build-to-order and mass customization will compete well against competitors both large and small because of a superior combination of speed, cost, and customization.

Without a superior business model, companies might have to compromise profits to enhance market share. In the worst case competitive positions, products revert to commodity status with purchasing decisions made solely on price.

BTO&MC company products can avoid commodity status with the differentiation aspects of build-to-order (delivery, cost, quality, etc.) and mass customizing products to better satisfy customers. Learning relationships result in greater customer loyalty with each order.

BTO&MC companies have the agility to expand business into adjacencies, niche markets, derivatives, and so forth. Finally, all the above advantages can create a reputation as a leader, which further improves sales, impresses investors, and attracts the best talent.

BOTTOM LINE ADVANTAGES OF BTO&MC

Lower total cost results in increased profits or lower selling price or both. Faster delivery, better quality, and lower cost can grow revenue and market share. Agility allows expansion into new markets. Additional value-added work offers high profit opportunities. With enough competitive advantage, premium prices may be possible.

SUPPLING RETAILERS

Manufacturers who resist the change to BTO&MC will make their first attempts to satisfy retailers from inventory made in forecasted batch production. But, this is will always be sub-optimal because inventory and its management adds cost and poses an endless dilemma, which will get worse as forecasts become less accurate:

- Trying to improve order-fulfillment rates raises inventory costs and obsolescence risk.
- Trying to lower inventory cost and risk lowers order-fulfillment rates

Many companies stubbornly hold on to the old mass production until it is catastrophically too late. Again quoting Finkelstein's study of business failures:

> "There is one blind spot that appears somewhere near the center of almost every major business disaster: a seriously inaccurate perception of reality among executives.".[6]

Savvy manufacturers will utilize BTO&MC principles to dominate their markets and deliver products on-demand without forecasts or inventory. This reduces or eliminates all the costs associated with inventory, obsolescence, and distribution in addition to easily providing rapid and frequent deliveries to store distribution centers or directly to stores.

To support manufacturers' BTO&MC programs, suppliers will also have to make parts on-demand even more quickly than their customers. If OEM's have 2-3 days to resupply a store (including shipping time), suppliers may only have one day *including shipping.* Sequential steps in the supply chain will be under even more time pressure.

GROWTH, SALES, and PROFITS FROM BTO&MC

Trying to Grow without a Viable Strategy

> Trying to grow without a viable strategy means "take all orders" & "dip lower into the barrel"

If companies don't use an effective growth strategy, like BTO&MC, then growth pressures will push them to "take all orders." Expanding sales by "dipping lower into the barrel" for low-volume, unusual, or marginal products may appear to satisfy growth goals, at least for a while, but in reality, it will have several unpleasant effects:

1. **Marginal *products* have marginal *profits.*** Marginal products usually have marginal or negative profits, thus lowering enterprise profitability. This is usually hard to realize since the vast majority of cost systems are totally inadequate at quantifying true profitability for individual products.

> Marginal *products* have marginal *profits*

2. **Production of low-volume, unusual products is inefficient.** Unless low-volume, unusual products are part of a well structured BTO&MC line, building them will be much less efficient if they are hard-to-build, require high overhead demands, involve a lot of firefighting, have excessive setups, or are incompatible with flexible strategies like lean production, build-to-order, and mass customization

3. **Unusual products raise overhead costs.** Bringing in many more low-volume, unusual products will raise overhead demands and thus overhead costs, although they are rarely quantified by conventional cost systems.

   Intel's Systems Group, in response to a downturn made the strategic decision to take in contract manufacturing jobs to "fill the factory." However, the first new customer *doubled* part variety in the plant and each new customer increased it again as much. Material overhead cost skyrocketed and the Manufacturing Engineering department *tripled* in a few months to generate machine tool programs, create manual instructions, and build new tooling. Ironically, the "filled" factory had to be closed later because – no surprise – operations were not competitive.

4. Cash-cows subsidize losers. These increased overhead demands, quantified or not, must be paid somehow. "Loser" products do not bring in enough money to pay for their own processing inefficiency, so the increased overhead costs must be paid by cash cow products as a "loser tax," which, in turn, will raise the overhead charges on the cash-cows. This will either make them less profitable (at the same price) or less competitive (if the price has to be raised).

Cash-cows subsidize losers and must pay a "loser tax"

5. More setups slow all production. Changing setups more frequently for low-volume products will slow production and deliveries of *all* products *including* the cash-cows. In addition, the increased amount of fire-drills needed to produce the unusual products may draw attention away from standard products and possibly lower *their* quality.

6. Discounting Erodes Profits. Unless you are the undisputed low-cost producer, slashing prices to grow revenue or market share will sacrifice profits in a vain attempt to gain market share. Without a real cost advantage, "growth" or market share protection will only be sustainable while the profit-lowering discounts are in place. But discounting is expensive and can cause significant losses. The cost of subsidized financing at Ford Motor Company was 16% of revenues in the third quarter of 2001 which became even higher in the fourth quarter. Ford President Nick Steele acknowledged that marketing cost should be 11% and that the extra promotional expenses cost Ford $7 million in 2001.[7]

MERGERS AND ACQUISITIONS

In Jim Collins' book on how good companies become great companies, appropriately titled, *Good to Great,* he states flatly that "you absolutely cannot buy your way to greatness."[8]

Yet many companies have tried to grow by mergers and acquisitions. A *Wall Street Journal* article cited problems of "serial acquirers" and questioned the strategy trying to grow by doing deals:

> "You absolutely cannot buy your way to greatness"
> - Jim Collins, *Good to Great*

> "The troubles of the serial acquirers raise new questions about the growth-by-acquisition strategy that enthralled so many companies and investors during the bull market." [9]

A Thomson Financial/First Call study commissioned by the Wall Street Journal study found that "stocks of the top 20 acquirers in the late 1990s have fallen nearly twice as much as the Dow Jones Industrial Average and the Standard & Poor's 500 Index."[10] An A.T. Kearney study of 115 global mergers in the mid 1990s showed that the total return to shareholders (relative to peer companies) was *minus 58%!*[11]

Growth by acquisition is hard to sustain. A *Business Week* cover story about General Electric summed up the problem:

> "About 15% of GE's earnings growth came from acquisitions last year. But because these are one-time events, it has to gobble more companies each year to keep up the pace."[12]

That same article reported what happened when one of the financial community's most admired figures, Bill Gross, accused GE of "inflating earnings through acquisitions and cheap debt rather than through organic growth;" GE's stock fell 6% in two days.[13]

The odds are not favorable for success in mergers and acquisitions. According to KPMG research, only one in five deals lives up to its expectations.[14] Time and cost estimates are usually unrealistic; Professor Thomas Lys of the Kellogg Graduate School of Management at Northwestern University, says: "It costs twice as much and takes twice as long as you planned."[15]

The Fallacy about Mergers Saving Cost

One of the reasons for the poor financial performance of mergers and acquisitions is that the often touted "synergies" come at *a significant cost* for downsizing, coordinating the integration, training, relocations, and other overhead costs. Dartmouth's Finkelstein, writing in his thorough study of business failures, offered the following guideline:

Economies of scale only work if all merged plants use the same parts, which is *very* unlikely

"As a rule of thumb, synergy realization costs often come in at two to three times the value of annual synergy benefits."[16]

Therefore, it would take two to three years to breakeven on the projected synergies, assuming that they really can be effective at all. And during that time, the merger or acquisition is diverting focus from other opportunities (see later section on M&A Distractions).

The *cost* of realizing merger synergies is 2 to 3 times annual benefits – Professor Sidney Finkelstein, author of book on business failures

But, despite the odds, many companies still try to acquire or merge with competitors to try to gain "pricing power" or "economies of scale" and are lured into this apparent panacea because they quantify parts cost but do not quantify all the total costs including the real cost of the merger and all the overhead costs involved in bulk purchases.

However, upon deeper analysis, merging two mass producers would provide a benefit *only if* the merged companies used the *same* parts and *same* processing equipment, which is highly unlikely. Even if some parts were the same, the pricing power would only be realized for centralized procurement, which still has all the overhead costs and inventory problems associated with purchasing and distributing parts, often to multiple sites. This approach is the opposite of the spontaneous supply chain principles of automatic resupply which arrange for local suppliers to keep the bins full (and bill the company at the end of the month) and thus avoid "purchasing" all together.

The biggest dangers from merging with or acquiring competitors is allowing dissimilar products to be built in the same plant. If there was a "Murphy's Law" of mergers, it would surely emphasize the unlikely probably of any similarity or commonality of parts, materials, or processes. Merging two dissimilar product lines into the same factory would probably double the variety of parts and raw materials, which would thwart any progress made toward standardization. It would also create massive efforts to translate documentation and create setbacks to setup up reduction efforts. In production, such a merger would present the dilemma of acquiring duplicate, dissimilar equipment and tooling or converting one product line's equipment, tooling, and procedures. In summary, merging two dissimilar product lines into the same factory would be a big step backward for build-to-order and mass customization.

> Merging dissimilar products into the same plant thwarts standardization and ultimately BTO&MC

Although the greater sales numbers may seem intoxicating, the hoped for synergies may prove more elusive than real, especially if the merged companies do not have advanced business models. Writing in the Wall Street Journal, Jim Collins summarized his philosophy is that "two big mediocrities never make one great company."[17]

M & A Distractions

A big problem with mergers and acquisitions, especially among competitors, is the *distractions* they cause. When General Mills acquired Pillsbury, it slashed promotions and new products to concentrate on integration. A half a year after the merger was completed, CEO Stephen Sanger admitted, "It's obvious that the focus on integration has taken its toll on our sales growth and our earnings growth."[18]

The most contentious deal of this decade involved Hewlett Packard and Compaq, which spent over one million man-hours in just integration *planning* – not including the actual integration.[19] To put this into perspective and do the proverbial math, one million man-hours is equivalent to 1,000 people working full time for half a year. How many companies have an extra million man-hours to spare? The obvious answer is that companies don't have *any* man-hours to spare, so what that means is that the merger integration has *distracted* them from what they were hired to do – important tasks like improving quality, product development, marketing, customer relations, and enhancing truly competitive business models that can *generate* growth, not just buy it.

One of the justifications for that merger was to become the largest PC maker and thus compete better with Dell Computer. But while Hewlett-Packard and Compaq were busy planning and merging, Dell's uninterrupted focus on growth soon surpassed the sales of the merged HP/Compaq, Dell starting marketing printers to challenge HP's cash cows, and, a year later, Dell was rated highest in customer satisfaction for desktop PCs (for the 12th time in 13 years) in PC magazine's user satisfaction survey of 18,000 subscribers. Dell received an "A+" grade, while both HP and Compaq got an "E" – the lowest possible grade.[20]

Even more distractions will come from layoffs that are usually part of the overall "cost savings" promised when pitching the deal. Chapter 13 presented several compelling reasons not to lay off people in general. Laying off people in the midst of the major upheaval of a merger or acquisition adds even *more* distractions. In the case of Hewlett-Packard's expected 15,000 job cuts, the above cited *Business Week* article referred to "a cloud of uncertainty that has left many of HP's 145, 000 workers in limbo." The Business Week article about the General Mills/Pillsbury acquisition said that: "Many Pillsbury salespeople had left, unsure of whether they would have jobs, and retailers were unhappy about the lack of support Pillsbury goods were getting."[21]

Unfortunately, companies distracted by M&A limbo and integration tasks are not focusing on improving their overall business model. One of the ambitious "idea practitioners" profiled in the book cited earlier on implementing business ideas, Lawrence Baxter, commented about the climate when his company, Wachovia Corporation, was merging with First Union Bark:

> Competitors not distracted by mergers will be innovating, improving quality, and gaining market share

> "We had a lot of very difficult and complex operational issues to address during the merger. There was a very clear focus on integrating the banks over a certain time period. It wasn't

a time to explore new ideas that might add risk to the organization."[22]

But, competitors that are *not* distracted by mergers keep on innovating. While HP was merging with Compaq, Dell gained market share with their more advanced assemble-to-order/ship-direct business model. And, while HP and Compaq were *integrating,* Dell was *innovating* and making plans to offer its own products for handheld computers and printers, which presents new competition in a key market for Compaq and in HP's most profitable product line.[23]

Some companies credit steering clear of mega-mergers as their formula of success. Peugeot resisted "merger mania" in the automobile industry and became one of the most profitable car manufacturers outside of Japan.[24] CEO Jean-Martin Folz argued that the key to success these days is "not amassing economies of scale with a merger, but producing innovative cars in rapid succession." Mr. Folz's views coincide with the theme of this section1: "Managers can't crank out that many products if they're struggling to integrated two companies."

Appropriate Use of M & A

Instead of disruptive M&A with competitors, look for supportive opportunities up and down the supply chain

Jim Collins said that great companies "used acquisitions as an accelerator of flywheel momentum, not a creator of it."[25] For BTO&MC, appropriate acquisitions, mergers, and alliances would be up and down the supply chain, as discussed in Chapter 13. Upstream acquisitions may be one solution to unresolvable part lead time problems; the other solution would be bringing production in-house, as discussed in Chapter 6. Upstream acquisitions may also be a way to gain better control of total cost, quality, and delivery in general.

Downstream acquisitions may be a way to capitalize on the ability to quickly make products on-demand, if existing distribution channels do not take advantage of such an opportunity.

Another category of appropriate mergers and acquisitions would be those that are *supportive,* for instance, complimentary product lines, marketing prowess, or distribution channels. Examples of complimentary products include combining production of out-of-season products like heaters with air conditioners and lawn mowers with snow blowers. This can help level the production loads and keep a stable workforce productive all year.

GROWTH FROM THE CORE

Despite the ambitious goals and attempts by many companies, few actually achieve *sustainable* growth. Adrian J. Slywotzky, author of *The Profit Zone* and *Profit Patterns,* concludes that "from 1990 to 2000, just 10% of publicly traded companies enjoyed eight or more years of double-digit growth in their top line." He and co-author Richard Wise also say that "tactics used in recent years to increase revenue are running out of steam and will no longer provide the foundation for long-term, double-digit growth."[26]

A thorough study by Bain & Company analyzed two hundred case studies and a database on almost 2000 companies. That study had similar results: 90% of companies failed to achieve sustained profitable growth over the past decade! The study analyzed the 10% that *did* grow and concluded that they owed their success to a strategy that is the title of the book, *Profit from the Core,*[27] meaning that they focused on their core products, capabilities, customers, and channels.

> Only 10% of companies get sustained growth; and they get it from their core business
> – Chris Zook,
> *Profit from the Core*

Build-to-order *enhances* and *grows* "the core" by building standard products on-demand at less cost without tying up working capital or incurring all the risks of forecasts and inventory. Further BTO adapts instantaneously to changing market conditions and enables the fastest introduction of new products. Mass customization *expands* the core to "adjacencies" (opportunities near the core) by offering variations of standard products and a vast range of customized products synergistic with core assets.

> Build-to-order *enhances and grows* the core. Mass customization *expands* the core.

BTO&MC companies can maximize sales with broader product lines, unbeatable responsiveness, and lower costs. They have the agility to quickly adjust to changing market conditions and be the first to respond while their slower competitors are still trying to react.

BTO&MC companies never miss a sale because products are built on-demand instead of being shipped from inventory based on forecasts that can never be 100% accurate. Whenever a product becomes an unexpected hit, the BTO&MC company can always deliver by shifting production to other flexible lines. Scarce production capacity and materials are used only for sold products, not for inventory that may not be selling.

BTO&MC companies can gain *new* customers by expanding into niche markets and mass-customizing products for various market segments, countries, regions, private labels, clubs, stores, and individual customers. Being able to manufacture products that exactly match customer needs can increase sales to consumers that would not normally buy less of a match. Pine and Gilmore, editors of a collection of Harvard Business Review articles related to mass customization, write in the Introduction:

> "In most, if not all, industries, customers refrain from spending money – by not buying at all, postponing repurchases, or buying a lower-priced offering – because existing options require them to comprise by accepting something other than what they want exactly." [28]

BTO&MC companies can better retain existing customers and expand offerings to them through learning relationships. Combining learning relations with lower cost and faster delivery keeps BTO&MC companies out of the commodity trap.

Expansion is also possible up and down the value chain. BTO&MC factories can expand into related businesses with customizations that can be manufactured just as efficiently, but which customers have been paying much more because they are being made "the hard way." For example, Hoffman Engineering uses the same laser cutters that cut sheet metal for its enclosures to cut custom holes and cut-outs that its customers need to mount their hardware.

Finally, BTO&MC companies may be able to command premium prices for this agility and customer satisfaction.

Achieving Growth Through BTO&MC

In Jim Collins' book, *Good to Great,* his research concluded that not one of "great" companies focused on an *obsession* for growth, and yet, paradoxically they *got* growth through good management.[29]

> None of the "great" companies were *obsessed* with growth, but got it through good management - Jim Collins, *Good to Great*

Build-to-order and mass customization have unique abilities to generate significant growth in four ways: (A) expansion of standard product sales; (B) new market expansion into niche market derivatives of standard products; (C) new market expansion into customized products; and (D) maintaining sales to existing customers by avoiding the commodity trap.

A. Expansion of *Standard* Product Sales

In contrast to the "take all orders" approach, the product portfolio planning aspect of BTO&MC encourages focusing all resources on the most promising products that can grow the most, rather than diluting focus over all products. And ironically, it is usually the *worst* products that demand most of the attention.

Build-to-order enables growth by expanding the sales of *standard* products to allow manufacturers to efficiently process smaller orders of compatible products. This enables the quick and efficient manufacture of standard products in any order quantity, even as low as one.

BTO&MC enables manufacturers to deliver 100% of orders, which will effectively expand sales compared to competitors with lower order fulfillment rates.

B. Niche Market Growth

Mass customization can open new markets for niche market derivatives of standard products. Niche markets offer a significant growth opportunity for the mass-customizer because many niche markets can be just many "variations-of-a-theme" which mass customization can handle quickly and efficiently. However, attempting an expansion into niche markets with batch-oriented mass production factories will have all the problems of bringing in many low-volume products mentioned above. Even if margins *appear* to be satisfactory, costs are bound to be higher than projected, good products will subsidize the losers, and other production will be disrupted.

C. New Markets for *Customized* Products

Custom products offer significant growth opportunities for new markets. These may be (1) entirely new markets or (2) may replace current customer solutions that are inadequate.

(1) For new markets, mass-customized capabilities could enable manufacturers to be the first to offer mass-customized products with the obvious potential for early market domination. Higher sales will result when customized products satisfy customers better than the current customer choices, which may not be optimal in all respects. Chemstation is an example of a customized product offering that opened new markets and locked in customers with *learning relationships.*[30]

(2) Mass-customized products that replace current customization practices have a huge potential for growth and profits, since the existing customizations are probably sub-optimal, either coming from craft-based local suppliers or inefficient activities in customers' factories or not-so-savvy suppliers. Mass-customized products have the dual benefit that they can be made much more efficiently *and* provide great value to customers by providing just what customers want and avoiding the cost, delays, and quality problems of their current practices. Thus, mass-customized products offer high value to the customer at a low production cost to Manufacturers: this is a recipe for high profits and growth.

One note of caution on customized products: Sometimes mass production companies try to achieve growth by expanding into custom products with the following consequences: Costs often exceed revenues because:

(a) Development costs are higher than expected on custom products if engineering is dealing with them on an *ad hoc* basis instead of through parametric templates where customer input is entered and product and process variables are automatically processed and sent to the factory.

(b) Production costs often exceed projections because of the inefficiencies dealing with unusual and unfamiliar, not to mention troublesome, products and procedures.

(c) The resulting losses must be paid for by cash-cow products, thus degrading their profitability or competitiveness. Such subsidies are common in conventional custom product programs.

(d) Factory operations are disrupted by all of the above resulting in delays and possible quality problems on not only the custom products but standard products as well.

D. Avoiding Commodity Status

Part of any growth strategy is to avoid situations in which growth programs will have to "swim upstream" to compensate for lost sales to existing customers as products fall into the commodity trap. The internet is enabling bidding to go on-line which may drive down prices, and probably profits, and allow industrial customers to easily switch suppliers. An ominous quote from a Business Week editorial, in summarizing the issue's focus on the 21st century corporation, predicts that the internet will commoditize all products:[31]

BTO&MC products can escape commoditization

> "'Everything gets cheaper forever,' according to John Chambers, CEO of Cisco Systems. The Net destroys corporate pricing power. It allows customers, suppliers, and partners to compare prices from 100 or 1000 sources, not just two or three, and erases market inefficiencies. It rapidly commoditizes all that is new, reducing prices fast. It quickly bids down prices toward marginal cost."

However, build-to-order and mass customization can *differentiate* products and avoid commodity status by providing the following:

1) **Lower *prices*** to customers by lowering *all costs*, including setup, inventory, and all other overhead costs.

2) **Faster delivery** approaching almost instantaneous response to store replenishment orders or OEM customer "pull signals" in support of *their* programs to build mass-customized or standard products to-order. Ultimately, this may even manifest as BTO&MC production lines at a customer's facility, which would result in long term relationships.

3) **100% order-fulfilment rate** by building products on-demand instead of shipping from forecasted inventory, thus eliminating out-of-stocks and expediting. A 100% order fulfilment rate will eliminate missed sales and bring sales up to 100% realization. It will also satisfy industrial customers and retailers, possibly providing a unique competitive advantage when competitors are disappointing, or even alienating, demanding customers.

4) **Development of supplier partnerships** to optimize *customers'* competitive positions, promote standardization, lower customers' costs, and implement build-to-order and mass customization throughout the supply chain. The value of these relationships to customers is far more important than any savings that come from the less effective approach of treating supplies as commodities and going with the low-bidder. Supplier partnerships are "learning relationships" that satisfy customers better with every transaction thus strengthening these relationships to keep customers forever. This, in turn, locks in customers since they will be several generations ahead than with any competitor.

BTO&MC AS A BUSINESS MODEL

The following is a summary of the advantages of the BTO&MC business model with respect to cost, responsiveness, customer satisfaction, competition, and the bottom line:

Cost advantages

- Product line rationalization, product portfolio planning, and total cost measurements eliminate low-margin and money-losing products
- Eliminating high-overhead products eliminates the "loser tax" on cash-cow products
- Shift to mass customization eliminates craft production and costly customizations
- Flexible fixturing reduces welding, machining, and assembly labor and setup costs
- Setup reduction helps maximize worker and equipment utilization
- CNC automation saves labor cost; flexible fixturing and setup reduction makes CNC even more efficient
- Greater productivity reduces overtime costs
- Lean activities eliminate waste
- Incoming JIT part deliveries minimize parts inventory carrying cost, which is a cost of 25% of value per year
- Flow manufacturing and setup/batch reduction efforts minimize work-in-process (WIP) inventory carrying costs
- The rapid feedback aspect of flow production eliminates the cost of recurring defects, which is more likely to happen with large batches
- Automatic resupply techniques reduce costs of scheduling, purchasing, expediting, material overhead, and MRP
- Concurrent engineering products for manufacturability generally results in significant cost savings, but these are more profound for BTO&MC companies because of the greater opportunities to *design out* several categories of overhead cost
- Learning relationships make repeat orders more efficient

- Better utilization of space delays or eliminates the need for more factory space for growth
- Total cost measurements will identify high costs and direct corrective efforts
- Advanced business model eliminates the need to discount or take low-margin sales

Responsiveness Advantages

- Delivery time can be reduced drastically for products entirely within the BTO&MC system and moderately for products mostly within the system
- Bid responses are quicker with configurators with less delays due to order-entry errors and customer-induced changes
- New products and options can be introduced faster as "variations on a theme"
- Ramping up on new products can be quicker on flexible equipment
- Optimal responsiveness is enhanced by internal production and supply chains that allow assemblers to *pull* parts quickly without lengthy lead-time delays, at least on the most common options and customizations
- Flexible operations result in a better match between production capacity and customer demand, thus resulting in less setup changes and inventory.
- Learning relationships make repeat orders quicker
- Postponement may allow dealers to order dealer-installed options as products are shipped

Customer Satisfaction Advantages

- Customers can save purchase costs
- Customers can receive products faster
- Customers can get a wide range of *expected* customizations
- Customers can be offered *unexpected* customization possibilities
- Customers can make better choices and do more "what ifs" with a good configurator

- Kaizen and the rapid feedback aspect of flow production improves quality
- Higher proportion of CNC work improves quality with more consistent tolerances
- Learning relationships result in better customer satisfaction with each order

Competitive Advantage

- In general, competition now is between business models
- Speed, cost, and customization advantages will compete well against competitors both large and small competitors
- Avoid commodity status with the differentiation aspects of MC
- Potential to expand business into adjacencies, niche markets, derivatives, etc.
- Learning relationships result in greater customer loyalty with each order
- Reputation as a leader improves retention and new sales
- Lower volume threshold for introducing new designs
- Protection of existing market share from competitive encroachment

Bottom Line

- Lower total cost
- Growth of revenue and market share
- Premium prices, or at least no need to discount, may be possible with improved reputation and enough competitive advantage
- Additional value-added work offer high profit opportunities
- Increased profits or lower selling price or both

CONCLUSIONS

BTO&MC is a business model that offers an unbeatable combination of responsiveness, cost, and what customers want when they want it. BTO&MC companies can achieve substantial cost advantages from eliminating inventory, forecasting, expediting, kitting, setup, and inefficient fire-drill efforts to customize products.

BTO&MC companies can substantially *simplify* supply chains – not just "manage" them – to the point where parts and materials can be *pulled* into production without forecasts, MRP, purchasing, waiting, or warehousing. Stores and dealers can be rapidly replenished at high order-fulfillment rates without the cost and risk of inventory.

BTO&MC enables manufacturers to be the first to market with new technologies and efficiently mass-customize products for niche markets, countries, regions, industries, and individual customers.

Sales and profits will grow from expanded sales for standard, customized, derivative, and niche market products, while enhancing sales to existing customers.

ENDNOTES/REFERENCES

1. James P. Womack and Daniel T. Jones, *Lean Thinking; Banish Waste and Create Wealth in Your Corporation,* (1996, Simon & Schuster), p. 27.

2. Jonathan Welsh, "A New Status Symbol: Overpaying for Your Minivan; Despite Discounts, More Cards Sell Above the Sticker Price," *Wall Street Journal,* July 23, 2003, p. B1.

3. Peter M. Senge, *The Fifth Discipline; The Art & Practice of the Learning Organization,* (1990, Doubleday Currency), Ch. 3.

4. Philip Siekman, "Where Build to Order Works Best," *Fortune,* April 26, 1999 (Industrial Management & Technology edition), p. 169J.

5. Sydney Finkelstein, *Why Smart Executives Fail and What You Can Learn from Their Mistakes,* (2003, Portfolio/Penguin), p. 138.

6. Ibid., p. 138.

7. Alex Taylor III, "The Fiasco at Ford," *Fortune,* February 4, 2003, pages 111 - 112.

8. Jim Collins, *Good to Great; Why Some Companies Make the Leap . . . and Others Don't,* (2001, Harper Business), see the section: "The Misguided Use of Acquisitions" pp. 180-181.

9. Robert Frank and Robin Sidel, "Firms that Lived by the Deal in the 90's, Nos Sink by the Dozens," *Wall Street Journal,* June 6, 2002, citing a Thomson/First Call study done for the Wall Street Journal.

10. Ibid.

11. Jeffery L. Hiday, "Most Mergers Fail to Add Value, Consultants Find," *Wall Street Journal,* October 12, 1998.

12. Diane Brady, "The Education of Jeff Immelt," *Business Week,* April 29, 2002; Cover story.

13. Ibid.

14. Michael Arndt, Emily Thornton, and Dean Foust, "Let's Talk Turkeys; Some mergers were never meant to be," *Business Week,* December 11, 2000, pp. 44 - 46.

15. Ibid.

16. Sydney Finkelstein, *Why Smart Executives Fail and What You Can Learn from Their Mistakes,* (2003, Portfolio/Penguin), p. 96.

17. Jim Collins, "Beware of the Self-Promoting CEO," *Wall Street Journal,* Editorial page, November 26, 2001.

18. Julie Forster, "General Malaise at General Mills; Rivals have been eating its lunch since the Pillsbury deal," *Business Week,* July 1, 2002, pp. 68 - 70.

19. Cliff Edwards with Andrew Park, "HP and Compaq; It's Showtime; The Cost-Cutting Looks Doable, But Other Synergies May be More Elusive than Expected," *Business Week,* June 17, 2002.

20. "HP, Compaq Fail PC User's Test," *Los Angeles Times,* July 10, 2003, p. C3.

21. Julie Forster, "General Malaise at General Mills; Rivals have been eating its lunch since the Pillsbury deal," *Business Week,* July 1, 2002, pp. 68 - 70.

22. Thomas H. Davenport and Laurence Prusak with H. James Wilson, *What's the Big Idea? Creating and Capitalizing on the Best Management Thinking,* (2003, Harvard Business School Press), page 133.

23. Andrew Park, Faith Keenan, and Cliff Edwards, "Whose Lunch Will Dell Eat Next?," *Business Week,* August 12, 2002, pp. 66-67.

24. Neal E. Boudette, "Peugeot's Formula for Success: Steering Clear of Megamergers," *Wall Street Journal,* August 4, 2003, front page, column 5.

25. Collins, *Good to Great,* p. 180.

26. Adrian J. Slywotzky and Richard Wise, "The Growth Crisis - and How to Escape it, *Harvard Business Review,* July 2002, pages 73 -83.

27. Chris Zook, *Profit from the Core; Growth Strategy in an Era of Turbulence,* (2001, Harvard Business School Press).

28. James H. Gilmore and B. Joseph Pine II, editors, *Markets of One; Creating Customer-Unique Value through Mass Customization,* (2000, Harvard Business School Press), a collection of articles that appeared from 1988 through 1999 in the Harvard Business Review; quote is from the Introduction (page xx) by Gilmore and Pine.

29. Jim Collins, *Good to Great; Why Some Companies Make the Leap . . . and Others Don't,* (2001, Harper Business), Chapter 5.

30. David M. Anderson, with an introduction by B. Joseph Pine II, *Agile Product Development for Mass Customization,* (1997, McGraw-Hill), Ch. 1, Introduction.

31. *Business Week,* August 28, 2000, Editorial Page (p. 278).

APPENDIX: RESOURCES

BOOKS

Design for Manufacturability & Concurrent Engineering; How to Design for Low Cost, Design in High Quality, Design for Lean Manufacture, and Design Quickly for Fast Production by David M. Anderson, (2008, CIM Press, 448 pages); for a description of the book, see *www.design4manufacturability.com/books.htm.*

Good to Great; Why Some Companies Make the Leap . . . and Others Don't by Jim Collins, (2001, Harper Business)

Why Smart Executives Fail, and What you Can Learn from their Mistakes by Sydney Finkelstein,(2003, Portfolio/Penguin Group)

The Goal, a novel by Eliyahu M. Goldratt,(1992, North River Press)

Activity-Based Costing; Making it Work at Small and Mid-Sized Companies, Second Edition, by Douglas T. Hicks, (1998, Wiley, 1998).

Toyota Production System, An Integrated Approach to Just-in-Time, **Second Edition,** by Yasuhiro Monden (1993, Industrial Engineering and Management Press, IIE)

The Toyota Product Development System by James Morgan & Jeffrey K. Liker, (2006, Productivity Press)

A Revolution in Manufacturing, The SMED System by Shigeo Shingo, (Portland, OR, Productivity Press, 1985).

The Machine that Changed the World*; *The Story of Lean Production, by James Womack, Daniel Jones, and Daniel Roos,(1991, paperback edition, Harper Perennial)

Lean Thinking; Banish Waste and Create Wealth in Your Corporation **by** James P. Womack and Daniel T. Jones,(1996, Simon & Schuster)

Profit from the Core; Growth Strategy in an Era of Turbulence by Chris Zook, (2001, Harvard Business School Press).

WEB-SITES by Dr. Anderson

www.HalfCostProducts.com

Home page: Eight-step half-cost reduction strategy (with links to related articles) followed by "How Not to Lower Cost" (with links to articles on bidding, offshoring, and cost reduction after design)

Statistics: Content equivalent to 250 page book; 700 hyperlinks

Articles:
- Build-to-Order
- Build-to-Order Future
- Cost of Quality
- Cost Reduction; How Not to
- Designing for Build-to-Order
- Designing for Lean
- Designing for Manufacturability
- Designing for Mass Customization
- Designing for Quality
- Low-Bidding
- Mass Customization
- Mass Production, End of
- Mergers & Acquisitions
- Lean Production
- Off-Shore Manufacture
- Outsourcing
- Rationalization
- Standardization
- SCM Cost Reduction
- Total Cost

www.build-to-order-consulting.com

Pages: Home Page (with summaries and links), Seminar Page (with comments from attendees), Consulting, Implementation, Articles, Books, Credentials, Client List, Site Map

Articles:
- Build-to-Order
- Mass Customization
- Shortcomings of Mass Production
- Business Model for BTO& MC
- Achieving Growth with BTO & MC
- On-Demand Lean Production
- Standardization
- Kanban Resupply
- Hoffman case study
- Rationalization
- Training for BTO&MC

www.design4manufacturability.com

Pages: Home page (with summaries and links), Seminar Page (with comments from attendees), Consulting, Implementation, Books, Credentials, Client List, Site Map

Articles:
- Design for Manufacturability
- Standardization
- Product Line Rationalization
- Build-to-Order
- Mass Customization

IN-HOUSE BTO&MC SEMINAR AGENDA

Introduction. The seminar will begin with discussions of the challenges and opportunities facing the company with respect to responsiveness, cost, product variety, growth, and profits.

Shortcomings of Mass Production. Mass production was the ideal way to make Model T's in the 1920s, but is not suitable for today's environment of increasing product variety and market volatility.

Supply Chain Simplification. Rather than just "managing" complex supply chains with an unnecessary proliferation of parts and suppliers, this seminar will show how to rationalize product lines, standardize parts, and establish a spontaneous supply chain that can *pull* in standard parts and materials automatically.

Outsourcing vs. Integration. Dr. Anderson will show how excessive outsourcing to far-flung supply chains hampers responsiveness while not really reducing cost on a total cost basis. Instead, he will show how the optimal level of integration enables manufacturers to quickly and cost-effectively build parts on-demand and them assemble them to-order.

On-demand lean production extends the proven principles of lean production, setup elimination, and flow production to enable operations to build *any* product *any* time in *any* quantity in a truly batch-size-of-one mode without forecasts or inventory.

Mass customization. The same operations and supply chain employed for standard products can efficiently *mass customize* a wide range of products for many niche markets, countries, regions, industries, stores, and individual customers.

Product development for BTO&MC. Dr. Anderson draws on two decades of experience in design for manufacturability (DFM) to show how to *concurrently engineer* families of products and versatile processes for build-to-order and mass customization.

Cost Reduction Strategies. BTO and mass customization offer many opportunities to substantially reduce total cost by eliminating all the costs of setup and inventory while minimizing overhead costs for customization, quality, distribution, and material overhead.

Implementation. Practical implementation strategies will be presented for several independently justifiable and self-supporting implementation steps. Scenarios for fabricated and electronic products will be illustrated with perspective factory drawings that show the production equipment and the flow of parts and information.

The Business Case. Finally, the seminar will present the business case for build-to-order and mass customization, itemizing all the advantages for cost, responsiveness, and customer satisfaction, including strategies for growth of sales and profits.

805-924-0100 e-mail: ***anderson@build-to-order-consulting.com***

Dr. Anderson's

RATIONALIZATION WORKSHOP

The rationalization workshop presents product line rationalization principles, identifies product lines to investigate, and analyzes Pareto plots to ascertain which product variations could be eliminated, outsourced, improved, or grouped into product family lines for BTO&MC.

The workshop group would then discuss and resolve issues related to overcoming inhibitions, fears, and resistance that were presented in Chapter 3. The group would start the process of developing profiles to "red flag" certain product variations for special scrutiny, adjusting responsibilities, incentives, and compensation, and finally assigning subsequent tasks and responsibilities for the rationalization effort.

STANDARDIZATION WORKSHOP

The standardization workshop starts with a presentation of standardization principles and then reviews Pareto plots first for all parts and then for specific categories. The workshop group will identify potential categories to standardize and vote to prioritize categories for implementation. The workshop would then map out an implementation plan, assign tasks to generate more Pareto plots, and designate people to do the above.

CONSULTING

Clients can benefit early from Dr. Anderson's expertise with on-site consulting after the seminar to help identify opportunities, set strategies, make decisions, and plan implementation. This can help maximize impact of training and workshops to ensure implementation success. After that, he can provide remote consulting by phone, e-mail, or teleconference calls.

Dr. Anderson's bio-sketch appears opposite page one in this book and his three web-sites are described on page 468. He can be reached at:

805-924-0100 e-mail: ***anderson@build-to-order-consulting.com***

INDEX